Mathematics
Third Edition

Harry

Ric Pimentel
Terry Wall

®IGCSE is the registered trademark of Cambridge International Examinations.

The Publishers would like to thank the following for permission to reproduce copyright material:

Photo credits
p.3 © Inter-University Centre for Astronomy and Astrophysics, Pune, India; **p.103** © dbimages/Alamy; **p.213** © Erica Guilane-Nachez – Fotolia; **p.275** © The Art Archive/Alamy; **p.315** © Georgios Kollidas – Fotolia; **p.345** © The Granger Collection, NYC/TopFoto; **p.381** © Stefano Bianchetti/Corbis; **p.441** © Bernard 63 – Fotolia; **p.465** © Jason Butcher/cultura/Corbis.
Every effort has been made to trace all copyright holders, but if any have been inadvertently overlooked the Publishers will be pleased to make the necessary arrangements at the first opportunity.

Although every effort has been made to ensure that website addresses are correct at time of going to press, Hodder Education cannot be held responsible for the content of any website mentioned in this book. It is sometimes possible to find a relocated web page by typing in the address of the home page for a website in the URL window of your browser.

Hachette UK's policy is to use papers that are natural, renewable and recyclable products and made from wood grown in sustainable forests. The logging and manufacturing processes are expected to conform to the environmental regulations of the country of origin.

Orders: please contact Bookpoint Ltd, 130 Milton Park, Abingdon, Oxon OX14 4SB.
Telephone: (44) 01235 827720. Fax: (44) 01235 400454. Lines are open 9.00–5.00, Monday to Saturday, with a 24-hour message answering service. Visit our website at www.hoddereducation.com

© Ric Pimentel and Terry Wall 1997, 2006, 2010, 2013
First edition published 1997
Second edition published 2006
Second edition updated with CD published 2010

This third edition published 2013 by
Hodder Education, an Hachette UK Company,
338 Euston Road
London NW1 3BH

Impression number 5 4 3
Year 2017 2016 2015 2014

All rights reserved. Apart from any use permitted under UK copyright law, no part of this publication may be reproduced or transmitted in any form or by any means, electronic or mechanical, including photocopying and recording, or held within any information storage and retrieval system, without permission in writing from the publisher or under licence from the Copyright Licensing Agency Limited. Further details of such licences (for reprographic reproduction) may be obtained from the Copyright Licensing Agency Limited, Saffron House, 6–10 Kirby Street, London EC1N 8TS.

Cover photo © senoldo – Fotolia
Illustrations by Pantek Media
Typeset in 10/12pt Times Ten by Pantek Media
Printed in Dubai

A catalogue record for this title is available from the British Library
ISBN 978 1 4441 91707

Contents

● Introduction — 1

● **Topic 1 Number** — 2
   Chapter 1 Number and language — 4
   Chapter 2 Accuracy — 12
   Chapter 3 Calculations and order — 24
   Chapter 4 Integers, fractions, decimals and percentages — 30
   Chapter 5 Further percentages — 43
   Chapter 6 Ratio and proportion — 52
   Chapter 7 Indices and standard form — 62
   Chapter 8 Money and finance — 75
   Chapter 9 Time — 85
   Chapter 10 Set notation and Venn diagrams — 87
   Topic 1 Mathematical investigations and ICT — 98

● **Topic 2 Algebra and graphs** — 102
   Chapter 11 Algebraic representation and manipulation — 104
   Chapter 12 Algebraic indices — 123
   Chapter 13 Equations and inequalities — 127
   Chapter 14 Linear programming — 149
   Chapter 15 Sequences — 155
   Chapter 16 Variation — 167
   Chapter 17 Graphs in practical situations — 174
   Chapter 18 Graphs of functions — 189
   Chapter 19 Functions — 203
   Topic 2 Mathematical investigations and ICT — 209

● **Topic 3 Geometry** — 212
   Chapter 20 Geometrical vocabulary — 214
   Chapter 21 Geometrical constructions and scale drawings — 221
   Chapter 22 Similarity — 231
   Chapter 23 Symmetry — 240
   Chapter 24 Angle properties — 245
   Chapter 25 Loci — 265
   Topic 3 Mathematical investigations and ICT — 271

● **Topic 4 Mensuration** — 274
   Chapter 26 Measures — 276
   Chapter 27 Perimeter, area and volume — 281
   Topic 4 Mathematical investigations and ICT — 311

Contents

- **Topic 5 Coordinate geometry** — **314**
  - Chapter 28 Straight-line graphs — 316
  - Topic 5 Mathematical investigations and ICT — 341

- **Topic 6 Trigonometry** — **344**
  - Chapter 29 Bearings — 346
  - Chapter 30 Trigonometry — 349
  - Chapter 31 Further trigonometry — 366
  - Topic 6 Mathematical investigations and ICT — 378

- **Topic 7 Matrices and transformations** — **380**
  - Chapter 32 Vectors — 382
  - Chapter 33 Matrices — 394
  - Chapter 34 Transformations — 408
  - Topic 7 Mathematical investigations and ICT — 436

- **Topic 8 Probability** — **440**
  - Chapter 35 Probability — 442
  - Chapter 36 Further probability — 452
  - Topic 8 Mathematical investigations and ICT — 460

- **Topic 9 Statistics** — **464**
  - Chapter 37 Mean, median, mode and range — 466
  - Chapter 38 Collecting and displaying data — 472
  - Chapter 39 Cumulative frequency — 491
  - Topic 9 Mathematical investigations and ICT — 499

- **Index** — **502**

Introduction

To the Student

This textbook has been written by two experienced mathematics teachers.

The book is written to cover every section of the Cambridge IGCSE® Mathematics (0580) syllabus (Core and Extended). The syllabus headings (Number, Algebra and graphs, Geometry, Mensuration, Coordinate geometry, Trigonometry, Matrices and transformations, Probability, Statistics) are mirrored in the textbook. Each major topic is divided into a number of chapters, and each chapter has its own discrete exercises and student assessments. The Core sections are identified with a green band and the Extended with a red band. Students using this book may follow either a Core or Extended curriculum.

The syllabus specifically refers to 'Applying mathematical techniques to solve problems' and this is fully integrated into the exercises and assessments. This book also includes a number of such problems so that students develop their skills in this area throughout the course. Ideas for ICT activities are also included, although this is not part of the examination.

The CD included with this book contains Personal Tutor audio-visual worked examples covering the main concepts.

The study of mathematics crosses all lands and cultures. A mathematician in Africa may be working with another in Japan to extend work done by a Brazilian in the USA; art, music, language and literature belong to the culture of the country of origin. Opera is European. Noh plays are Japanese. It is not likely that people from different cultures could work together on a piece of Indian art for example. But all people in all cultures have tried to understand their world, and mathematics has been a common way of furthering that understanding, even in cultures which have left no written records. Each Topic in this textbook has an introduction which tries to show how, over a period of thousands of years, mathematical ideas have been passed from one culture to another.

The Ishango Bone from Stone-Age Africa has marks suggesting it was a tally stick. It was the start of arithmetic. 4500 years ago in ancient Mesopotamia, clay tablets show multiplication and division problems. An early abacus may have been used at this time. 3600 years ago what is now called The Rhind Papyrus was found in Egypt. It shows simple algebra and fractions. The Moscow Papyrus shows how to find the volume of a pyramid. The Egyptians advanced our knowledge of geometry. The Babylonians worked with arithmetic. 3000 years ago in India the great wise men advanced mathematics and their knowledge travelled to Egypt and later to Greece, then to the rest of Europe when great Arab mathematicians took their knowledge with them to Spain.

Europeans and later Americans made mathematical discoveries from the fifteenth century. It is likely that, with the re-emergence of China and India as major world powers, these countries will again provide great mathematicians and the cycle will be completed. So when you are studying from this textbook remember that you are following in the footsteps of earlier mathematicians who were excited by the discoveries they had made. These discoveries changed our world.

You may find some of the questions in this book difficult. It is easy when this happens to ask the teacher for help. Remember though that mathematics is intended to stretch the mind. If you are trying to get physically fit, you do not stop as soon as things get hard. It is the same with mental fitness. Think logically, try harder. You can solve that difficult problem and get the feeling of satisfaction that comes with learning something new.

Ric Pimentel
Terry Wall

Topic 1: Number

◯ Syllabus

E1.1

Identify and use natural numbers, integers (positive, negative and zero), prime numbers, square numbers, common factors and common multiples, rational and irrational numbers (e.g. π, $\sqrt{2}$), real numbers.

E1.2

Use language, notation and Venn diagrams to describe sets and represent relationships between sets as follows:
Definition of sets
e.g. $A = \{x: x \text{ is a natural number}\}$
$B = \{(x, y): y = mx + c\}$
$C = \{x: a \leq x \leq b\}$
$D = \{a, b, c, \ldots\}$

E1.3

Calculate squares, square roots, cubes and cube roots of numbers.

E1.4

Use directed numbers in practical situations.

E1.5

Use the language and notation of simple vulgar and decimal fractions and percentages in appropriate contexts.
Recognise equivalence and convert between these forms.

E1.6

Order quantities by magnitude and demonstrate familiarity with the symbols $=, \neq, >, <, \geq, \leq$

E1.7

Understand the meaning and rules of indices. Use the standard form $A \times 10^n$ where n is a positive or negative integer, and $1 \leq A < 10$.

E1.8

Use the four rules for calculations with whole numbers, decimals and vulgar (and mixed) fractions, including correct ordering of operations and use of brackets.

E1.9

Make estimates of numbers, quantities and lengths, give approximations to specified numbers of significant figures and decimal places and round off answers to reasonable accuracy in the context of a given problem.

E1.10

Give appropriate upper and lower bounds for data given to a specified accuracy.
Obtain appropriate upper and lower bounds to solutions of simple problems given data to a specified accuracy.

E1.11

Demonstrate an understanding of ratio and proportion.
Increase and decrease a quantity by a given ratio.
Use common measures of rate.
Calculate average speed.

E1.12

Calculate a given percentage of a quantity.
Express one quantity as a percentage of another.
Calculate percentage increase or decrease.
Carry out calculations involving reverse percentages.

E1.13
Use a calculator efficiently.
Apply appropriate checks of accuracy.

E1.14
Calculate times in terms of the 24-hour and 12-hour clock.
Read clocks, dials and timetables.

E1.15
Calculate using money and convert from one currency to another.

E1.16
Use given data to solve problems on personal and household finance involving earnings, simple interest and compound interest.
Extract data from tables and charts.

E1.17
Use exponential growth and decay in relation to population and finance.

Contents

Chapter 1 Number and language (E1.1, E1.3, E1.4)
Chapter 2 Accuracy (E1.9, E1.10)
Chapter 3 Calculations and order (E1.6, E1.13)
Chapter 4 Integers, fractions, decimals and percentages (E1.5, E1.8)
Chapter 5 Further percentages (E1.12)
Chapter 6 Ratio and proportion (E1.11)
Chapter 7 Indices and standard form (E1.7)
Chapter 8 Money and finance (E1.15, E1.16, E1.17)
Chapter 9 Time (E1.14)
Chapter 10 Set notation and Venn diagrams (E1.2)

Hindu mathematicians

In 1300BCE a Hindu teacher named Laghada used geometry and trigonometry for his astronomical calculations. At around this time, other Indian mathematicians solved quadratic and simultaneous equations.

Much later, in about AD500, another Indian teacher, Aryabhata, worked on approximations for π (pi) and on the trigonometry of the sphere. He realised that not only did the planets go round the Sun but that their paths were elliptic.

Brahmagupta, a Hindu, was the first to treat zero as a number in its own right. This helped to develop the decimal system of numbers.

One of the greatest mathematicians of all time was Bhascara who, in the twelfth century, worked in algebra and trigonometry. He discovered that:

$$\sin(A + B) = \sin A \cos B + \cos A \sin B$$

His work was taken to Arabia and later to Europe.

Aryabhata (476–550)

1 Number and language

● Vocabulary for sets of numbers

A square can be classified in many different ways. It is a quadrilateral but it is also a polygon and a two-dimensional shape. Just as shapes can be classified in many different ways, so can numbers. Below is a description of some of the more common types of numbers.

● Natural numbers

A child learns to count 'one, two, three, four, …'. These are sometimes called the counting numbers or whole numbers.

The child will say 'I am three', or 'I live at number 73'.

If we include the number 0, then we have the set of numbers called the **natural numbers**.

The set of natural numbers $\mathbb{N} = \{0, 1, 2, 3, 4, …\}$.

● Integers

On a cold day, the temperature may be $4\,°C$ at 10 p.m. If the temperature drops by a further $6\,°C$, then the temperature is 'below zero'; it is $-2\,°C$.

If you are overdrawn at the bank by $200, this might be shown as $-$200.

The set of **integers** $\mathbb{Z} = \{…, -3, -2, -1, 0, 1, 2, 3, …\}$.

$\mathbb{Z}$ is therefore an extension of $\mathbb{N}$. Every natural number is an integer.

● Rational numbers

A child may say 'I am three'; she may also say 'I am three and a half', or even 'three and a quarter'. $3\frac{1}{2}$ and $3\frac{1}{4}$ are **rational numbers**. All rational numbers can be written as a fraction whose denominator is not zero. All terminating and recurring decimals are rational numbers as they can also be written as fractions, e.g.

$$0.2 = \tfrac{1}{5} \quad 0.3 = \tfrac{3}{10} \quad 7 = \tfrac{7}{1} \quad 1.53 = \tfrac{153}{100} \quad 0.\dot{2} = \tfrac{2}{9}$$

The set of rational numbers $\mathbb{Q}$ is an extension of the set of integers.

● Real numbers

Numbers which cannot be expressed as a fraction are not rational numbers; they are **irrational numbers**.

Using Pythagoras' rule in the diagram to the left, the length of the hypotenuse AC is found as:

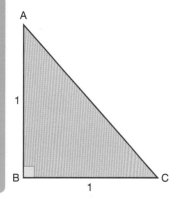

$$AC^2 = 1^2 + 1^2 = 2$$
$$AC = \sqrt{2}$$

$\sqrt{2} = 1.41421356\ldots$. The digits in this number do not recur or repeat in a pattern. This is a property of all irrational numbers. Another example of an irrational number you will come across is π (pi). It is the ratio of the circumference of a circle to the length of its diameter. Although it is often rounded to 3.142, the digits continue indefinitely never repeating themselves in any particular pattern.

The set of rational and irrational numbers together form the set of **real numbers** $\mathbb{R}$.

● Prime numbers

A prime number is one whose only factors are 1 and itself. (Note that 1 is not a prime number.)

Exercise 1.1

1. In a 10 by 10 square, write the numbers 1 to 100.
   Cross out number 1.
   Cross out all the even numbers after 2 (these have 2 as a factor).
   Cross out every third number after 3 (these have 3 as a factor).
   Continue with 5, 7, 11 and 13, then list all the prime numbers less than 100.

● Square numbers

The number 1 can be written as 1×1 or 1^2.
The number 4 can be written as 2×2 or 2^2.
9 can be written as 3×3 or 3^2.
16 can be written as 4×4 or 4^2.

When an integer (whole number) is multiplied by itself, the result is a square number. In the examples above, 1, 4, 9 and 16 are all square numbers.

● Cube numbers

The number 1 can be written as $1 \times 1 \times 1$ or 1^3.
The number 8 can be written as $2 \times 2 \times 2$ or 2^3.
27 can be written as $3 \times 3 \times 3$ or 3^3.
64 can be written as $4 \times 4 \times 4$ or 4^3.

When an integer is multiplied by itself and then by itself again, the result is a cube number. In the examples above 1, 8, 27 and 64 are all cube numbers.

● Factors

The factors of 12 are all the numbers which will divide exactly into 12, i.e. 1, 2, 3, 4, 6 and 12.

Number

Exercise 1.2 1. List all the factors of the following numbers:

a) 6 b) 9 c) 7 d) 15 e) 24
f) 36 g) 35 h) 25 i) 42 j) 100

● **Prime factors**

The factors of 12 are 1, 2, 3, 4, 6 and 12.
 Of these, 2 and 3 are prime numbers, so 2 and 3 are the prime factors of 12.

Exercise 1.3 1. List the prime factors of the following numbers:

a) 15 b) 18 c) 24 d) 16 e) 20
f) 13 g) 33 h) 35 i) 70 j) 56

An easy way to find prime factors is to divide by the prime numbers in order, smallest first.

Worked examples a) Find the prime factors of 18 and express it as a product of prime numbers:

	18
2	9
3	3
3	1

$18 = 2 \times 3 \times 3$ or 2×3^2

b) Find the prime factors of 24 and express it as a product of prime numbers:

	24
2	12
2	6
2	3
3	1

$24 = 2 \times 2 \times 2 \times 3$ or $2^3 \times 3$

c) Find the prime factors of 75 and express it as a product of prime numbers:

	75
3	25
5	5
5	1

$75 = 3 \times 5 \times 5$ or 3×5^2

Number and language

Exercise 1.4 1. Find the prime factors of the following numbers and express them as a product of prime numbers:

a) 12 b) 32 c) 36 d) 40 e) 44
f) 56 g) 45 h) 39 i) 231 j) 63

● **Highest common factor**

The prime factors of 12 are $2 \times 2 \times 3$.
The prime factors of 18 are $2 \times 3 \times 3$.
So the highest common factor (HCF) can be seen by inspection to be 2×3, i.e. 6.

● **Multiples**

Multiples of 5 are 5, 10, 15, 20, etc.
The lowest common multiple (LCM) of 2 and 3 is 6, since 6 is the smallest number divisible by 2 and 3.
The LCM of 3 and 5 is 15.
The LCM of 6 and 10 is 30.

Exercise 1.5 1. Find the HCF of the following numbers:

a) 8, 12 b) 10, 25 c) 12, 18, 24
d) 15, 21, 27 e) 36, 63, 108 f) 22, 110
g) 32, 56, 72 h) 39, 52 i) 34, 51, 68
j) 60, 144

2. Find the LCM of the following:

a) 6, 14 b) 4, 15 c) 2, 7, 10 d) 3, 9, 10
e) 6, 8, 20 f) 3, 5, 7 g) 4, 5, 10 h) 3, 7, 11
i) 6, 10, 16 j) 25, 40, 100

● **Rational and irrational numbers**

A **rational number** is any number which can be expressed as a fraction. Examples of some rational numbers and how they can be expressed as a fraction are shown below:

$0.2 = \frac{1}{5}$ $0.3 = \frac{3}{10}$ $7 = \frac{7}{1}$ $1.53 = \frac{153}{100}$ $0.\dot{2} = \frac{2}{9}$

An **irrational number** cannot be expressed as a fraction. Examples of irrational numbers include:

$\sqrt{2}, \ \sqrt{5}, \ 6 - \sqrt{3}, \ \pi$

In summary:
Rational numbers include:
- whole numbers,
- fractions,
- recurring decimals,
- terminating decimals.

Number

Irrational numbers include:
- the square root of any number other than square numbers,
- a decimal which neither repeats nor terminates (e.g. π).

Exercise 1.6

1. For each of the numbers shown below state whether it is rational or irrational:

   a) 1.3   b) $0.\dot{6}$ c) $\sqrt{3}$
   d) $-2\frac{3}{5}$ e) $\sqrt{25}$ f) $\sqrt[3]{8}$
   g) $\sqrt{7}$ h) 0.625   i) $0.1\dot{1}$

2. For each of the numbers shown below state whether it is rational or irrational:

   a) $\sqrt{2} \times \sqrt{3}$ b) $\sqrt{2} + \sqrt{3}$ c) $(\sqrt{2} \times \sqrt{3})^2$
   d) $\dfrac{\sqrt{8}}{\sqrt{2}}$ e) $\dfrac{2\sqrt{5}}{2\sqrt{20}}$ f) $4 + (\sqrt{9} - 4)$

3. In each of the following decide whether the quantity required is rational or irrational. Give reasons for your answer.

a)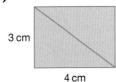
3 cm, 4 cm
The length of the diagonal

b)
4 cm
The circumference of the circle

c)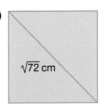
$\sqrt{72}$ cm
The side length of the square

d)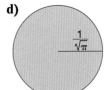
$\dfrac{1}{\sqrt{\pi}}$
The area of the circle

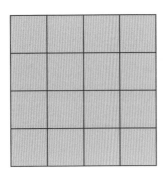

● **Square roots**

The square on the left contains 16 squares. It has sides of length 4 units.

So the square root of 16 is 4.
This can be written as $\sqrt{16} = 4$.

Note that 4×4 is 16 so 4 is the square root of 16.

However, -4×-4 is also 16 so -4 is also the square root of 16.

By convention, $\sqrt{16}$ means 'the positive square root of 16' so $\sqrt{16} = 4$ but the square root of 16 is ±4 i.e. +4 or −4.

Note that −16 has no square root since any integer squared is positive.

Exercise 1.7

1. Find the following:
   a) $\sqrt{25}$ b) $\sqrt{9}$ c) $\sqrt{49}$ d) $\sqrt{100}$
   e) $\sqrt{121}$ f) $\sqrt{169}$ g) $\sqrt{0.01}$ h) $\sqrt{0.04}$
   i) $\sqrt{0.09}$ j) $\sqrt{0.25}$

2. Use the $\sqrt{}$ key on your calculator to check your answers to question 1.

3. Calculate the following:
   a) $\sqrt{\frac{1}{9}}$ b) $\sqrt{\frac{1}{16}}$ c) $\sqrt{\frac{1}{25}}$ d) $\sqrt{\frac{1}{49}}$
   e) $\sqrt{\frac{1}{100}}$ f) $\sqrt{\frac{4}{9}}$ g) $\sqrt{\frac{9}{100}}$ h) $\sqrt{\frac{49}{81}}$
   i) $\sqrt{2\frac{7}{9}}$ j) $\sqrt{6\frac{1}{4}}$

Exercise 1.8

● **Using a graph**

1. Copy and complete the table below for the equation $y = \sqrt{x}$.

x	0	1	4	9	16	25	36	49	64	81	100
y											

   Plot the graph of $y = \sqrt{x}$. Use your graph to find the approximate values of the following:
   a) $\sqrt{35}$ b) $\sqrt{45}$ c) $\sqrt{55}$ d) $\sqrt{60}$ e) $\sqrt{2}$

2. Check your answers to question 1 above by using the $\sqrt{}$ key on a calculator.

● **Cube roots**

The cube below has sides of 2 units and occupies 8 cubic units of space. (That is, $2 \times 2 \times 2$.)

So the cube root of 8 is 2.

This can be written as $\sqrt[3]{8} = 2$.

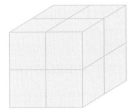

$\sqrt[3]{}$ is read as 'the cube root of …'.

$\sqrt[3]{64}$ is 4, since $4 \times 4 \times 4 = 64$.

Note that $\sqrt[3]{64}$ is not -4

since $-4 \times -4 \times -4 = -64$

but $\sqrt[3]{-64}$ is -4.

Number

Exercise 1.9 1. Find the following cube roots:
   a) $\sqrt[3]{8}$ b) $\sqrt[3]{125}$ c) $\sqrt[3]{27}$ d) $\sqrt[3]{0.001}$
   e) $\sqrt[3]{0.027}$ f) $\sqrt[3]{216}$ g) $\sqrt[3]{1000}$ h) $\sqrt[3]{1000000}$
   i) $\sqrt[3]{-8}$ j) $\sqrt[3]{-27}$ k) $\sqrt[3]{-1000}$ l) $\sqrt[3]{-1}$

● **Directed numbers**

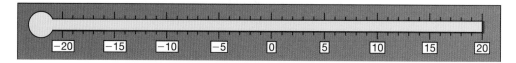

Worked example   The diagram above shows the scale of a thermometer. The temperature at 04 00 was −3 °C. By 09 00 the temperature had risen by 8 °C. What was the temperature at 09 00?

$$(-3)° + (8)° = (5)°$$

Exercise 1.10 1. The highest temperature ever recorded was in Libya. It was 58 °C. The lowest temperature ever recorded was −88 °C in Antarctica. What is the temperature difference?

2. My bank account shows a credit balance of $105. Describe my balance as a positive or negative number after each of these transactions is made in sequence:
   a) rent $140
   b) car insurance $283
   c) 1 week's salary $230
   d) food bill $72
   e) credit transfer $250

3. The roof of an apartment block is 130 m above ground level. The car park beneath the apartment is 35 m below ground level. How high is the roof above the floor of the car park?

4. A submarine is at a depth of 165 m. If the ocean floor is 860 m from the surface, how far is the submarine from the ocean floor?

Student assessment 1

1. State whether the following numbers are rational or irrational:
   a) 1.5
   b) $\sqrt{7}$
   c) $0.\dot{7}$
   d) $0.\dot{7}\dot{3}$
   e) $\sqrt{121}$
   f) π

2. Show, by expressing them as fractions or whole numbers, that the following numbers are rational:
   a) 0.625
   b) $\sqrt[3]{27}$
   c) 0.44

3. Find the value of:
   a) 9^2
   b) 15^2
   c) $(0.2)^2$
   d) $(0.7)^2$

4. Calculate:
   a) $(3.5)^2$
   b) $(4.1)^2$
   c) $(0.15)^2$

5. Without using a calculator, find:
   a) $\sqrt{225}$
   b) $\sqrt{0.01}$
   c) $\sqrt{0.81}$
   d) $\sqrt{\frac{9}{25}}$
   e) $\sqrt{5\frac{4}{9}}$
   f) $\sqrt{2\frac{23}{49}}$

6. Without using a calculator, find:
   a) 4^3
   b) $(0.1)^3$
   c) $(\frac{2}{3})^3$

7. Without using a calculator, find:
   a) $\sqrt[3]{27}$
   b) $\sqrt[3]{1\,000\,000}$
   c) $\sqrt[3]{\frac{64}{125}}$

8. My bank statement for seven days in October is shown below:

Date	Payments ($)	Receipts ($)	Balance ($)
01/10			200
02/10	284		(a)
03/10		175	(b)
04/01	(c)		46
05/10		(d)	120
06/10	163		(e)
07/10		28	(f)

Copy and complete the statement by entering the amounts (a) to (f).

2 Accuracy

● Approximation

In many instances exact numbers are not necessary or even desirable. In those circumstances approximations are given. The approximations can take several forms. The common types of approximation are dealt with below.

● Rounding

If 28 617 people attend a gymnastics competition, this figure can be reported to various levels of accuracy.

> To the nearest 10 000 this figure would be rounded up to 30 000.
> To the nearest 1000 the figure would be rounded up to 29 000.
> To the nearest 100 the figure would be rounded down to 28 600.

In this type of situation it is unlikely that the exact number would be reported.

Exercise 2.1

1. Round the following numbers to the nearest 1000:
   a) 68 786   b) 74 245   c) 89 000
   d) 4020   e) 99 500   f) 999 999

2. Round the following numbers to the nearest 100:
   a) 78 540   b) 6858   c) 14 099
   d) 8084   e) 950   f) 2984

3. Round the following numbers to the nearest 10:
   a) 485   b) 692   c) 8847
   d) 83   e) 4   f) 997

Decimal places

A number can also be approximated to a given number of decimal places (d.p.). This refers to the number of digits written after a decimal point.

Worked examples

a) Write 7.864 to 1 d.p.
   The answer needs to be written with one digit after the decimal point. However, to do this, the second digit after the decimal point also needs to be considered. If it is 5 or more then the first digit is rounded up.

   i.e. 7.864 is written as 7.9 to 1 d.p.

b) Write 5.574 to 2 d.p.
The answer here is to be given with two digits after the decimal point. In this case the third digit after the decimal point needs to be considered. As the third digit after the decimal point is less than 5, the second digit is not rounded up.

i.e. 5.574 is written as 5.57 to 2 d.p.

Exercise 2.2 **1.** Give the following to 1 d.p.
a) 5.58   b) 0.73   c) 11.86
d) 157.39 e) 4.04   f) 15.045
g) 2.95   h) 0.98   i) 12.049

2. Give the following to 2 d.p.
a) 6.473   b) 9.587   c) 16.476
d) 0.088   e) 0.014   f) 9.3048
g) 99.996 h) 0.0048 i) 3.0037

Significant figures

Numbers can also be approximated to a given number of significant figures (s.f.). In the number 43.25 the 4 is the most significant figure as it has a value of 40. In contrast, the 5 is the least significant as it only has a value of 5 hundredths.

Worked examples **a)** Write 43.25 to 3 s.f.
Only the three most significant digits are written, however the fourth digit needs to be considered to see whether the third digit is to be rounded up or not.

i.e. 43.25 is written as 43.3 to 3 s.f.

b) Write 0.0043 to 1 s.f.
In this example only two digits have any significance, the 4 and the 3. The 4 is the most significant and therefore is the only one of the two to be written in the answer.

i.e. 0.0043 is written as 0.004 to 1 s.f.

Exercise 2.3 **1.** Write the following to the number of significant figures written in brackets:
a) 48 599 (1 s.f.)   b) 48 599 (3 s.f.)   c) 6841 (1 s.f.)
d) 7538 (2 s.f.)   e) 483.7 (1 s.f.)   f) 2.5728 (3 s.f.)
g) 990 (1 s.f.)   h) 2045 (2 s.f.)   i) 14.952 (3 s.f.)

2. Write the following to the number of significant figures written in brackets:
a) 0.085 62 (1 s.f.) b) 0.5932 (1 s.f.)   c) 0.942 (2 s.f.)
d) 0.954 (1 s.f.)   e) 0.954 (2 s.f.)   f) 0.003 05 (1 s.f.)
g) 0.003 05 (2 s.f.) h) 0.009 73 (2 s.f.) i) 0.009 73 (1 s.f.)

Appropriate accuracy

In many instances calculations carried out using a calculator produce answers which are not whole numbers. A calculator will give the answer to as many decimal places as will fit on its screen. In most cases this degree of accuracy is neither desirable nor necessary. Unless specifically asked for, answers should not be given to more than two decimal places. Indeed, one decimal place is usually sufficient. In the examination, you will usually be asked to give your answers exactly or correct to three significant figures as appropriate; answers in degrees to be given to one decimal place.

Worked example Calculate $4.64 \div 2.3$ giving your answer to an appropriate degree of accuracy.

The calculator will give the answer to $4.64 \div 2.3$ as 2.0173913. However the answer given to 1 d.p. is sufficient.
Therefore $4.64 \div 2.3 = 2.0$ (1 d.p.).

Estimating answers to calculations

Even though many calculations can be done quickly and effectively on a calculator, often an estimate for an answer can be a useful check. This is done by rounding each of the numbers in such a way that the calculation becomes relatively straightforward.

Worked examples **a)** Estimate the answer to 57×246.

Here are two possibilities:
i) $60 \times 200 = 12\,000$,
ii) $50 \times 250 = 12\,500$.

b) Estimate the answer to $6386 \div 27$.

$6000 \div 30 = 200$.

Exercise 2.4

1. Calculate the following, giving your answer to an appropriate degree of accuracy:
   a) 23.456×17.89 b) 0.4×12.62 c) 18×9.24
   d) $76.24 \div 3.2$ e) 7.6^2 f) 16.42^3
   g) $\dfrac{2.3 \times 3.37}{4}$ h) $\dfrac{8.31}{2.02}$ i) $9.2 \div 4^2$

2. Without using a calculator, estimate the answers to the following:
   a) 62×19 b) 270×12 c) 55×60
   d) 4950×28 e) 0.8×0.95 f) 0.184×475

3. Without using a calculator, estimate the answers to the following:
   a) $3946 \div 18$ b) $8287 \div 42$ c) $906 \div 27$
   d) $5520 \div 13$ e) $48 \div 0.12$ f) $610 \div 0.22$

4. Without using a calculator, estimate the answers to the following:
   a) $78.45 + 51.02$ b) $168.3 - 87.09$ c) 2.93×3.14
   d) $84.2 \div 19.5$ e) $\dfrac{4.3 \times 752}{15.6}$ f) $\dfrac{(9.8)^3}{(2.2)^2}$

5. Using estimation, identify which of the following are definitely incorrect. Explain your reasoning clearly.
   a) $95 \times 212 = 20\,140$ b) $44 \times 17 = 748$
   c) $689 \times 413 = 28\,457$ d) $142\,656 \div 8 = 17\,832$
   e) $77.9 \times 22.6 = 2512.54$
   f) $\dfrac{8.42 \times 46}{0.2} = 19\,366$

6. Estimate the shaded areas of the following shapes. Do *not* work out an exact answer.

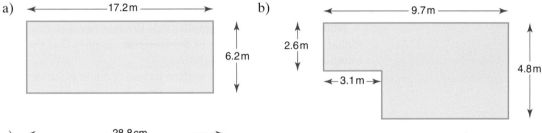

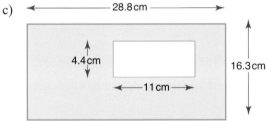

7. Estimate the volume of each of the solids below. Do *not* work out an exact answer.

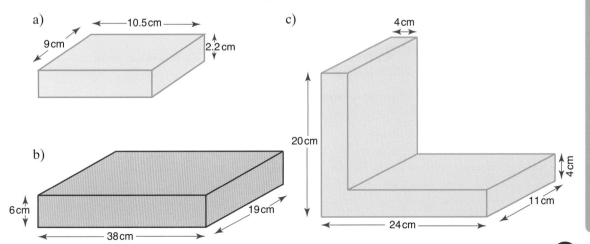

Upper and lower bounds

Numbers can be written to different degrees of accuracy. For example 4.5, 4.50 and 4.500, although appearing to represent the same number, do not. This is because they are written to different degrees of accuracy.

4.5 is rounded to one decimal place and therefore any number from 4.45 up to but not including 4.55 would be rounded to 4.5. On a number line this would be represented as:

As an inequality where x represents the number it would be expressed as

$$4.45 \leqslant x < 4.55$$

4.45 is known as the **lower bound** of 4.5, whilst 4.55 is known as the **upper bound**.

Note that ○→ implies that the number is not included in the solution whilst ●→ implies that the number is included in the solution.

4.50 on the other hand is written to two decimal places and therefore only numbers from 4.495 up to but not including 4.505 would be rounded to 4.50. This therefore represents a much smaller range of numbers than that being rounded to 4.5. Similarly the range of numbers being rounded to 4.500 would be even smaller.

Worked example A girl's height is given as 162 cm to the nearest centimetre.

i) Work out the lower and upper bounds within which her height can lie.

Lower bound = 161.5 cm
Upper bound = 162.5 cm

ii) Represent this range of numbers on a number line.

iii) If the girl's height is h cm, express this range as an inequality.

$$161.5 \leqslant h < 162.5$$

Exercise 2.5

1. Each of the following numbers is expressed to the nearest whole number.
   i) Give the upper and lower bounds of each.
   ii) Using x as the number, express the range in which the number lies as an inequality.
   a) 6   b) 83   c) 152
   d) 1000   e) 100

2. Each of the following numbers is correct to one decimal place.
   i) Give the upper and lower bounds of each.
   ii) Using x as the number, express the range in which the number lies as an inequality.
   a) 3.8   b) 15.6   c) 1.0
   d) 10.0   e) 0.3

3. Each of the following numbers is correct to two significant figures.
   i) Give the upper and lower bounds of each.
   ii) Using x as the number, express the range in which the number lies as an inequality.
   a) 4.2   b) 0.84   c) 420
   d) 5000   e) 0.045   f) 25 000

4. The mass of a sack of vegetables is given as 5.4 kg.
   a) Illustrate the lower and upper bounds of the mass on a number line.
   b) Using M kg for the mass, express the range of values in which M must lie, as an inequality.

5. At a school sports day, the winning time for the 100 m race was given as 11.8 seconds.
   a) Illustrate the lower and upper bounds of the time on a number line.
   b) Using T seconds for the time, express the range of values in which T must lie, as an inequality.

6. The capacity of a swimming pool is given as 620 m³ correct to two significant figures.
   a) Calculate the lower and upper bounds of the pool's capacity.
   b) Using x cubic metres for the capacity, express the range of values in which x must lie, as an inequality.

7. A farmer measures the dimensions of his rectangular field to the nearest 10 m. The length is recorded as 630 m and the width is recorded as 400 m.
   a) Calculate the lower and upper bounds of the length.
   b) Using W metres for the width, express the range of values in which W must lie, as an inequality.

Number

● Calculating with upper and lower bounds

When numbers are written to a specific degree of accuracy, calculations involving those numbers also give a range of possible answers.

Worked examples a) Calculate the upper and lower bounds for the following calculation, given that each number is given to the nearest whole number.

$$34 \times 65$$

34 lies in the range $33.5 \leqslant x < 34.5$.
65 lies in the range $64.5 \leqslant x < 65.5$.

The lower bound of the calculation is obtained by multiplying together the two lower bounds. Therefore the minimum product is 33.5×64.5, i.e. 2160.75.
 The upper bound of the calculation is obtained by multiplying together the two upper bounds. Therefore the maximum product is 34.5×65.5, i.e. 2259.75.

b) Calculate the upper and lower bounds to $\frac{33.5}{22.0}$ given that each of the numbers is accurate to 1 d.p.

33.5 lies in the range $33.45 \leqslant x < 33.55$.
22.0 lies in the range $21.95 \leqslant x < 22.05$.

 The lower bound of the calculation is obtained by dividing the lower bound of the numerator by the *upper* bound of the denominator. So the minimum value is $33.45 \div 22.05$, i.e. 1.52 (2 d.p.).
 The upper bound of the calculation is obtained by dividing the upper bound of the numerator by the *lower* bound of the denominator. So the maximum value is $33.55 \div 21.95$, i.e. 1.53 (2 d.p.).

Exercise 2.6 1. Calculate lower and upper bounds for the following calculations, if each of the numbers is given to the nearest whole number.

a) 14×20 b) 135×25 c) 100×50
d) $\frac{40}{10}$ e) $\frac{33}{11}$ f) $\frac{125}{15}$
g) $\frac{12 \times 65}{16}$ h) $\frac{101 \times 28}{69}$ i) $\frac{250 \times 7}{100}$
j) $\frac{44}{3^2}$ k) $\frac{578}{17 \times 22}$ l) $\frac{1000}{4 \times (3+8)}$

2. Calculate lower and upper bounds for the following calculations, if each of the numbers is given to 1 d.p.
   a) $2.1 + 4.7$
   b) 6.3×4.8
   c) 10.0×14.9
   d) $17.6 - 4.2$
   e) $\dfrac{8.5 + 3.6}{6.8}$
   f) $\dfrac{7.7 - 6.2}{3.5}$
   g) $\dfrac{(16.4)^2}{(3.0 - 0.3)^2}$
   h) $\dfrac{100.0}{(50.0 - 40.0)^2}$
   i) $(0.1 - 0.2)^2$

3. Calculate lower and upper bounds for the following calculations, if each of the numbers is given to 2 s.f.
   a) 64×320
   b) 1.7×0.65
   c) 4800×240
   d) $\dfrac{54\,000}{600}$
   e) $\dfrac{4.2}{0.031}$
   f) $\dfrac{100}{5.2}$
   g) $\dfrac{6.8 \times 42}{120}$
   h) $\dfrac{100}{(4.5 \times 6.0)}$
   i) $\dfrac{180}{(7.3 - 4.5)}$

Exercise 2.7

1. The masses to the nearest 0.5 kg of two parcels are 1.5 kg and 2.5 kg. Calculate the lower and upper bounds of their combined mass.

2. Calculate upper and lower bounds for the perimeter of the rectangle shown (below), if its dimensions are correct to 1 d.p.

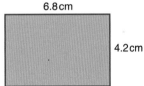

3. Calculate upper and lower bounds for the perimeter of the rectangle shown (below), whose dimensions are accurate to 2 d.p.

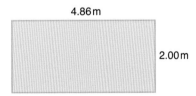

4. Calculate upper and lower bounds for the area of the rectangle shown (below), if its dimensions are accurate to 1 d.p.

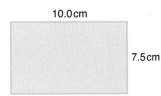

5. Calculate upper and lower bounds for the area of the rectangle shown (below), whose dimensions are correct to 2 s.f.

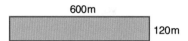

6. Calculate upper and lower bounds for the length marked x cm in the rectangle (below). The area and length are both given to 1 d.p.

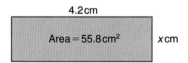

7. Calculate the upper and lower bounds for the length marked x cm in the rectangle (below). The area and length are both accurate to 2 s.f.

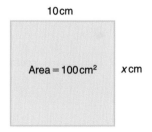

8. The radius of the circle shown (below) is given to 1 d.p. Calculate the upper and lower bounds of:
a) the circumference,
b) the area.

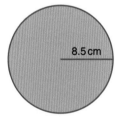

9. The area of the circle shown (below) is given to 2 s.f. Calculate the upper and lower bounds of:
a) the radius,
b) the circumference.

10. The mass of a cube of side 2 cm is given as 100 g. The side is accurate to the nearest millimetre and the mass accurate to the nearest gram. Calculate the maximum and minimum possible values for the density of the material (density = mass ÷ volume).

11. The distance to the nearest 100 000 km from Earth to the moon is given as 400 000 km. The average speed to the nearest 500 km/h of a rocket to the moon is given as 3500 km/h. Calculate the greatest and least time it could take the rocket to reach the moon.

Student assessment 1

1. Round the following numbers to the degree of accuracy shown in brackets:
   a) 2841 (nearest 100)   b) 7286 (nearest 10)
   c) 48 756 (nearest 1000) d) 951 (nearest 100)

2. Round the following numbers to the number of decimal places shown in brackets:
   a) 3.84 (1 d.p.)   b) 6.792 (1 d.p.)
   c) 0.8526 (2 d.p.) d) 1.5849 (2 d.p.)
   e) 9.954 (1 d.p.)   f) 0.0077 (3 d.p.)

3. Round the following numbers to the number of significant figures shown in brackets:
   a) 3.84 (1 s.f.)   b) 6.792 (2 s.f.)
   c) 0.7765 (1 s.f.) d) 9.624 (1 s.f.)
   e) 834.97 (2 s.f.) f) 0.004 51 (1 s.f.)

4. 1 mile is 1760 yards. Estimate the number of yards in 11.5 miles.

5. Estimate the shaded area of the figure below:

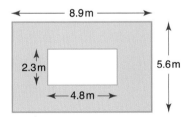

6. Estimate the answers to the following. Do *not* work out an exact answer.

   a) $\dfrac{5.3 \times 11.2}{2.1}$ b) $\dfrac{(9.8)^2}{(4.7)^2}$ c) $\dfrac{18.8 \times (7.1)^2}{(3.1)^2 \times (4.9)^2}$

7. A cuboid's dimensions are given as 12.32 cm by 1.8 cm by 4.16 cm. Calculate its volume, giving your answer to an appropriate degree of accuracy.

Student assessment 2

1. The following numbers are expressed to the nearest whole number. Illustrate on a number line the range in which each must lie.
   a) 7
   b) 40
   c) 300
   d) 2000

2. The following numbers are expressed correct to two significant figures. Representing each number by the letter x, express the range in which each must lie, using an inequality.
   a) 210
   b) 64
   c) 3.0
   d) 0.88

3. A school measures the dimensions of its rectangular playing field to the nearest metre. The length was recorded as 350 m and the width as 200 m. Express the range in which the length and width lie using inequalities.

4. A boy's mass was measured to the nearest 0.1 kg. If his mass was recorded as 58.9 kg, illustrate on a number line the range within which it must lie.

5. An electronic clock is accurate to $\frac{1}{1000}$ of a second. The duration of a flash from a camera is timed at 0.004 seconds. Express the upper and lower bounds of the duration of the flash using inequalities.

6. The following numbers are rounded to the degree of accuracy shown in brackets. Express the lower and upper bounds of these numbers as an inequality.
   a) $x = 4.83$ (2 d.p.)
   b) $y = 5.05$ (2 d.p.)
   c) $z = 10.0$ (1 d.p.)
   d) $p = 100.00$ (2 d.p.)

Student assessment 3

1. Calculate the upper and lower bounds of the following calculations given that each number is written to the nearest whole number.
   a) 20×50
   b) 100×63
   c) $\frac{500}{80}$
   d) $\frac{14 \times 73}{20}$
   e) $\frac{17-7}{4+6}$
   f) $\frac{8 \times (3+6)}{10^2}$

2. In the rectangle (below) both dimensions are given to 1 d.p. Calculate the upper and lower bounds for the area.

16.2 cm

7.4 cm

3. An equilateral triangle has sides of length 4 cm correct to the nearest whole number. Calculate the upper and lower bounds for the perimeter of the triangle.

4. The height to 1 d.p. of a room is given as 3.0 m. A door to the room has a height to 1 d.p. of 2.1 m. Write as an inequality the upper and lower bounds for the gap between the top of the door and the ceiling.

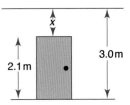

5. The mass of 85 oranges is given as 40 kg correct to 2 s.f. Calculate the lower and upper bounds for the average mass of one orange.

Student assessment 4

1. Five boys have a mass, given to the nearest 10 kg, of: 40 kg, 50 kg, 50 kg, 60 kg and 80 kg. Calculate the least possible total mass.

2. A water tank measures 30 cm by 50 cm by 20 cm. If each of these measurements is given to the nearest centimetre, calculate the largest possible volume of the tank.

3. The volume of a cube is given as 125 cm^3 to the nearest whole number.
   a) Express as an inequality the upper and lower bounds of the cube's volume.
   b) Express as an inequality the upper and lower bounds of the length of each of the cube's edges.

4. The radius of a circle is given as 4.00 cm to 2 d.p. Express as an inequality the upper and lower bounds for:
   a) the circumference of the circle,
   b) the area of the circle.

5. A cylindrical water tank has a volume of 6000 cm^3 correct to 1 s.f. A full cup of water from the tank has a volume of 300 cm^3 correct to 2 s.f. Calculate the maximum number of full cups of water that can be drawn from the tank.

6. A match measures 5 cm to the nearest centimetre. 100 matches end to end measure 5.43 m correct to 3 s.f.
   a) Calculate the upper and lower limits of the length of one match.
   b) How can the limits of the length of a match be found to 2 d.p.?

3 Calculations and order

● Ordering

The following symbols have a specific meaning in mathematics:

$=$ is equal to
$\neq$ is not equal to
$>$ is greater than
$\geq$ is greater than or equal to
$<$ is less than
$\leq$ is less than or equal to

$x \geq 3$ implies that x is greater than or equal to 3, i.e. x can be 3, 4, 4.2, 5, 5.6, etc.
$3 \leq x$ implies that 3 is less than or equal to x, i.e. x is still 3, 4, 4.2, 5, 5.6, etc.

Therefore:

$5 > x$ can be rewritten as $x < 5$, i.e. x can be 4, 3.2, 3, 2.8, 2, 1, etc.
$-7 \leq x$ can be rewritten as $x \geq -7$, i.e. x can be $-7, -6, -5$, etc.

These inequalities can also be represented on a number line:

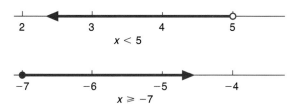

Note that ○——▶ implies that the number is not included in the solution whilst ●——▶ implies that the number is included in the solution.

Worked examples **a)** The maximum number of players from one football team allowed on the pitch at any one time is eleven. Represent this information:
i) as an inequality,
ii) on a number line.

i) Let the number of players be represented by the letter n. n must be less than or equal to 11. Therefore $n \leq 11$.

ii) ◀————————●
       8   9   10   11

b) The maximum number of players from one football team allowed on the pitch at any one time is eleven. The minimum allowed is seven players. Represent this information:

i) as an inequality,
ii) on a number line.

i) Let the number of players be represented by the letter n. n must be greater than or equal to 7, but less than or equal to 11.

Therefore $7 \leqslant n \leqslant 11$.

ii)

Exercise 3.1

1. Copy each of the following statements, and insert one of the symbols $=, >, <$ into the space to make the statement correct:
   a) $7 \times 2 \ldots 8 + 7$
   b) $6^2 \ldots 9 \times 4$
   c) $5 \times 10 \ldots 7^2$
   d) 80 cm ... 1 m
   e) 1000 litres ... 1 m³
   f) $48 \div 6 \ldots 54 \div 9$

2. Represent each of the following inequalities on a number line, where x is a real number:
   a) $x < 2$
   b) $x \geqslant 3$
   c) $x \leqslant -4$
   d) $x \geqslant -2$
   e) $2 < x < 5$
   f) $-3 < x < 0$
   g) $-2 \leqslant x < 2$
   h) $2 \geqslant x \geqslant -1$

3. Write down the inequalities which correspond to the following number lines:

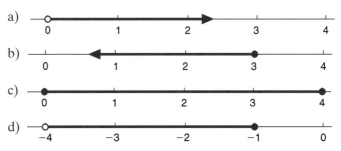

4. Write the following sentences using inequality signs.
   a) The maximum capacity of an athletics stadium is 20 000 people.
   b) In a class the tallest student is 180 cm and the shortest is 135 cm.
   c) Five times a number plus 3 is less than 20.
   d) The maximum temperature in May was 25 °C.
   e) A farmer has between 350 and 400 apples on each tree in his orchard.
   f) In December temperatures in Kenya were between 11 °C and 28 °C.

Number

Exercise 3.2

1. Write the following decimals in order of magnitude, starting with the smallest:

    6.0   0.6   0.66   0.606   0.06   6.6   6.606

2. Write the following fractions in order of magnitude, starting with the largest:

    $\frac{1}{2}$ $\frac{1}{3}$ $\frac{6}{13}$ $\frac{4}{5}$ $\frac{7}{18}$ $\frac{2}{19}$

3. Write the following lengths in order of magnitude, starting with the smallest:

    2 m   60 cm   800 mm   180 cm   0.75 m

4. Write the following masses in order of magnitude, starting with the largest:

    4 kg   3500 g $\frac{3}{4}$ kg   700 g   1 kg

5. Write the following volumes in order of magnitude, starting with the smallest:

    1 l   430 ml   800 cm^3 120 cl   150 cm^3

● The order of operations

When carrying out calculations, care must be taken to ensure that they are carried out in the correct order.

Worked examples

a) Use a scientific calculator to work out the answer to the following:

$$2 + 3 \times 4 =$$

| 2 | + | 3 | × | 4 | = | 14

b) Use a scientific calculator to work out the answer to the following:

$$(2 + 3) \times 4 =$$

| (| 2 | + | 3 |) | × | 4 | = | 20

The reason why different answers are obtained is because, by convention, the operations have different priorities. These are as follows:

(1) brackets
(2) multiplication/division
(3) addition/subtraction

Therefore in **Worked example a)** 3×4 is evaluated first, and then the 2 is added, whilst in **Worked example b)** $(2 + 3)$ is evaluated first, followed by multiplication by 4.

Calculations and order

Exercise 3.3 In the following questions, evaluate the answers:
  i) in your head,
  ii) using a scientific calculator.

1. a) $8 \times 3 + 2$ b) $4 \div 2 + 8$
   c) $12 \times 4 - 6$ d) $4 + 6 \times 2$
   e) $10 - 6 \div 3$ f) $6 - 3 \times 4$

2. a) $7 \times 2 + 3 \times 2$ b) $12 \div 3 + 6 \times 5$
   c) $9 + 3 \times 8 - 1$ d) $36 - 9 \div 3 - 2$
   e) $14 \times 2 - 16 \div 2$ f) $4 + 3 \times 7 - 6 \div 3$

3. a) $(4 + 5) \times 3$ b) $8 \times (12 - 4)$
   c) $3 \times (8 + 3) - 3$ d) $(4 + 11) \div (7 - 2)$
   e) $4 \times 3 \times (7 + 5)$ f) $24 \div 3 \div (10 - 5)$

Exercise 3.4 In each of the following questions:
  i) Copy the calculation and put in any brackets which are needed to make it correct.
  ii) Check your answer using a scientific calculator.

1. a) $6 \times 2 + 1 = 18$ b) $1 + 3 \times 5 = 16$
   c) $8 + 6 \div 2 = 7$ d) $9 + 2 \times 4 = 44$
   e) $9 \div 3 \times 4 + 1 = 13$ f) $3 + 2 \times 4 - 1 = 15$

2. a) $12 \div 4 - 2 + 6 = 7$ b) $12 \div 4 - 2 + 6 = 12$
   c) $12 \div 4 - 2 + 6 = -5$ d) $12 \div 4 - 2 + 6 = 1.5$
   e) $4 + 5 \times 6 - 1 = 33$ f) $4 + 5 \times 6 - 1 = 29$
   g) $4 + 5 \times 6 - 1 = 53$ h) $4 + 5 \times 6 - 1 = 45$

It is important to use brackets when dealing with more complex calculations.

Worked examples a) Evaluate the following using a scientific calculator:

$$\frac{12+9}{10-3} =$$

 3

b) Evaluate the following using a scientific calculator:

$$\frac{20+12}{4^2} =$$

 2

Number

c) Evaluate the following using a scientific calculator:

$$\frac{90+38}{4^3} =$$

(9 0 + 3 8) ÷ 4 x^y 3 = 2

Note: different types of calculator have different 'to the power of' buttons.

Exercise 3.5 Using a scientific calculator, evaluate the following:

1. a) $\dfrac{9+3}{6}$ b) $\dfrac{30-6}{5+3}$

   c) $\dfrac{40+9}{12-5}$ d) $\dfrac{15\times 2}{7+8}+2$

   e) $\dfrac{100+21}{11}+4\times 3$ f) $\dfrac{7+2\times 4}{7-2}-3$

2. a) $\dfrac{4^2-6}{2+8}$ b) $\dfrac{3^2+4^2}{5}$

   c) $\dfrac{6^3-4^2}{4\times 25}$ d) $\dfrac{3^3\times 4^4}{12^2}+2$

   e) $\dfrac{3+3^3}{5}+\dfrac{4^2-2^3}{8}$ f) $\dfrac{(6+3)\times 4}{2^3}-2\times 3$

Student assessment 1

1. Write the information on the following number lines as inequalities:

   a)

   b)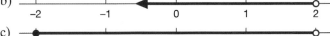

   c)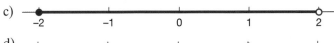

   d)

2. Illustrate each of the following inequalities on a number line:
   a) $x \geqslant 3$
   b) $x < 4$
   c) $0 < x < 4$
   d) $-3 \leqslant x < 1$

3. Write the following fractions in order of magnitude, starting with the smallest:

   $$\frac{4}{7} \quad \frac{3}{14} \quad \frac{9}{10} \quad \frac{1}{2} \quad \frac{2}{5}$$

Student assessment 2

1. Evaluate the following:
   a) $6 \times 8 - 4$
   b) $3 + 5 \times 2$
   c) $3 \times 3 + 4 \times 4$
   d) $3 + 3 \times 4 + 4$
   e) $(5 + 2) \times 7$
   f) $18 \div 2 \div (5 - 2)$

2. Copy the following, if necessary putting in brackets to make the statement correct:
   a) $7 - 4 \times 2 = 6$
   b) $12 + 3 \times 3 + 4 = 33$
   c) $5 + 5 \times 6 - 4 = 20$
   d) $5 + 5 \times 6 - 4 = 56$

3. Evaluate the following using a calculator:
   a) $\dfrac{2^4 - 3^2}{2}$
   b) $\dfrac{(8-3) \times 3}{5} + 7$

Student assessment 3

1. Evaluate the following:
   a) $3 \times 9 - 7$
   b) $12 + 6 \div 2$
   c) $3 + 4 \div 2 \times 4$
   d) $6 + 3 \times 4 - 5$
   e) $(5 + 2) \div 7$
   f) $14 \times 2 \div (9 - 2)$

2. Copy the following, if necessary putting in brackets to make the statement correct:
   a) $7 - 5 \times 3 = 6$
   b) $16 + 4 \times 2 + 4 = 40$
   c) $4 + 5 \times 6 - 1 = 45$
   d) $1 + 5 \times 6 - 6 = 30$

3. Using a calculator, evaluate the following:
   a) $\dfrac{3^3 - 4^2}{2}$
   b) $\dfrac{(15-3) \div 3}{2} + 7$

Integers, fractions, decimals and percentages

● **Fractions**

A single unit can be broken into equal parts called fractions, e.g. $\frac{1}{2}, \frac{1}{3}, \frac{1}{6}$.

If, for example, the unit is broken into ten equal parts and three parts are then taken, the fraction is written as $\frac{3}{10}$. That is, three parts out of ten parts.

In the fraction $\frac{3}{10}$:
The ten is called the **denominator**.
The three is called the **numerator**.
A **proper fraction** has its numerator less than its denominator, e.g. $\frac{3}{4}$.
An **improper fraction** has its numerator more than its denominator, e.g. $\frac{9}{2}$.
A **mixed number** is made up of a whole number and a proper fraction, e.g. $4\frac{1}{5}$.

Worked examples a) Find $\frac{1}{5}$ of 35.

This means 'divide 35 into 5 equal parts'.

$\frac{1}{5}$ of 35 is $35 \div 5 = 7$.

b) Find $\frac{3}{5}$ of 35.

Since $\frac{1}{5}$ of 35 is 7, $\frac{3}{5}$ of 35 is 7×3.

That is, 21.

Exercise 4.1 1. Evaluate the following:

a) $\frac{3}{4}$ of 12   b) $\frac{4}{5}$ of 20   c) $\frac{4}{9}$ of 45   d) $\frac{5}{8}$ of 64

e) $\frac{3}{11}$ of 66   f) $\frac{9}{10}$ of 80   g) $\frac{5}{7}$ of 42   h) $\frac{8}{9}$ of 54

i) $\frac{7}{8}$ of 240   j) $\frac{4}{5}$ of 65

Changing a mixed number to an improper fraction

Worked examples a) Change $3\frac{5}{8}$ into an improper fraction.

$$3\frac{5}{8} = \frac{24}{8} + \frac{5}{8}$$
$$= \frac{24+5}{8}$$
$$= \frac{29}{8}$$

b) Change $\frac{27}{4}$ into a mixed number.

$$\frac{27}{4} = \frac{24+3}{4}$$
$$= \frac{24}{4} + \frac{3}{4}$$
$$= 6\frac{3}{4}$$

4 Integers, fractions, decimals and percentages

Exercise 4.2
1. Change the following mixed numbers into improper fractions:
   a) $4\frac{2}{3}$ b) $3\frac{3}{5}$ c) $5\frac{7}{8}$ d) $2\frac{5}{6}$
   e) $8\frac{1}{2}$ f) $9\frac{5}{7}$ g) $6\frac{4}{9}$ h) $4\frac{1}{4}$
   i) $5\frac{4}{11}$ j) $7\frac{6}{7}$ k) $4\frac{3}{10}$ l) $11\frac{3}{13}$

2. Change the following improper fractions into mixed numbers.
   a) $\frac{29}{4}$ b) $\frac{33}{5}$ c) $\frac{41}{6}$ d) $\frac{53}{8}$
   e) $\frac{49}{9}$ f) $\frac{17}{12}$ g) $\frac{66}{7}$ h) $\frac{33}{10}$
   i) $\frac{19}{2}$ j) $\frac{73}{12}$

● **Decimals**

H	T	U.	$\frac{1}{10}$	$\frac{1}{100}$	$\frac{1}{1000}$
		3.	2	7	
		0.	0	3	8

3.27 is 3 units, 2 tenths and 7 hundredths

i.e. $3.27 = 3 + \frac{2}{10} + \frac{7}{100}$

0.038 is 3 hundredths and 8 thousandths

i.e. $0.038 = \frac{3}{100} + \frac{8}{1000}$

Note that 2 tenths and 7 hundredths is equivalent to 27 hundredths

i.e. $\frac{2}{10} + \frac{7}{100} = \frac{27}{100}$

and that 3 hundredths and 8 thousandths is equivalent to 38 thousandths

i.e. $\frac{3}{100} + \frac{8}{1000} = \frac{38}{1000}$

Exercise 4.3
1. Write the following fractions as decimals:
   a) $4\frac{5}{10}$ b) $6\frac{3}{10}$ c) $17\frac{8}{10}$ d) $3\frac{7}{100}$
   e) $9\frac{27}{100}$ f) $11\frac{36}{100}$ g) $4\frac{6}{1000}$ h) $5\frac{27}{1000}$
   i) $4\frac{356}{1000}$ j) $9\frac{204}{1000}$

Number

2. Evaluate the following without using a calculator:

   a) $2.7 + 0.35 + 16.09$
   b) $1.44 + 0.072 + 82.3$
   c) $23.8 - 17.2$
   d) $16.9 - 5.74$
   e) $121.3 - 85.49$
   f) $6.03 + 0.5 - 1.21$
   g) $72.5 - 9.08 + 3.72$
   h) $100 - 32.74 - 61.2$
   i) $16.0 - 9.24 - 5.36$
   j) $1.1 - 0.92 - 0.005$

● Percentages

A fraction whose denominator is 100 can be expressed as a percentage.

$\frac{29}{100}$ can be written as 29%

$\frac{45}{100}$ can be written as 45%

By using equivalent fractions to change the denominator to 100, other fractions can be written as percentages.

Worked example Change $\frac{3}{5}$ to a percentage.

$\frac{3}{5} = \frac{3}{5} \times \frac{20}{20} = \frac{60}{100}$

$\frac{60}{100}$ can be written as 60%

Exercise 4.4

1. Express each of the following as a fraction with denominator 100, then write them as percentages:

   a) $\frac{29}{50}$ b) $\frac{17}{25}$ c) $\frac{11}{20}$ d) $\frac{3}{10}$

   e) $\frac{23}{25}$ f) $\frac{19}{50}$ g) $\frac{3}{4}$ h) $\frac{2}{5}$

2. Copy and complete the table of equivalents below.

Fraction	Decimal	Percentage
$\frac{1}{10}$		
	0.2	
		30%
$\frac{4}{10}$		
	0.5	
		60%
	0.7	
$\frac{4}{5}$		
	0.9	
$\frac{1}{4}$		
		75%

● The four rules

Addition, subtraction, multiplication and division are mathematical operations.

Long multiplication

When carrying out long multiplication, it is important to remember place value.

Worked example 184×37

$$\begin{array}{r} 184 \\ \times 37 \\ \hline 1288 \\ 5520 \\ \hline 6808 \end{array} \begin{array}{l} (184 \times 7) \\ (184 \times 30) \\ (184 \times 37) \end{array}$$

Short division

Worked example $453 \div 6$

$$6\overline{)4\,5\,^33} \quad 75\,\text{r}3$$

It is usual, however, to give the final answer in decimal form rather than with a remainder. The division should therefore be continued:

$453 \div 6$

$$6\overline{)4\,5\,^33\,.\,^30} \quad 75.5$$

Long division

Worked example Calculate $7184 \div 23$ to one decimal place (1 d.p.).

$$\begin{array}{r} 312.4 \\ 23\overline{)7184.00} \\ \underline{69} \\ 28 \\ \underline{23} \\ 54 \\ \underline{46} \\ 80 \\ \underline{69} \\ 110 \\ \underline{92} \\ 18 \end{array}$$

Therefore $7184 \div 23 = 312.3$ to 1 d.p.

Mixed operations

When a calculation involves a mixture of operations, the order of the operations is important. Multiplications and divisions are done first, whilst additions and subtractions are done afterwards. To override this, brackets need to be used.

Number

Worked examples

a) $3 + 7 \times 2 - 4$
$= 3 + 14 - 4$
$= 13$

b) $(3 + 7) \times 2 - 4$
$= 10 \times 2 - 4$
$= 20 - 4$
$= 16$

c) $3 + 7 \times (2 - 4)$
$= 3 + 7 \times (-2)$
$= 3 - 14$
$= -11$

d) $(3 + 7) \times (2 - 4)$
$= 10 \times (-2)$
$= -20$

Exercise 4.5

1. Evaluate the answer to each of the following:
   a) $3 + 5 \times 2 - 4$
   b) $12 \div 8 + 6 \div 4$

2. Copy these equations and put brackets in the correct places to make them correct:
   a) $6 \times 4 + 6 \div 3 = 20$
   b) $9 - 3 \times 7 + 2 = 54$

3. Without using a calculator, work out the solutions to the following multiplications:
   a) 785×38
   b) 164×253

4. Work out the remainders in the following divisions:
   a) $72 \div 7$
   b) $430 \div 9$

5. a) A length of rail track is 9 m long. How many complete lengths will be needed to lay 1 km of track?
   b) How many 35 cent stamps can be bought for 10 dollars?

6. Work out the following long divisions to 1 d.p.
   a) $7892 \div 7$
   b) $7892 \div 15$

● Calculations with fractions

Equivalent fractions

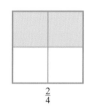
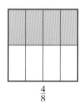

$\frac{1}{2}$ $\frac{2}{4}$ $\frac{4}{8}$

It should be apparent that $\frac{1}{2}, \frac{2}{4}$ and $\frac{4}{8}$ are equivalent fractions.

Similarly, $\frac{1}{3}, \frac{2}{6}, \frac{3}{9}$ and $\frac{4}{12}$ are equivalent, as are $\frac{1}{5}, \frac{10}{50}$ and $\frac{20}{100}$. Equivalent fractions are mathematically the same as each other. In the diagrams above $\frac{1}{2}$ is mathematically the same as $\frac{4}{8}$. However $\frac{1}{2}$ is a simplified form of $\frac{4}{8}$.

When carrying out calculations involving fractions it is usual to give your answer in its **simplest form**. Another way of saying 'simplest form' is '**lowest terms**'.

Integers, fractions, decimals and percentages

Worked examples

a) Write $\frac{4}{22}$ in its simplest form.

Divide both the numerator and the denominator by their highest common factor.

The highest common factor of both 4 and 22 is 2.

Dividing both 4 and 22 by 2 gives $\frac{2}{11}$.

Therefore $\frac{2}{11}$ is $\frac{4}{22}$ written in its simplest form.

b) Write $\frac{12}{40}$ in its lowest terms.

Divide both the numerator and the denominator by their highest common factor.

The highest common factor of both 12 and 40 is 4.

Dividing both 12 and 40 by 4 gives $\frac{3}{10}$.

Therefore $\frac{3}{10}$ is $\frac{12}{40}$ written in its lowest terms.

Exercise 4.6

1. Express the following fractions in their lowest terms.

   e.g. $\frac{12}{16} = \frac{3}{4}$

   a) $\frac{5}{10}$ b) $\frac{7}{21}$ c) $\frac{8}{12}$

   d) $\frac{16}{36}$ e) $\frac{75}{100}$ f) $\frac{81}{90}$

Addition and subtraction of fractions

For fractions to be either added or subtracted, the denominators need to be the same.

Worked examples

a) $\frac{3}{11} + \frac{5}{11} = \frac{8}{11}$

b) $\frac{7}{8} + \frac{5}{8} = \frac{12}{8} = 1\frac{1}{2}$

c) $\frac{1}{2} + \frac{1}{3}$

$= \frac{3}{6} + \frac{2}{6} = \frac{5}{6}$

d) $\frac{4}{5} - \frac{1}{3}$

$= \frac{12}{15} - \frac{5}{15} = \frac{7}{15}$

When dealing with calculations involving mixed numbers, it is sometimes easier to change them to improper fractions first.

Worked examples

a) $5\frac{3}{4} - 2\frac{5}{8}$

$= \frac{23}{4} - \frac{21}{8}$

$= \frac{46}{8} - \frac{21}{8}$

$= \frac{25}{8} = 3\frac{1}{8}$

b) $1\frac{4}{7} + 3\frac{3}{4}$

$= \frac{11}{7} + \frac{15}{4}$

$= \frac{44}{28} + \frac{105}{28}$

$= \frac{149}{28} = 5\frac{9}{28}$

Exercise 4.7

Evaluate each of the following and write the answer as a fraction in its simplest form:

1. a) $\frac{3}{5} + \frac{4}{5}$ b) $\frac{3}{11} + \frac{7}{11}$

   c) $\frac{2}{3} + \frac{1}{4}$ d) $\frac{3}{5} + \frac{4}{9}$

   e) $\frac{8}{13} + \frac{2}{5}$ f) $\frac{1}{2} + \frac{2}{3} + \frac{3}{4}$

Number

2. a) $\frac{1}{8} + \frac{3}{8} + \frac{5}{8}$ b) $\frac{3}{7} + \frac{5}{7} + \frac{4}{7}$
   c) $\frac{1}{3} + \frac{1}{2} + \frac{1}{4}$ d) $\frac{1}{5} + \frac{1}{3} + \frac{1}{4}$
   e) $\frac{3}{8} + \frac{3}{5} + \frac{3}{4}$ f) $\frac{3}{13} + \frac{1}{4} + \frac{1}{2}$

3. a) $\frac{3}{7} - \frac{2}{7}$ b) $\frac{4}{5} - \frac{7}{10}$
   c) $\frac{8}{9} - \frac{1}{3}$ d) $\frac{7}{12} - \frac{1}{2}$
   e) $\frac{5}{8} - \frac{2}{5}$ f) $\frac{3}{4} - \frac{2}{5} + \frac{7}{10}$

4. a) $\frac{3}{4} + \frac{1}{5} - \frac{2}{3}$ b) $\frac{3}{8} + \frac{7}{11} - \frac{1}{2}$
   c) $\frac{4}{5} - \frac{3}{10} + \frac{7}{20}$ d) $\frac{9}{13} + \frac{1}{3} - \frac{4}{5}$
   e) $\frac{9}{10} - \frac{1}{5} - \frac{1}{4}$ f) $\frac{8}{9} - \frac{1}{3} - \frac{1}{2}$

5. a) $2\frac{1}{2} + 3\frac{1}{4}$ b) $3\frac{3}{5} + 1\frac{7}{10}$
   c) $6\frac{1}{2} - 3\frac{2}{5}$ d) $8\frac{5}{8} - 2\frac{1}{3}$
   e) $5\frac{7}{8} - 4\frac{3}{4}$ f) $3\frac{1}{4} - 2\frac{5}{9}$

6. a) $2\frac{1}{2} + 1\frac{1}{4} + 1\frac{3}{8}$ b) $2\frac{4}{5} + 3\frac{1}{8} + 1\frac{3}{10}$
   c) $4\frac{1}{2} - 1\frac{1}{4} - 3\frac{5}{8}$ d) $6\frac{1}{2} - 2\frac{3}{4} - 3\frac{2}{5}$
   e) $2\frac{4}{7} - 3\frac{1}{4} - 1\frac{3}{5}$ f) $4\frac{7}{20} - 5\frac{1}{2} + 2\frac{2}{5}$

Multiplication and division of fractions

Worked examples   a) $\frac{3}{4} \times \frac{2}{3}$ b) $3\frac{1}{2} \times 4\frac{4}{7}$
$= \frac{6}{12}$ $= \frac{7}{2} \times \frac{32}{7}$
$= \frac{1}{2}$ $= \frac{224}{14}$
                                                                    $= 16$

The **reciprocal** of a number is obtained when 1 is divided by that number. The reciprocal of 5 is $\frac{1}{5}$, the reciprocal of $\frac{2}{5}$ is $\frac{5}{2}$, etc.

Dividing fractions is the same as multiplying by the reciprocal.

Worked examples   a) $\frac{3}{8} \div \frac{3}{4}$ b) $5\frac{1}{2} \div 3\frac{2}{3}$
$= \frac{3}{8} \times \frac{4}{3}$ $= \frac{11}{2} \div \frac{11}{3}$
$= \frac{12}{24}$ $= \frac{11}{2} \times \frac{3}{11}$
$= \frac{1}{2}$ $= \frac{3}{2}$

Integers, fractions, decimals and percentages

Exercise 4.8 1. Write the reciprocal of each of the following:
   a) $\frac{1}{8}$ b) $\frac{7}{12}$ c) $\frac{3}{5}$
   d) $1\frac{1}{2}$ e) $3\frac{3}{4}$ f) 6

2. Evaluate the following:
   a) $\frac{3}{8} \times \frac{4}{9}$ b) $\frac{2}{3} \times \frac{9}{10}$ c) $\frac{5}{7} \times \frac{4}{15}$
   d) $\frac{3}{4}$ of $\frac{8}{9}$ e) $\frac{5}{6}$ of $\frac{3}{10}$ f) $\frac{7}{8}$ of $\frac{2}{5}$

3. Evaluate the following:
   a) $\frac{5}{8} \div \frac{3}{4}$ b) $\frac{5}{6} \div \frac{1}{3}$ c) $\frac{4}{5} \div \frac{7}{10}$
   d) $1\frac{2}{3} \div \frac{2}{5}$ e) $\frac{3}{7} \div 2\frac{1}{7}$ f) $1\frac{1}{4} \div 1\frac{7}{8}$

4. Evaluate the following:
   a) $\frac{3}{4} \times \frac{4}{5}$ b) $\frac{7}{8} \times \frac{2}{3}$
   c) $\frac{3}{4} \times \frac{4}{7} \times \frac{3}{10}$ d) $(\frac{4}{5} \div \frac{2}{3}) \times \frac{7}{10}$
   e) $\frac{1}{2}$ of $\frac{3}{4}$ f) $4\frac{1}{2} \div 3\frac{1}{9}$

5. Evaluate the following:
   a) $(\frac{3}{8} \times \frac{4}{5}) + (\frac{1}{2}$ of $\frac{3}{5})$ b) $(1\frac{1}{2} \times 3\frac{3}{4}) - (2\frac{3}{5} \div 1\frac{1}{2})$
   c) $(\frac{3}{5}$ of $\frac{4}{9}) + (\frac{4}{9}$ of $\frac{3}{5})$ d) $(1\frac{1}{3} \times 2\frac{5}{8})^2$

● **Changing a fraction to a decimal**
To change a fraction to a decimal, divide the numerator by the denominator.

Worked examples a) Change $\frac{5}{8}$ to a decimal.

$$\begin{array}{r} 0.6\ 2\ 5 \\ 8\ \overline{)5.0\ ^20\ ^40} \end{array}$$

b) Change $2\frac{3}{5}$ to a decimal.
   This can be represented as $2 + \frac{3}{5}$.

$$\begin{array}{r} 0.6 \\ 5\ \overline{)3.0} \end{array}$$

Therefore $2\frac{3}{5} = 2.6$

Exercise 4.9 1. Change the following fractions to decimals:
   a) $\frac{3}{4}$ b) $\frac{4}{5}$ c) $\frac{9}{20}$
   d) $\frac{17}{50}$ e) $\frac{1}{3}$ f) $\frac{3}{8}$
   g) $\frac{7}{16}$ h) $\frac{2}{9}$ i) $\frac{7}{11}$

Number

2. Change the following mixed numbers to decimals:
   a) $2\frac{3}{4}$
   b) $3\frac{3}{5}$
   c) $4\frac{7}{20}$
   d) $6\frac{11}{50}$
   e) $5\frac{2}{3}$
   f) $6\frac{7}{8}$
   g) $5\frac{9}{16}$
   h) $4\frac{2}{9}$
   i) $5\frac{3}{7}$

● Changing a decimal to a fraction

Changing a decimal to a fraction is done by knowing the 'value' of each of the digits in any decimal.

Worked examples a) Change 0.45 from a decimal to a fraction.

units	.	tenths	hundredths
0	.	4	5

0.45 is therefore equivalent to 4 tenths and 5 hundredths, which in turn is the same as 45 hundredths.
Therefore $0.45 = \frac{45}{100} = \frac{9}{20}$

b) Change 2.325 from a decimal to a fraction.

units	.	tenths	hundredths	thousandths
2	.	3	2	5

Therefore $2.325 = 2\frac{325}{1000} = 2\frac{13}{40}$

Exercise 4.10

1. Change the following decimals to fractions:
   a) 0.5
   b) 0.7
   c) 0.6
   d) 0.75
   e) 0.825
   f) 0.05
   g) 0.050
   h) 0.402
   i) 0.0002

2. Change the following decimals to mixed numbers:
   a) 2.4
   b) 6.5
   c) 8.2
   d) 3.75
   e) 10.55
   f) 9.204
   g) 15.455
   h) 30.001
   i) 1.0205

● Recurring decimals

In Chapter 1 the definition of a rational number was given as any number that can be written as a fraction. These include integers, terminating decimals and recurring decimals. The examples given were:

$$0.2 = \frac{1}{5} \quad 0.3 = \frac{3}{10} \quad 7 = \frac{7}{1} \quad 1.53 = \frac{153}{100} \quad \text{and} \quad 0.\dot{2} = \frac{2}{9}$$

The first four examples are the more straightforward to understand regarding how the number can be expressed as a fraction. The fifth example shows a recurring decimal also written as a fraction. Any recurring decimal can be written as a fraction as any recurring decimal is also a rational number.

Changing a recurring decimal to a fraction

A recurring decimal is one in which the numbers after the decimal point repeat themselves infinitely. Which numbers are repeated is indicated by a dot above them.

$0.\dot{2}$ implies 0.222 222 222 222 222 …

$0.\dot{4}\dot{5}$ implies 0.454 545 454 545 …

$0.\dot{6}0\dot{2}\dot{4}$ implies 0.602 460 246 024 …

Note, the last example is usually written as $0.\dot{6}02\dot{4}$ where a dot only appears above the first and last numbers to be repeated.

Entering $\frac{4}{9}$ into a calculator will produce 0.444 444 444…

Therefore $0.\dot{4} = \frac{4}{9}$.

The example below will prove this.

Worked examples   **a)** Convert $0.\dot{4}$ to a fraction.

Let $x = 0.\dot{4}$ i.e. $x = 0.444\ 444\ 444\ 444\ …$
$10x = 4.\dot{4}$ i.e. $10x = 4.444\ 444\ 444\ 444\ …$

Subtracting x from $10x$ gives:

$$
\begin{aligned}
10x &= 4.444\ 444\ 444\ 444\ …\\
- \quad x &= 0.444\ 444\ 444\ 444\ …\\
\hline
9x &= 4
\end{aligned}
$$

Rearranging gives $x = \frac{4}{9}$

But $x = 0.\dot{4}$

Therefore $0.\dot{4} = \frac{4}{9}$

b) Convert $0.\dot{6}\dot{8}$ to a fraction.

Let $x = 0.\dot{6}\dot{8}$ i.e. $x = 0.686\ 868\ 686\ 868\ 686…$
$100x = 68.\dot{6}\dot{8}$ i.e. $100x = 68.686\ 868\ 686\ 868\ 686…$

Subtracting x from $100x$ gives:

$$
\begin{aligned}
100x &= 68.686\ 868\ 686\ 868\ 686…\\
- \quad x &= 0.686\ 868\ 686\ 868\ 686…\\
\hline
99x &= 68
\end{aligned}
$$

Rearranging gives $x = \frac{68}{99}$

But $x = 0.\dot{6}\dot{8}$

Therefore $0.\dot{6}\dot{8} = \frac{68}{99}$

Number

c) Convert $0.0\dot{3}\dot{1}$ to a fraction.

Let $x = 0.0\dot{3}\dot{1}$ i.e. $x = 0.031\,313\,131\,313\,131...$
$100x = 3.\dot{1}\dot{3}$ i.e. $100x = 3.131\,313\,131\,313\,131...$

Subtracting x from $100x$ gives:

$$
\begin{array}{rcl}
100x & = & 3.131\,313\,131\,313\,131...\\
-\quad x & = & 0.031\,313\,131\,313\,131...\\
\hline
99x & = & 3.1
\end{array}
$$

Multiplying both sides of the equation by 10 eliminates the decimal to give:

$990x = 31$

Rearranging gives $x = \frac{31}{990}$

But $x = 0.0\dot{3}\dot{1}$

Therefore $0.0\dot{3}\dot{1} = \frac{31}{990}$

The method is therefore to let the recurring decimal equal x and then to multiply this by a multiple of 10 so that when one is subtracted from the other either an integer (whole number) or terminating decimal (a decimal that has an end point) is left.

d) Convert $2.0\dot{4}0\dot{6}$ to a fraction.

Let $x = 2.0\dot{4}0\dot{6}$ i.e. $x = 2.040\,640\,640\,640...$
$1000x = 2040.\dot{6}4\dot{0}$ i.e. $1000x = 2040.640\,640\,640\,640...$

Subtracting x from $1000x$ gives:

$$
\begin{array}{rcl}
1000x & = & 2040.640\,640\,640\,640...\\
-\quad x & = & 2.040\,640\,640\,640...\\
\hline
999x & = & 2038.6
\end{array}
$$

Multiplying both sides of the equation by 10 eliminates the decimal to give:

$9990x = 20\,386$

Rearranging gives $x = \frac{20\,386}{9990} = 2\frac{406}{9990}$ which simplifies further to $2\frac{203}{4995}$.

But $x = 2.0\dot{4}0\dot{6}$

Therefore $2.0\dot{4}0\dot{6} = 2\frac{203}{4995}$

Exercise 4.11

1. Convert each of the following recurring decimals to fractions in their simplest form:

   a) $0.\dot{3}$ b) $0.\dot{7}$
   c) $0.\dot{4}\dot{2}$ d) $0.\dot{6}\dot{5}$

2. Convert each of the following recurring decimals to fractions in their simplest form:
   a) $0.0\dot{5}$
   b) $0.0\dot{6}\dot{2}$
   c) $1.0\dot{2}$
   d) $4.00\dot{3}\dot{8}$

3. Without using a calculator work out the sum $0.1\dot{5} + 0.0\dot{4}$ by converting each decimal to a fraction first. Give your answer as a fraction in its simplest form.

4. Without using a calculator evaluate $0.\dot{2}\dot{7} - 0.1\dot{0}\dot{6}$ by converting each decimal to a fraction first. Give your answer as a fraction in its simplest form.

Student assessment 1

1. Evaluate the following:
   a) $\frac{1}{5}$ of 60
   b) $\frac{3}{5}$ of 55
   c) $\frac{2}{7}$ of 21
   d) $\frac{3}{4}$ of 120

2. Write the following as percentages:
   a) $\frac{3}{10}$
   b) $\frac{29}{100}$
   c) $\frac{1}{2}$
   d) $\frac{7}{10}$
   e) $\frac{4}{5}$
   f) $2\frac{19}{100}$
   g) $\frac{6}{100}$
   h) $\frac{3}{4}$
   i) 0.31
   j) 0.07
   k) 3.4
   l) 2

3. Evaluate the following:
   a) $5 + 8 \times 3 - 6$
   b) $15 + 45 \div 3 - 12$

4. Work out 851×27.

5. Work out $6843 \div 19$ giving your answer to 1 d.p.

6. Evaluate the following:
   a) $3\frac{3}{4} - 1\frac{11}{16}$
   b) $4\frac{4}{5} \div \frac{8}{15}$

7. Change the following fractions to decimals:
   a) $\frac{2}{5}$
   b) $1\frac{3}{4}$
   c) $\frac{9}{11}$
   d) $1\frac{2}{3}$

8. Change the following decimals to fractions. Give each fraction in its simplest form.
   a) 4.2
   b) 0.06
   c) 1.85
   d) 2.005

9. Convert the following decimals to fractions, giving your answer in its simplest form:
   a) $0.\dot{3}\dot{7}$
   b) $0.0\dot{8}\dot{0}$
   c) $1.2\dot{1}$

10. Work out $0.62\dot{5} + 0.09\dot{6}$ by first converting each decimal to a fraction. Give your answer in its simplest form.

Number

Student assessment 2

1. Evaluate the following:
   a) $\frac{1}{3}$ of 63   b) $\frac{3}{8}$ of 72   c) $\frac{2}{5}$ of 55   d) $\frac{3}{13}$ of 169

2. Write the following as percentages:
   a) $\frac{3}{5}$ b) $\frac{49}{100}$ c) $\frac{1}{4}$ d) $\frac{9}{10}$
   e) $1\frac{1}{2}$ f) $3\frac{27}{100}$ g) $\frac{5}{100}$ h) $\frac{7}{20}$
   i) 0.77   j) 0.03   k) 2.9   l) 4

3. Evaluate the following:
   a) $6 \times 4 - 3 \times 8$ b) $15 \div 3 + 2 \times 7$

4. Work out 368×49.

5. Work out $7835 \div 23$ giving your answer to 1 d.p.

6. Evaluate the following:
   a) $2\frac{1}{2} - \frac{4}{5}$ b) $3\frac{1}{2} \times \frac{4}{7}$

7. Change the following fractions to decimals:
   a) $\frac{7}{8}$ b) $1\frac{2}{5}$ c) $\frac{8}{9}$ d) $3\frac{2}{7}$

8. Change the following decimals to fractions. Give each fraction in its simplest form.
   a) 6.5   b) 0.04   c) 3.65   d) 3.008

9. Convert the following decimals to fractions, giving your answer in its simplest form:
   a) $0.0\dot{7}$ b) $0.000\dot{9}$ c) $3.0\dot{2}\dot{0}$

10. Work out $1.02\dot{5} - 0.80\dot{5}$ by first converting each decimal to a fraction. Give your answer in its simplest form.

5 Further percentages

You should already be familiar with the percentage equivalents of simple fractions and decimals as outlined in the table below.

Fraction	Decimal	Percentage
$\frac{1}{2}$	0.5	50%
$\frac{1}{4}$	0.25	25%
$\frac{3}{4}$	0.75	75%
$\frac{1}{8}$	0.125	12.5%
$\frac{3}{8}$	0.375	37.5%
$\frac{5}{8}$	0.625	62.5%
$\frac{7}{8}$	0.875	87.5%
$\frac{1}{10}$	0.1	10%
$\frac{2}{10}$ or $\frac{1}{5}$	0.2	20%
$\frac{3}{10}$	0.3	30%
$\frac{4}{10}$ or $\frac{2}{5}$	0.4	40%
$\frac{6}{10}$ or $\frac{3}{5}$	0.6	60%
$\frac{7}{10}$	0.7	70%
$\frac{8}{10}$ or $\frac{4}{5}$	0.8	80%
$\frac{9}{10}$	0.9	90%

● **Simple percentages**

Worked examples a) Of 100 sheep in a field, 88 are ewes.

   i) What percentage of the sheep are ewes?
   88 out of 100 are ewes
   = 88%

   ii) What percentage are not ewes?
   12 out of 100
   = 12%

   b) Convert the following percentages into fractions and decimals:
   i) 27%   ii) 5%

   $$\frac{27}{100} = 0.27 \qquad \frac{5}{100} = \frac{1}{20} = 0.05$$

c) Convert $\frac{3}{16}$ to a percentage:
This example is more complicated as 16 is not a factor of 100.
Convert $\frac{3}{16}$ to a decimal first.

$$3 \div 16 = 0.1875$$

Convert the decimal to a percentage by multiplying by 100.

$$0.1875 \times 100 = 18.75$$

Therefore $\frac{3}{16} = 18.75\%$.

Exercise 5.1

1. There are 200 birds in a flock. 120 of them are female. What percentage of the flock are:
   a) female?   b) male?

2. Write these fractions as percentages:
   a) $\frac{7}{8}$ b) $\frac{11}{15}$ c) $\frac{7}{24}$ d) $\frac{1}{7}$

3. Convert the following percentages to decimals:
   a) 39%   b) 47%   c) 83%
   d) 7%   e) 2%   f) 20%

4. Convert the following decimals to percentages:
   a) 0.31   b) 0.67   c) 0.09
   d) 0.05   e) 0.2   f) 0.75

● **Calculating a percentage of a quantity**

Worked examples **a)** Find 25% of 300 m.

25% can be written as 0.25.
0.25 × 300 m = 75 m.

b) Find 35% of 280 m.

35% can be written as 0.35.
0.35 × 280 m = 98 m.

Exercise 5.2

1. Write the percentage equivalent of the following fractions:
   a) $\frac{1}{4}$ b) $\frac{2}{3}$ c) $\frac{5}{8}$
   d) $1\frac{4}{5}$ e) $4\frac{9}{10}$ f) $3\frac{7}{8}$

2. Write the decimal equivalent of the following:
   a) $\frac{3}{4}$ b) 80%   c) $\frac{1}{5}$
   d) 7%   e) $1\frac{7}{8}$ f) $\frac{1}{6}$

3. Evaluate the following:
   a) 25% of 80   b) 80% of 125   c) 62.5% of 80
   d) 30% of 120   e) 90% of 5   f) 25% of 30

4. Evaluate the following:
   a) 17% of 50
   b) 50% of 17
   c) 65% of 80
   d) 80% of 65
   e) 7% of 250
   f) 250% of 7

5. In a class of 30 students, 20% have black hair, 10% have blonde hair and 70% have brown hair. Calculate the number of students with
   a) black hair,
   b) blonde hair,
   c) brown hair.

6. A survey conducted among 120 school children looked at which type of meat they preferred. 55% said they preferred lamb, 20% said they preferred chicken, 15% preferred duck and 10% turkey. Calculate the number of children in each category.

7. A survey was carried out in a school to see what nationality its students were. Of the 220 students in the school, 65% were Australian, 20% were Pakistani, 5% were Greek and 10% belonged to other nationalities. Calculate the number of students of each nationality.

8. A shopkeeper keeps a record of the number of items he sells in one day. Of the 150 items he sold, 46% were newspapers, 24% were pens, 12% were books whilst the remaining 18% were other items. Calculate the number of each item he sold.

● **Expressing one quantity as a percentage of another**

To express one quantity as a percentage of another, first write the first quantity as a fraction of the second and then multiply by 100.

Worked example In an examination a girl obtains 69 marks out of 75. Express this result as a percentage.

$$\frac{69}{75} \times 100\% = 92\%$$

Exercise 5.3
1. Express the first quantity as a percentage of the second.
   a) 24 out of 50
   b) 46 out of 125
   c) 7 out of 20
   d) 45 out of 90
   e) 9 out of 20
   f) 16 out of 40
   g) 13 out of 39
   h) 20 out of 35

2. A hockey team plays 42 matches. It wins 21, draws 14 and loses the rest. Express each of these results as a percentage of the total number of games played.

3. Four candidates stood in an election:

   A received 24 500 votes
   B received 18 200 votes
   C received 16 300 votes
   D received 12 000 votes

   Express each of these as a percentage of the total votes cast.

4. A car manufacturer produces 155 000 cars a year. The cars are available for sale in six different colours. The numbers sold of each colour were:

   Red 55 000
   Blue 48 000
   White 27 500
   Silver 10 200
   Green 9300
   Black 5000

   Express each of these as a percentage of the total number of cars produced. Give your answers to 1 d.p.

● **Percentage increases and decreases**

Worked examples a) A shop assistant has a salary of $16 000 per month. If his salary increases by 8%, calculate:

   i) the amount extra he receives a month,
   ii) his new monthly salary.

   i) Increase = 8% of $16 000
                = 0.08 × $16 000 = $1280
   ii) New salary = old salary + increase
                  = $16 000 + $1280 per month
                  = $17 280 per month

b) A garage increases the price of a truck by 12%. If the original price was $14 500, calculate its new price.
   The original price represents 100%, therefore the increased price can be represented as 112%.

   New price = 112% of $14 500
             = 1.12 × $14 500
             = $16 240

c) A shop is having a sale. It sells a set of tools costing $130 at a 15% discount. Calculate the sale price of the tools.
   The old price represents 100%, therefore the new price can be represented as (100 − 15)% = 85%.

   85% of $130 = 0.85 × $130
              = $110.50

Exercise 5.4

1. Increase the following by the given percentage:
  a) 150 by 25%   b) 230 by 40%   c) 7000 by 2%
  d) 70 by 250%   e) 80 by 12.5%   f) 75 by 62%

2. Decrease the following by the given percentage:
  a) 120 by 25%   b) 40 by 5%   c) 90 by 90%
  d) 1000 by 10%   e) 80 by 37.5%   f) 75 by 42%

3. In the following questions the first number is increased to become the second number. Calculate the percentage increase in each case.
  a) $50 \to 60$ b) $75 \to 135$ c) $40 \to 84$
  d) $30 \to 31.5$ e) $18 \to 33.3$ f) $4 \to 13$

4. In the following questions the first number is decreased to become the second number. Calculate the percentage decrease in each case.
  a) $50 \to 25$ b) $80 \to 56$ c) $150 \to 142.5$
  d) $3 \to 0$ e) $550 \to 352$ f) $20 \to 19$

5. A farmer increases the yield on his farm by 15%. If his previous yield was 6500 tonnes, what is his present yield?

6. The cost of a computer in a computer store is reduced by 12.5% in a sale. If the computer was priced at $7800, what is its price in the sale?

7. A winter coat is priced at $100. In the sale its price is reduced by 25%.
  a) Calculate the sale price of the coat.
  b) After the sale its price is increased by 25% again. Calculate the coat's price after the sale.

8. A farmer takes 250 chickens to be sold at a market. In the first hour he sells 8% of his chickens. In the second hour he sells 10% of those that were left.
  a) How many chickens has he sold in total?
  b) What percentage of the original number did he manage to sell in the two hours?

9. The number of fish on a fish farm increases by approximately 10% each month. If there were originally 350 fish, calculate to the nearest 100 how many fish there would be after 12 months.

Reverse percentages

Worked examples **a)** In a test Ahmed answered 92% of the questions correctly. If he answered 23 questions correctly, how many had he got wrong?

92% of the marks is equivalent to 23 questions.
1% of the marks therefore is equivalent to $\frac{23}{92}$ questions.
So 100% is equivalent to $\frac{23}{92} \times 100 = 25$ questions.
Ahmed got 2 questions wrong.

b) A boat is sold for $15 360. This represents a profit of 28% to the seller. What did the boat originally cost the seller?

The selling price is 128% of the original cost to the seller.
128% of the original cost is $15 360.

1% of the original cost is $\frac{\$15\,360}{128}$.

100% of the original cost is $\frac{\$15\,360}{128} \times 100$, i.e. $12 000.

Exercise 5.5

1. Calculate the value of X in each of the following:

   a) 40% of X is 240
   b) 24% of X is 84
   c) 85% of X is 765
   d) 4% of X is 10
   e) 15% of X is 18.75
   f) 7% of X is 0.105

2. Calculate the value of Y in each of the following:

   a) 125% of Y is 70
   b) 140% of Y is 91
   c) 210% of Y is 189
   d) 340% of Y is 68
   e) 150% of Y is 0.375
   f) 144% of Y is -54.72

3. In a geography text book, 35% of the pages are coloured. If there are 98 coloured pages, how many pages are there in the whole book?

4. A town has 3500 families who own a car. If this represents 28% of the families in the town, how many families are there in total?

5. In a test Isabel scored 88%. If she got three questions incorrect, how many did she get correct?

6. Water expands when it freezes. Ice is less dense than water so it floats. If the increase in volume is 4%, what volume of water will make an iceberg of 12 700 000 m³? Give your answer to three significant figures.

Student assessment 1

1. Find 40% of 1600 m.

2. A shop increases the price of a television set by 8%. If the present price is $320 what is the new price?

3. A car loses 55% of its value after four years. If it cost $22 500 when new, what is its value after the four years?

4. Express the first quantity as a percentage of the second.
   a) 40 cm, 2 m
   b) 25 mins, 1 hour
   c) 450 g, 2 kg
   d) 3 m, 3.5 m
   e) 70 kg, 1 tonne
   f) 75 cl, 2.5 l

5. A house is bought for $75 000 and then resold for $87 000. Calculate the percentage profit.

6. A pair of shoes is priced at $45. During a sale the price is reduced by 20%.
   a) Calculate the sale price of the shoes.
   b) What is the percentage increase in the price if after the sale it is once again restored to $45?

Student assessment 2

1. Find 30% of 2500 m.

2. In a sale a shop reduces its prices by 12.5%. What is the sale price of a desk previously costing $600?

3. In the last six years the value of a house has increased by 35%. If it cost $72 000 six years ago, what is its value now?

4. Express the first quantity as a percentage of the second.
   a) 35 mins, 2 hours
   b) 650 g, 3 kg
   c) 5 m, 4 m
   d) 15 s, 3 mins
   e) 600 kg, 3 tonnes
   f) 35 cl, 3.5 l

5. Shares in a company are bought for $600. After a year the same shares are sold for $550. Calculate the percentage depreciation.

6. In a sale the price of a jacket originally costing $1700 is reduced by $400. Any item not sold by the last day of the sale is reduced by a further 50%. If the jacket is sold on the last day of the sale:
   a) calculate the price it is finally sold for,
   b) calculate the overall percentage reduction in price.

Student assessment 3

1. Calculate the original price for each of the following:

Selling price	Profit
$3780	8%
$14 880	24%
$3.50	250%
$56.56	1%

2. Calculate the original price for each of the following:

Selling price	Loss
$350	30%
$200	20%
$8000	60%
$27 500	80%

3. In a test Ben gained 90% by answering 135 questions correctly. How many questions did he answer incorrectly?

4. A one-year-old car is worth $11 250. If its value has depreciated by 25% in that first year, calculate its price when new.

5. This year a farmer's crop yielded 50 000 tonnes. If this represents a 25% increase on last year, what was the yield last year?

Student assessment 4

1. Calculate the original price for each of the following:

Selling price	Profit
$224	12%
$62.50	150%
$660.24	26%
$38.50	285%

2. Calculate the original price for each of the following:

Selling price	Loss
$392.70	15%
$2480	38%
$3937.50	12.5%
$4675	15%

3. In an examination Sarah obtained 87.5% by gaining 105 marks. How many marks did she lose?

4. At the end of a year a factory has produced 38 500 television sets. If this represents a 10% increase in productivity on last year, calculate the number of sets that were made last year.

5. A computer manufacturer is expected to have produced 24 000 units by the end of this year. If this represents a 4% decrease on last year's output, calculate the number of units produced last year.

6. A company increased its productivity by 10% each year for the last two years. If it produced 56 265 units this year, how many units did it produce two years ago?

6 Ratio and proportion

● Direct proportion

Workers in a pottery factory are paid according to how many plates they produce. The wage paid to them is said to be in **direct proportion** to the number of plates made. As the number of plates made increases so does their wage. Other workers are paid for the number of hours worked. For them the wage paid is in **direct proportion** to the number of hours worked. There are two main methods for solving problems involving direct proportion: the ratio method and the unitary method.

Worked example A bottling machine fills 500 bottles in 15 minutes. How many bottles will it fill in $1\frac{1}{2}$ hours?

Note: The time units must be the same, so for either method the $1\frac{1}{2}$ hours must be changed to 90 minutes.

The ratio method
Let x be the number of bottles filled. Then:

$$\frac{x}{90} = \frac{500}{15}$$

so $x = \dfrac{500 \times 90}{15} = 3000$

3000 bottles are filled in $1\frac{1}{2}$ hours.

The unitary method
In 15 minutes 500 bottles are filled.

Therefore in 1 minute $\frac{500}{15}$ bottles are filled.

So in 90 minutes $90 \times \frac{500}{15}$ bottles are filled.

In $1\frac{1}{2}$ hours, 3000 bottles are filled.

Exercise 6.1 Use either the ratio method or the unitary method to solve the problems below.

1. A machine prints four books in 10 minutes. How many will it print in 2 hours?

2. A farmer plants five apple trees in 25 minutes. If he continues to work at a constant rate, how long will it take him to plant 200 trees?

3. A television set uses 3 units of electricity in 2 hours. How many units will it use in 7 hours? Give your answer to the nearest unit.

4. A bricklayer lays 1500 bricks in an 8-hour day. Assuming he continues to work at the same rate, calculate:
   a) how many bricks he would expect to lay in a five-day week,
   b) how long to the nearest hour it would take him to lay 10 000 bricks.

5. A machine used to paint white lines on a road uses 250 litres of paint for each 8 km of road marked. Calculate:
   a) how many litres of paint would be needed for 200 km of road,
   b) what length of road could be marked with 4000 litres of paint.

6. An aircraft is cruising at 720 km/h and covers 1000 km. How far would it travel in the same period of time if the speed increased to 800 km/h?

7. A production line travelling at 2 m/s labels 150 tins. In the same period of time how many will it label at:
   a) 6 m/s   b) 1 m/s   c) 1.6 m/s?

8. A car travels at an average speed of 80 km/h for 6 hours.
   a) How far will it travel in the 6 hours?
   b) What average speed will it need to travel at in order to cover the same distance in 5 hours?

If the information is given in the form of a ratio, the method of solution is the same.

Worked example Tin and copper are mixed in the ratio 8 : 3. How much tin is needed to mix with 36 g of copper?

The ratio method
Let x grams be the mass of tin needed.

$$\frac{x}{36} = \frac{8}{3}$$

Therefore $x = \dfrac{8 \times 36}{3}$
$= 96$

So 96 g of tin is needed.

The unitary method
3 g of copper mixes with 8 g of tin.
1 g of copper mixes with $\frac{8}{3}$ g of tin.
So 36 g of copper mixes with $36 \times \frac{8}{3}$ g of tin.
Therefore 36 g of copper mixes with 96 g of tin.

Number

Exercise 6.2

1. Sand and gravel are mixed in the ratio 5 : 3 to form ballast.
   a) How much gravel is mixed with 750 kg of sand?
   b) How much sand is mixed with 750 kg of gravel?

2. A recipe uses 150 g butter, 500 g flour, 50 g sugar and 100 g currants to make 18 small cakes.
   a) How much of each ingredient will be needed to make 72 cakes?
   b) How many whole cakes could be made with 1 kg of butter?

3. A paint mix uses red and white paint in a ratio of 1 : 12.
   a) How much white paint will be needed to mix with 1.4 litres of red paint?
   b) If a total of 15.5 litres of paint is mixed, calculate the amount of white paint and the amount of red paint used. Give your answers to the nearest 0.1 litre.

4. A tulip farmer sells sacks of mixed bulbs to local people. The bulbs develop into two different colours of tulips, red and yellow. The colours are packaged in a ratio of 8 : 5 respectively.
   a) If a sack contains 200 red bulbs, calculate the number of yellow bulbs.
   b) If a sack contains 351 bulbs in total, how many of each colour would you expect to find?
   c) One sack is packaged with a bulb mixture in the ratio 7 : 5 by mistake. If the sack contains 624 bulbs, how many more yellow bulbs would you expect to have compared with a normal sack of 624 bulbs?

5. A pure fruit juice is made by mixing the juices of oranges and mangoes in the ratio of 9 : 2.
   a) If 189 litres of orange juice are used, calculate the number of litres of mango juice needed.
   b) If 605 litres of the juice are made, calculate the number of litres of orange juice and mango juice used.

● Divide a quantity in a given ratio

Worked examples a) Divide 20 m in the ratio 3 : 2.

The ratio method
3 : 2 gives 5 parts.

$\frac{3}{5} \times 20\,\text{m} = 12\,\text{m}$

$\frac{2}{5} \times 20\,\text{m} = 8\,\text{m}$

20 m divided in the ratio 3 : 2 is 12 m : 8 m.

Ratio and proportion

The unitary method
3 : 2 gives 5 parts.
5 parts is equivalent to 20 m.
1 part is equivalent to $\frac{20}{5}$ m.
Therefore 3 parts is $3 \times \frac{20}{5}$ m; that is 12 m.
Therefore 2 parts is $2 \times \frac{20}{5}$ m; that is 8 m.

b) A factory produces cars in red, blue, white and green in the ratio 7 : 5 : 3 : 1. Out of a production of 48 000 cars how many are white?

7 + 5 + 3 + 1 gives a total of 16 parts.
Therefore the total number of white cars = $\frac{3}{16} \times 48\,000 = 9000$.

Exercise 6.3

1. Divide 150 in the ratio 2 : 3.
2. Divide 72 in the ratio 2 : 3 : 4.
3. Divide 5 kg in the ratio 13 : 7.
4. Divide 45 minutes in the ratio 2 : 3.
5. Divide 1 hour in the ratio 1 : 5.
6. $\frac{7}{8}$ of a can of drink is water, the rest is syrup. What is the ratio of water to syrup?
7. $\frac{5}{9}$ of a litre carton of orange is pure orange juice, the rest is water. How many millilitres of each are in the carton?
8. 55% of students in a school are boys.
    a) What is the ratio of boys to girls?
    b) How many boys and how many girls are there if the school has 800 students?
9. A piece of wood is cut in the ratio 2 : 3. What fraction of the length is the longer piece?
10. If the piece of wood in question 9 is 80 cm long, how long is the shorter piece?
11. A gas pipe is 7 km long. A valve is positioned in such a way that it divides the length of the pipe in the ratio 4 : 3. Calculate the distance of the valve from each end of the pipe.
12. The size of the angles of a quadrilateral are in the ratio 1 : 2 : 3 : 3. Calculate the size of each angle.
13. The angles of a triangle are in the ratio 3 : 5 : 4. Calculate the size of each angle.

Number

14. A millionaire leaves 1.4 million dollars in his will to be shared between his three children in the ratio of their ages. If they are 24, 28 and 32 years old, calculate to the nearest dollar the amount they will each receive.

15. A small company makes a profit of $8000. This is divided between the directors in the ratio of their initial investments. If Alex put $20 000 into the firm, Maria $35 000 and Ahmet $25 000, calculate the amount of the profit they will each receive.

● Inverse proportion

Sometimes an increase in one quantity causes a decrease in another quantity. For example, if fruit is to be picked by hand, the more people there are picking the fruit, the less time it will take.

Worked examples

a) If 8 people can pick the apples from the trees in 6 days, how long will it take 12 people?

8 people take 6 days.
1 person will take 6 × 8 days.

Therefore 12 people will take $\frac{6 \times 8}{12}$ days, i.e. 4 days.

b) A cyclist averages a speed of 27 km/h for 4 hours. At what average speed would she need to cycle to cover the same distance in 3 hours?

Completing it in 1 hour would require cycling at 27 × 4 km/h. Completing it in 3 hours requires cycling at

$\frac{27 \times 4}{3}$ km/h; that is 36 km/h.

Exercise 6.4

1. A teacher shares sweets among 8 students so that they get 6 each. How many sweets would they each have got had there been 12 students?

2. The table below represents the relationship between the speed and the time taken for a train to travel between two stations.

Speed (km/h)	60			120	90	50	10
Time (h)	2	3	4				

Copy and complete the table.

3. Six people can dig a trench in 8 hours.
   a) How long would it take:
      i) 4 people ii) 12 people iii) 1 person?
   b) How many people would it take to dig the trench in:
      i) 3 hours ii) 16 hours iii) 1 hour?

4. Chairs in a hall are arranged in 35 rows of 18.
   a) How many rows would there be with 21 chairs to a row?
   b) How many chairs would there be in each row if there were 15 rows?

5. A train travelling at 100 km/h takes 4 hours for a journey. How long would it take a train travelling at 60 km/h?

6. A worker in a sugar factory packs 24 cardboard boxes with 15 bags of sugar in each. If he had boxes which held 18 bags of sugar each, how many fewer boxes would be needed?

7. A swimming pool is filled in 30 hours by two identical pumps. How much quicker would it be filled if five similar pumps were used instead?

● Increase and decrease by a given ratio

Worked examples a) A photograph is 12 cm wide and 8 cm tall. It is enlarged in the ratio 3 : 2. What are the dimensions of the enlarged photograph?

3 : 2 is an enlargement of $\frac{3}{2}$. Therefore the enlarged width is 12 cm $\times \frac{3}{2}$; that is 18 cm.
The enlarged height is 8 cm $\times \frac{3}{2}$; that is 12 cm.

b) A photographic transparency 5 cm wide and 3 cm tall is projected onto a screen. If the image is 1.5 m wide:
   i) calculate the ratio of the enlargement,
   ii) calculate the height of the image.

   i) 5 cm width is enlarged to become 150 cm.
      So 1 cm width becomes $\frac{150}{5}$ cm; that is 30 cm.
      Therefore the enlargement ratio is 30 : 1.
   ii) The height of the image = 3 cm $\times$ 30 = 90 cm.

Exercise 6.5

1. Increase 100 by the following ratios:
   a) 8 : 5   b) 5 : 2   c) 7 : 4
   d) 11 : 10   e) 9 : 4   f) 32 : 25

2. Increase 70 by the following ratios:
   a) 4 : 3   b) 5 : 3   c) 8 : 7
   d) 9 : 4   e) 11 : 5   f) 17 : 14

3. Decrease 60 by the following ratios:
   a) 2 : 3   b) 5 : 6   c) 7 : 12
   d) 3 : 5   e) 1 : 4   f) 13 : 15

Number

4. Decrease 30 by the following ratios:
   a) 3 : 4 b) 2 : 9 c) 7 : 12
   d) 3 : 16 e) 5 : 8 f) 9 : 20

5. Increase 40 by a ratio of 5 : 4.

6. Decrease 40 by a ratio of 4 : 5.

7. Increase 150 by a ratio of 7 : 5.

8. Decrease 210 by a ratio of 3 : 7.

Exercise 6.6

1. A photograph measuring 8 cm by 6 cm is enlarged by a ratio of 11 : 4. What are the dimensions of the new print?

2. A photocopier enlarges in the ratio 7 : 4. What would be the new size of a diagram measuring 16 cm by 12 cm?

3. A drawing measuring 10 cm by 16 cm needs to be enlarged. The dimensions of the enlargement need to be 25 cm by 40 cm. Calculate the enlargement needed and express it as a ratio.

4. A banner needs to be enlarged from its original format. The dimensions of the original are 4 cm tall by 25 cm wide. The enlarged banner needs to be at least 8 m wide but no more than 1.4 m tall. Calculate the minimum and maximum ratios of enlargement possible.

5. A rectangle measuring 7 cm by 4 cm is enlarged by a ratio of 2 : 1.
   a) What is the area of:
      i) the original rectangle?
      ii) the enlarged rectangle?
   b) By what ratio has the area been enlarged?

6. A square of side length 3 cm is enlarged by a ratio of 3 : 1.
   a) What is the area of:
      i) the original square?
      ii) the enlarged square?
   b) By what ratio has the area been enlarged?

7. A cuboid measuring 3 cm by 5 cm by 2 cm is enlarged by a ratio of 2 : 1.
   a) What is the volume of:
      i) the original cuboid?
      ii) the enlarged cuboid?
   b) By what ratio has the volume been increased?

8. A cube of side 4 cm is enlarged by a ratio of 3 : 1.
   a) What is the volume of:
      i) the original cube?
      ii) the enlarged cube?
   b) By what ratio has the volume been increased?

9. The triangle is to be reduced by a ratio of 1 : 2.

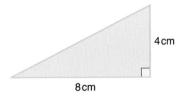

   a) Calculate the area of the original triangle.
   b) Calculate the area of the reduced triangle.
   c) Calculate the ratio by which the area of the triangle has been reduced.

10. From questions 5−9 can you conclude what happens to two- and three-dimensional figures when they are either enlarged or reduced?

Student assessment I

1. A boat travels at an average speed of 15 km/h for 1 hour.
   a) Calculate the distance it travels in one hour.
   b) What average speed will the boat need to travel at in order to cover the same distance in $2\frac{1}{2}$ hours?

2. A ruler 30 cm long is broken into two parts in the ratio 8 : 7. How long are the two parts?

3. A recipe needs 400 g of flour to make 8 cakes. How much flour would be needed in order to make 24 cakes?

4. To make 6 jam tarts, 120 g of jam is needed. How much jam is needed to make 10 tarts?

5. The scale of a map is 1 : 25 000.
   a) Two villages are 8 cm apart on the map. How far apart are they in real life? Give your answer in kilometres.
   b) The distance from a village to the edge of a lake is 12 km in real life. How far apart would they be on the map? Give your answer in centimetres.

6. A motorbike uses petrol and oil mixed in the ratio 13 : 2.
   a) How much of each is there in 30 litres of mixture?
   b) How much petrol would be mixed with 500 ml of oil?

7. a) A model car is a $\frac{1}{40}$ scale model. Express this as a ratio.
   b) If the length of the real car is 5.5 m, what is the length of the model car?

8. An aunt gives a brother and sister $2000 to be divided in the ratio of their ages. If the girl is 13 years old and the boy 12 years old, how much will each get?

9. The angles of a triangle are in the ratio 2 : 5 : 8. Find the size of each of the angles.

Number

10. A photocopying machine is capable of making 50 copies each minute.
    a) If four identical copiers are used simultaneously how long would it take to make a total of 50 copies?
    b) How many copiers would be needed to make 6000 copies in 15 minutes?

11. It takes 16 hours for three bricklayers to build a wall. Calculate how long it would take for eight bricklayers to build a similar wall.

12. A photocopier enlarges by a ratio of 7 : 4. A picture measures 6 cm by 4 cm. How many consecutive enlargements can be made so that the largest possible picture will fit on a sheet measuring 30 cm by 20 cm?

Student assessment 2

1. A cyclist travels at an average speed of 20 km/h for 1.5 hours.
   a) Calculate the distance she travels in 1.5 hours.
   b) What average speed will the cyclist need to travel in order to cover the same distance in 1 hour?

2. A piece of wood is cut in the ratio 3 : 7.
   a) What fraction of the whole is the longer piece?
   b) If the wood is 1.5 m long, how long is the shorter piece?

3. A recipe for two people requires $\frac{1}{4}$ kg of rice to 150 g of meat.
   a) How much meat would be needed for five people?
   b) How much rice would there be in 1 kg of the final dish?

4. The scale of a map is 1 : 10 000.
   a) Two rivers are 4.5 cm apart on the map, how far apart are they in real life? Give your answer in metres.
   b) Two towns are 8 km apart in real life. How far apart are they on the map? Give your answer in centimetres.

5. a) A model train is a $\frac{1}{25}$ scale model. Express this as a ratio.
   b) If the length of the model engine is 7 cm, what is the true length of the engine?

6. Divide 3 tonnes in the ratio 2 : 5 : 13.

7. The ratio of the angles of a quadrilateral is 2 : 3 : 3 : 4. Calculate the size of each of the angles.

8. The ratio of the interior angles of a pentagon is 2 : 3 : 4 : 4 : 5. Calculate the size of the largest angle.

NB: All diagrams are not drawn to scale.

9. A large swimming pool takes 36 hours to fill using three identical pumps.
   a) How long would it take to fill using eight identical pumps?
   b) If the pool needs to be filled in 9 hours, how many pumps will be needed?

10. The first triangle is an enlargement of the second. Calculate the size of the missing sides and angles.

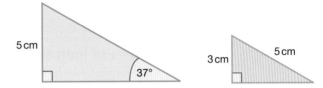

11. A tap issuing water at a rate of 1.2 litres per minute fills a container in 4 minutes.
    a) How long would it take to fill the same container if the rate was decreased to 1 litre per minute? Give your answer in minutes and seconds.
    b) If the container is to be filled in 3 minutes, calculate the rate at which the water should flow.

12. A map measuring 60 cm by 25 cm is reduced twice in the ratio 3 : 5. Calculate the final dimensions of the map.

7 Indices and standard form

The index refers to the power to which a number is raised. In the example 5^3 the number 5 is raised to the power 3. The 3 is known as the **index**. Indices is the plural of index.

Worked examples
a) $5^3 = 5 \times 5 \times 5$
$ = 125$

b) $7^4 = 7 \times 7 \times 7 \times 7$
$ = 2401$

c) $3^1 = 3$

● Laws of indices

When working with numbers involving indices there are three basic laws which can be applied. These are:

(1) $a^m \times a^n = a^{m+n}$

(2) $a^m \div a^n$ or $\dfrac{a^m}{a^n} = a^{m-n}$

(3) $(a^m)^n = a^{mn}$

● Positive indices

Worked examples
a) Simplify $4^3 \times 4^2$.
$4^3 \times 4^2 = 4^{(3+2)}$
$ = 4^5$

b) Simplify $2^5 \div 2^3$.
$2^5 \div 2^3 = 2^{(5-3)}$
$ = 2^2$

c) Evaluate $3^3 \times 3^4$.
$3^3 \times 3^4 = 3^{(3+4)}$
$ = 3^7$
$ = 2187$

d) Evaluate $(4^2)^3$.
$(4^2)^3 = 4^{(2 \times 3)}$
$ = 4^6$
$ = 4096$

Exercise 7.1

1. Using indices, simplify the following expressions:
   a) $3 \times 3 \times 3$
   b) $2 \times 2 \times 2 \times 2 \times 2$
   c) 4×4
   d) $6 \times 6 \times 6 \times 6$
   e) $8 \times 8 \times 8 \times 8 \times 8 \times 8$
   f) 5

2. Simplify the following using indices:
   a) $2 \times 2 \times 2 \times 3 \times 3$
   b) $4 \times 4 \times 4 \times 4 \times 4 \times 5 \times 5$
   c) $3 \times 3 \times 4 \times 4 \times 4 \times 5 \times 5$
   d) $2 \times 7 \times 7 \times 7 \times 7$
   e) $1 \times 1 \times 6 \times 6$
   f) $3 \times 3 \times 3 \times 4 \times 4 \times 6 \times 6 \times 6 \times 6 \times 6$

Indices and standard form

3. Write out the following in full:
   a) 4^2
   b) 5^7
   c) 3^5
   d) $4^3 \times 6^3$
   e) $7^2 \times 2^7$
   f) $3^2 \times 4^3 \times 2^4$

4. Without a calculator work out the value of the following:
   a) 2^5
   b) 3^4
   c) 8^2
   d) 6^3
   e) 10^6
   f) 4^4
   g) $2^3 \times 3^2$
   h) $10^3 \times 5^3$

Exercise 7.2

1. Simplify the following using indices:
   a) $3^2 \times 3^4$
   b) $8^5 \times 8^2$
   c) $5^2 \times 5^4 \times 5^3$
   d) $4^3 \times 4^5 \times 4^2$
   e) $2^1 \times 2^3$
   f) $6^2 \times 3^2 \times 3^3 \times 6^4$
   g) $4^5 \times 4^3 \times 5^5 \times 5^4 \times 6^2$
   h) $2^4 \times 5^7 \times 5^3 \times 6^2 \times 6^6$

2. Simplify the following:
   a) $4^6 \div 4^2$
   b) $5^7 \div 5^4$
   c) $2^5 \div 2^4$
   d) $6^5 \div 6^2$
   e) $\dfrac{6^5}{6^2}$
   f) $\dfrac{8^6}{8^5}$
   g) $\dfrac{4^8}{4^5}$
   h) $\dfrac{3^9}{3^2}$

3. Simplify the following:
   a) $(5^2)^2$
   b) $(4^3)^4$
   c) $(10^2)^5$
   d) $(3^3)^5$
   e) $(6^2)^4$
   f) $(8^2)^3$

4. Simplify the following:
   a) $\dfrac{2^2 \times 2^4}{2^3}$
   b) $\dfrac{3^4 \times 3^2}{3^5}$
   c) $\dfrac{5^6 \times 5^7}{5^2 \times 5^8}$
   d) $\dfrac{(4^2)^5 \times 4^2}{4^7}$
   e) $\dfrac{4^4 \times 2^5 \times 4^2}{4^3 \times 2^3}$
   f) $\dfrac{6^3 \times 6^3 \times 8^5 \times 8^6}{8^6 \times 6^2}$
   g) $\dfrac{(5^5)^2 \times (4^4)^3}{5^8 \times 4^9}$
   h) $\dfrac{(6^3)^4 \times 6^3 \times 4^9}{6^8 \times 4^8}$

The zero index

The zero index indicates that a number is raised to the power 0. A number raised to the power 0 is equal to 1. This can be explained by applying the laws of indices.

$$a^m \div a^n = a^{m-n} \quad \text{therefore} \quad \frac{a^m}{a^m} = a^{m-m}$$
$$= a^0$$

However, $\quad \dfrac{a^m}{a^m} = 1$

therefore $a^0 = 1$

Negative indices

A negative index indicates that a number is being raised to a negative power: e.g. 4^{-3}.

Another law of indices states that $a^{-m} = \dfrac{1}{a^m}$. This can be proved as follows.

$$a^{-m} = a^{0-m}$$
$$= \frac{a^0}{a^m} \quad \text{(from the second law of indices)}$$
$$= \frac{1}{a^m}$$

therefore $a^{-m} = \dfrac{1}{a^m}$

Exercise 7.3 Without using a calculator, evaluate the following:

1. a) $2^3 \times 2^0$ b) $5^2 \div 6^0$
   c) $5^2 \times 5^{-2}$ d) $6^3 \times 6^{-3}$
   e) $(4^0)^2$ f) $4^0 \div 2^2$

2. a) 4^{-1} b) 3^{-2}
   c) 6×10^{-2} d) 5×10^{-3}
   e) 100×10^{-2} f) 10^{-3}

3. a) 9×3^{-2} b) 16×2^{-3}
   c) 64×2^{-4} d) 4×2^{-3}
   e) 36×6^{-3} f) 100×10^{-1}

4. a) $\dfrac{3}{2^{-2}}$ b) $\dfrac{4}{2^{-3}}$
   c) $\dfrac{9}{5^{-2}}$ d) $\dfrac{5}{4^{-2}}$
   e) $\dfrac{7^{-3}}{7^{-4}}$ f) $\dfrac{8^{-6}}{8^{-8}}$

7 Indices and standard form

● Exponential equations
Equations that involve indices as unknowns are known as **exponential equations**.

Worked examples
a) Find the value of x if $2^x = 32$.

32 can be expressed as a power of 2,
$32 = 2^5$.

Therefore $2^x = 2^5$

$x = 5$

b) Find the value of m if $3^{(m-1)} = 81$.

81 can be expressed as a power of 3,
$81 = 3^4$.

Therefore $3^{(m-1)} = 3^4$

$m - 1 = 4$

$m = 5$

Exercise 7.4

1. Find the value of x in each of the following:
   a) $2^x = 4$
   b) $2^x = 16$
   c) $4^x = 64$
   d) $10^x = 1000$
   e) $5^x = 625$
   f) $3^x = 1$

2. Find the value of z in each of the following:
   a) $2^{(z-1)} = 8$
   b) $3^{(z+2)} = 27$
   c) $4^{2z} = 64$
   d) $10^{(z+1)} = 1$
   e) $3^z = 9^{(z-1)}$
   f) $5^z = 125^z$

3. Find the value of n in each of the following:
   a) $(\frac{1}{2})^n = 8$
   b) $(\frac{1}{3})^n = 81$
   c) $(\frac{1}{2})^n = 32$
   d) $(\frac{1}{2})^n = 4^{(n+1)}$
   e) $(\frac{1}{2})^{(n+1)} = 2$
   f) $(\frac{1}{16})^n = 4$

4. Find the value of x in each of the following:
   a) $3^{-x} = 27$
   b) $2^{-x} = 128$
   c) $2^{(-x+3)} = 64$
   d) $4^{-x} = \frac{1}{16}$
   e) $2^{-x} = \frac{1}{256}$
   f) $3^{(-x+1)} = \frac{1}{81}$

● Standard form
Standard form is also known as standard index form or sometimes as scientific notation. It involves writing large numbers or very small numbers in terms of powers of 10.

Number

● Positive indices and large numbers

$$100 = 1 \times 10^2$$
$$1000 = 1 \times 10^3$$
$$10\,000 = 1 \times 10^4$$
$$3000 = 3 \times 10^3$$

For a number to be in standard form it must take the form $A \times 10^n$ where the index n is a positive or negative integer and A must lie in the range $1 \leq A < 10$.

e.g. 3100 can be written in many different ways:

$$3.1 \times 10^3 \quad 31 \times 10^2 \quad 0.31 \times 10^4 \quad \text{etc.}$$

However, only 3.1×10^3 satisfies the above conditions and therefore is the only one which is written in standard form.

Worked examples

a) Write 72 000 in standard form.

$$7.2 \times 10^4$$

b) Write 4×10^4 as an ordinary number.

$$4 \times 10^4 = 4 \times 10\,000$$
$$= 40\,000$$

c) Multiply the following and write your answer in standard form:

$$600 \times 4000$$
$$= 2\,400\,000$$
$$= 2.4 \times 10^6$$

d) Multiply the following and write your answer in standard form:

$$(2.4 \times 10^4) \times (5 \times 10^7)$$
$$= 12 \times 10^{11}$$
$$= 1.2 \times 10^{12} \text{ when written in standard form}$$

e) Divide the following and write your answer in standard form:

$$(6.4 \times 10^7) \div (1.6 \times 10^3)$$
$$= 4 \times 10^4$$

f) Add the following and write your answer in standard form:

$$(3.8 \times 10^6) + (8.7 \times 10^4)$$

Changing the indices to the same value gives the sum:

$$(380 \times 10^4) + (8.7 \times 10^4)$$
$$= 388.7 \times 10^4$$
$$= 3.887 \times 10^6 \text{ when written in standard form}$$

Indices and standard form

g) Subtract the following and write your answer in standard form:

$(6.5 \times 10^7) - (9.2 \times 10^5)$

Changing the indices to the same value gives
$(650 \times 10^5) - (9.2 \times 10^5)$
$= 640.8 \times 10^5$
$= 6.408 \times 10^7$ when written in standard form

Exercise 7.5

1. Which of the following are not in standard form?
   a) 6.2×10^5
   b) 7.834×10^{16}
   c) 8.0×10^5
   d) 0.46×10^7
   e) 82.3×10^6
   f) 6.75×10^1

2. Write the following numbers in standard form:
   a) 600 000
   b) 48 000 000
   c) 784 000 000 000
   d) 534 000
   e) 7 million
   f) 8.5 million

3. Write the following in standard form:
   a) 68×10^5
   b) 720×10^6
   c) 8×10^5
   d) 0.75×10^8
   e) 0.4×10^{10}
   f) 50×10^6

4. Write the following as ordinary numbers:
   a) 3.8×10^3
   b) 4.25×10^6
   c) 9.003×10^7
   d) 1.01×10^5

5. Multiply the following and write your answers in standard form:
   a) 200×3000
   b) 6000×4000
   c) 7 million $\times$ 20
   d) 500×6 million
   e) 3 million $\times$ 4 million
   f) 4500×4000

6. Light from the Sun takes approximately 8 minutes to reach Earth. If light travels at a speed of 3×10^8 m/s, calculate to three significant figures (s.f.) the distance from the Sun to the Earth.

7. Find the value of the following and write your answers in standard form:
   a) $(4.4 \times 10^3) \times (2 \times 10^5)$
   b) $(6.8 \times 10^7) \times (3 \times 10^3)$
   c) $(4 \times 10^5) \times (8.3 \times 10^5)$
   d) $(5 \times 10^9) \times (8.4 \times 10^{12})$
   e) $(8.5 \times 10^6) \times (6 \times 10^{15})$
   f) $(5.0 \times 10^{12})^2$

8. Find the value of the following and write your answers in standard form:
   a) $(3.8 \times 10^8) \div (1.9 \times 10^6)$
   b) $(6.75 \times 10^9) \div (2.25 \times 10^4)$
   c) $(9.6 \times 10^{11}) \div (2.4 \times 10^5)$
   d) $(1.8 \times 10^{12}) \div (9.0 \times 10^7)$
   e) $(2.3 \times 10^{11}) \div (9.2 \times 10^4)$
   f) $(2.4 \times 10^8) \div (6.0 \times 10^3)$

Number

9. Find the value of the following and write your answers in standard form:
   a) $(3.8 \times 10^5) + (4.6 \times 10^4)$ b) $(7.9 \times 10^9) + (5.8 \times 10^8)$
   c) $(6.3 \times 10^7) + (8.8 \times 10^5)$ d) $(3.15 \times 10^9) + (7.0 \times 10^6)$
   e) $(5.3 \times 10^8) - (8.0 \times 10^7)$ f) $(6.5 \times 10^7) - (4.9 \times 10^6)$
   g) $(8.93 \times 10^{10}) - (7.8 \times 10^9)$ h) $(4.07 \times 10^7) - (5.1 \times 10^6)$

● **Negative indices and small numbers**

A negative index is used when writing a number between 0 and 1 in standard form.

e.g. $100 = 1 \times 10^2$
$10 = 1 \times 10^1$
$1 = 1 \times 10^0$
$0.1 = 1 \times 10^{-1}$
$0.01 = 1 \times 10^{-2}$
$0.001 = 1 \times 10^{-3}$
$0.0001 = 1 \times 10^{-4}$

Note that A must still lie within the range $1 \leq A < 10$.

Worked examples

a) Write 0.0032 in standard form.

3.2×10^{-3}

b) Write 1.8×10^{-4} as an ordinary number.

$1.8 \times 10^{-4} = 1.8 \div 10^4$
$= 1.8 \div 10000$
$= 0.00018$

c) Write the following numbers in order of magnitude, starting with the largest:

3.6×10^{-3} 5.2×10^{-5} 1×10^{-2} 8.35×10^{-2} 6.08×10^{-8}

8.35×10^{-2} 1×10^{-2} 3.6×10^{-3} 5.2×10^{-5} 6.08×10^{-8}

Exercise 7.6

1. Write the following numbers in standard form:
   a) 0.0006   b) 0.000 053
   c) 0.000 864   d) 0.000 000 088
   e) 0.000 000 7   f) 0.000 414 5

2. Write the following numbers in standard form:
   a) 68×10^{-5} b) 750×10^{-9}
   c) 42×10^{-11} d) 0.08×10^{-7}
   e) 0.057×10^{-9} f) 0.4×10^{-10}

3. Write the following as ordinary numbers:
   a) 8×10^{-3} b) 4.2×10^{-4}
   c) 9.03×10^{-2} d) 1.01×10^{-5}

4. Deduce the value of n in each of the following cases:
a) $0.00025 = 2.5 \times 10^n$
b) $0.00357 = 3.57 \times 10^n$
c) $0.00000006 = 6 \times 10^n$
d) $0.004^2 = 1.6 \times 10^n$
e) $0.00065^2 = 4.225 \times 10^n$
f) $0.0002^n = 8 \times 10^{-12}$

5. Write these numbers in order of magnitude, starting with the largest:

3.2×10^{-4} 6.8×10^5 5.57×10^{-9} 6.2×10^3
5.8×10^{-7} 6.741×10^{-4} 8.414×10^2

● Fractional indices

$16^{\frac{1}{2}}$ can be written as $(4^2)^{\frac{1}{2}}$.

$$(4^2)^{\frac{1}{2}} = 4^{(2 \times \frac{1}{2})}$$
$$= 4^1$$
$$= 4$$

Therefore $16^{\frac{1}{2}} = 4$

but $\sqrt{16} = 4$

therefore $16^{\frac{1}{2}} = \sqrt{16}$

Similarly:

$27^{\frac{1}{3}}$ can be written as $(3^3)^{\frac{1}{3}}$

$$(3^3)^{\frac{1}{3}} = 3^{(3 \times \frac{1}{3})}$$
$$= 3^1$$
$$= 3$$

Therefore $27^{\frac{1}{3}} = 3$

but $\sqrt[3]{27} = 3$

therefore $27^{\frac{1}{3}} = \sqrt[3]{27}$

In general:

$$a^{\frac{1}{n}} = \sqrt[n]{a}$$

$$a^{\frac{m}{n}} = \sqrt[n]{(a^m)} \text{ or } (\sqrt[n]{a})^m$$

Worked examples   **a)** Evaluate $16^{\frac{1}{4}}$ without the use of a calculator.

$16^{\frac{1}{4}} = \sqrt[4]{16}$ Alternatively: $16^{\frac{1}{4}} = (2^4)^{\frac{1}{4}}$
$\phantom{16^{\frac{1}{4}}} = \sqrt[4]{(2^4)}$ $\phantom{Alternatively: 16^{\frac{1}{4}}} = 2^1$
$\phantom{16^{\frac{1}{4}}} = 2$ $\phantom{Alternatively: 16^{\frac{1}{4}}} = 2$

b) Evaluate $25^{\frac{3}{2}}$ without the use of a calculator.

Number

$$25^{\frac{3}{2}} = (25^{\frac{1}{2}})^3 \quad \text{Alternatively:} \quad 25^{\frac{3}{2}} = (5^2)^{\frac{3}{2}}$$
$$= (\sqrt{25})^3 \qquad \qquad \qquad \qquad = 5^3$$
$$= 5^3 \qquad \qquad \qquad \qquad \qquad = 125$$
$$= 125$$

c) Solve $32^x = 2$

32 is 2^5 so $\sqrt[5]{32} = 2$
or $\quad 32^{\frac{1}{5}} = 2$
therefore $x = \frac{1}{5}$

d) Solve $125^x = 5$

125 is 5^3 so $\sqrt[3]{125} = 5$
or $\quad 125^{\frac{1}{3}} = 5$
therefore $x = \frac{1}{3}$

Exercise 7.7 Evaluate the following without the use of a calculator:

1. a) $16^{\frac{1}{2}}$ b) $25^{\frac{1}{2}}$ c) $100^{\frac{1}{2}}$
   d) $27^{\frac{1}{3}}$ e) $81^{\frac{1}{2}}$ f) $1000^{\frac{1}{3}}$

2. a) $16^{\frac{1}{4}}$ b) $81^{\frac{1}{4}}$ c) $32^{\frac{1}{5}}$
   d) $64^{\frac{1}{6}}$ e) $216^{\frac{1}{3}}$ f) $256^{\frac{1}{4}}$

3. a) $4^{\frac{3}{2}}$ b) $4^{\frac{5}{2}}$ c) $9^{\frac{3}{2}}$
   d) $16^{\frac{3}{2}}$ e) $1^{\frac{5}{2}}$ f) $27^{\frac{2}{3}}$

4. a) $125^{\frac{2}{3}}$ b) $32^{\frac{3}{5}}$ c) $64^{\frac{5}{6}}$
   d) $1000^{\frac{2}{3}}$ e) $16^{\frac{5}{4}}$ f) $81^{\frac{3}{4}}$

5. a) solve $16^x = 4$ b) solve $8^x = 2$
   c) solve $9^x = 3$ d) solve $27^x = 3$
   e) solve $100^x = 10$ f) solve $64^x = 2$

6. a) solve $1000^x = 10$ b) solve $49^x = 7$
   c) solve $81^x = 3$ d) solve $343^x = 7$
   e) solve $1\,000\,000^x = 10$ f) solve $216^x = 6$

Exercise 7.8 Evaluate the following without the use of a calculator:

1. a) $\dfrac{27^{\frac{2}{3}}}{3^2}$ b) $\dfrac{7^{\frac{3}{2}}}{\sqrt{7}}$ c) $\dfrac{4^{\frac{5}{2}}}{4^2}$

   d) $\dfrac{16^{\frac{3}{2}}}{2^6}$ e) $\dfrac{27^{\frac{5}{3}}}{\sqrt{9}}$ f) $\dfrac{6^{\frac{4}{3}}}{6^{\frac{1}{3}}}$

2. a) $5^{\frac{2}{3}} \times 5^{\frac{4}{3}}$ b) $4^{\frac{1}{4}} \times 4^{\frac{1}{4}}$ c) 8×2^{-2}
   d) $3^{\frac{4}{3}} \times 3^{\frac{5}{3}}$ e) $2^{-2} \times 16$ f) $8^{\frac{2}{3}} \times 8^{-\frac{4}{3}}$

3. a) $\dfrac{2^{\frac{1}{2}} \times 2^{\frac{5}{2}}}{2}$ b) $\dfrac{4^{\frac{5}{6}} \times 4^{\frac{1}{6}}}{4^{\frac{1}{2}}}$ c) $\dfrac{2^3 \times 8^{\frac{3}{2}}}{\sqrt{8}}$

   d) $\dfrac{(3^2)^{\frac{3}{2}} \times 3^{-\frac{1}{2}}}{3^{\frac{1}{2}}}$ e) $\dfrac{8^{\frac{1}{3}}+7}{27^{\frac{1}{3}}}$ f) $\dfrac{9^{\frac{1}{2}} \times 3^{\frac{5}{2}}}{3^{\frac{2}{3}} \times 3^{-\frac{1}{6}}}$

Student assessment 1

1. Using indices, simplify the following:
   a) $2 \times 2 \times 2 \times 5 \times 5$ b) $2 \times 2 \times 3 \times 3 \times 3 \times 3 \times 3$

2. Write the following out in full:
   a) 4^3 b) 6^4

3. Work out the value of the following without using a calculator:
   a) $2^3 \times 10^2$ b) $1^4 \times 3^3$

4. Simplify the following using indices:
   a) $3^4 \times 3^3$ b) $6^3 \times 6^2 \times 3^4 \times 3^5$
   c) $\dfrac{4^5}{2^3}$ d) $\dfrac{(6^2)^3}{6^5}$
   e) $\dfrac{3^5 \times 4^2}{3^3 \times 4^0}$ f) $\dfrac{4^{-2} \times 2^6}{2^2}$

5. Without using a calculator, evaluate the following:
   a) $2^4 \times 2^{-2}$ b) $\dfrac{3^5}{3^3}$
   c) $\dfrac{5^{-5}}{5^{-6}}$ d) $\dfrac{2^5 \times 4^{-3}}{2^{-1}}$

6. Find the value of x in each of the following:
   a) $2^{(x-2)} = 32$ b) $\dfrac{1}{4^x} = 16$
   c) $5^{(-x+2)} = 125$ d) $8^{-x} = \dfrac{1}{2}$

Student assessment 2

1. Using indices, simplify the following:
   a) $3 \times 2 \times 2 \times 3 \times 27$
   b) $2 \times 2 \times 4 \times 4 \times 4 \times 2 \times 32$

2. Write the following out in full:
   a) 6^5 b) 2^{-5}

3. Work out the value of the following without using a calculator:
   a) $3^3 \times 10^3$ b) $1^{-4} \times 5^3$

Number

4. Simplify the following using indices:
   a) $2^4 \times 2^3$
   b) $7^5 \times 7^2 \times 3^4 \times 3^8$
   c) $\dfrac{4^8}{2^{10}}$
   d) $\dfrac{(3^3)^4}{27^3}$
   e) $\dfrac{7^6 \times 4^2}{4^3 \times 7^6}$
   f) $\dfrac{8^{-2} \times 2^6}{2^{-2}}$

5. Without using a calculator, evaluate the following:
   a) $5^2 \times 5^{-1}$
   b) $\dfrac{4^5}{4^3}$
   c) $\dfrac{7^{-5}}{7^{-7}}$
   d) $\dfrac{3^{-5} \times 4^2}{3^{-6}}$

6. Find the value of x in each of the following:
   a) $2^{(2x+2)} = 128$
   b) $\dfrac{1}{4^{-x}} = \dfrac{1}{2}$
   c) $3^{(-x+4)} = 81$
   d) $8^{-3x} = \dfrac{1}{4}$

Student assessment 3

1. Write the following numbers in standard form:
   a) 8 million
   b) 0.000 72
   c) 75 000 000 000
   d) 0.0004
   e) 4.75 billion
   f) 0.000 000 64

2. Write the following as ordinary numbers:
   a) 2.07×10^4
   b) 1.45×10^{-3}
   c) 5.23×10^{-2}

3. Write the following numbers in order of magnitude, starting with the smallest:

   $6.2 \times 10^7 \quad 5.5 \times 10^{-3} \quad 4.21 \times 10^7 \quad 4.9 \times 10^8 \quad 3.6 \times 10^{-5}$
   7.41×10^{-9}

4. Write the following numbers:
   a) in standard form,
   b) in order of magnitude, starting with the largest.

   6 million   820 000   0.0044   0.8   52 000

5. Deduce the value of n in each of the following:
   a) $620 = 6.2 \times 10^n$
   b) $555\,000\,000 = 5.55 \times 10^n$
   c) $0.000\,45 = 4.5 \times 10^n$
   d) $500^2 = 2.5 \times 10^n$
   e) $0.0035^2 = 1.225 \times 10^n$
   f) $0.04^3 = 6.4 \times 10^n$

6. Write the answers to the following calculations in standard form:
   a) $4000 \times 30\,000$
   b) $(2.8 \times 10^5) \times (2.0 \times 10^3)$
   c) $(3.2 \times 10^9) \div (1.6 \times 10^4)$
   d) $(2.4 \times 10^8) \div (9.6 \times 10^2)$

7. The speed of light is 3×10^8 m/s. Venus is 108 million km from the Sun. Calculate the number of minutes it takes for sunlight to reach Venus.

8. A star system is 500 light years away from Earth. The speed of light is 3×10^5 km/s. Calculate the distance the star system is from Earth. Give your answer in kilometres and written in standard form.

Student assessment 4

1. Write the following numbers in standard form:
   a) 6 million
   b) 0.0045
   c) 3 800 000 000
   d) 0.000 000 361
   e) 460 million
   f) 3

2. Write the following as ordinary numbers:
   a) 8.112×10^6
   b) 4.4×10^5
   c) 3.05×10^{-4}

3. Write the following numbers in order of magnitude, starting with the largest:

   3.6×10^2 2.1×10^{-3} 9×10^1 4.05×10^8 1.5×10^{-2} 7.2×10^{-3}

4. Write the following numbers:
   a) in standard form,
   b) in order of magnitude, starting with the smallest.

   15 million   430 000   0.000 435   4.8   0.0085

5. Deduce the value of n in each of the following:
   a) $4750 = 4.75 \times 10^n$
   b) $6\,440\,000\,000 = 6.44 \times 10^n$
   c) $0.0040 = 4.0 \times 10^n$
   d) $1000^2 = 1 \times 10^n$
   e) $0.9^3 = 7.29 \times 10^n$
   f) $800^3 = 5.12 \times 10^n$

6. Write the answers to the following calculations in standard form:
   a) $50\,000 \times 2400$
   b) $(3.7 \times 10^6) \times (4.0 \times 10^4)$
   c) $(5.8 \times 10^7) + (9.3 \times 10^6)$
   d) $(4.7 \times 10^6) - (8.2 \times 10^5)$

7. The speed of light is 3×10^8 m/s. Jupiter is 778 million km from the Sun. Calculate the number of minutes it takes for sunlight to reach Jupiter.

8. A star is 300 light years away from Earth. The speed of light is 3×10^5 km/s. Calculate the distance from the star to Earth. Give your answer in kilometres and written in standard form.

Number

Student assessment 5

1. Evaluate the following without the use of a calculator:
   a) $81^{\frac{1}{2}}$
   b) $27^{\frac{1}{3}}$
   c) $9^{\frac{1}{2}}$
   d) $625^{\frac{3}{4}}$
   e) $343^{\frac{2}{3}}$
   f) $16^{-\frac{1}{4}}$
   g) $\dfrac{1}{25^{-\frac{1}{2}}}$
   h) $\dfrac{2}{16^{-\frac{3}{4}}}$

2. Evaluate the following without the use of a calculator:
   a) $\dfrac{16^{\frac{1}{2}}}{2^2}$
   b) $\dfrac{9^{\frac{5}{2}}}{3^3}$
   c) $\dfrac{8^{\frac{4}{3}}}{8^{\frac{2}{3}}}$
   d) $5^{\frac{6}{5}} \times 5^{\frac{4}{5}}$
   e) $4^{\frac{3}{2}} \times 2^{-2}$
   f) $\dfrac{27^{\frac{2}{3}} \times 3^{-2}}{4^{-\frac{3}{2}}}$
   g) $\dfrac{(4^3)^{-\frac{1}{2}} \times 2^{\frac{3}{2}}}{2^{-\frac{3}{2}}}$
   h) $\dfrac{(5^3)^{\frac{1}{2}} \times 5^{\frac{2}{3}}}{3^{-2}}$

3. Draw a pair of axes with x from -4 to 4 and y from 0 to 10.
   a) Plot a graph of $y = 3^{\frac{x}{2}}$.
   b) Use your graph to estimate when $3^{\frac{x}{2}} = 5$.

Student assessment 6

1. Evaluate the following without the use of a calculator:
   a) $64^{\frac{1}{6}}$
   b) $27^{\frac{4}{3}}$
   c) $9^{-\frac{1}{2}}$
   d) $512^{\frac{2}{9}}$
   e) $\sqrt[3]{27}$
   f) $\sqrt[4]{16}$
   g) $\dfrac{1}{36^{-\frac{1}{2}}}$
   h) $\dfrac{2}{64^{-\frac{2}{3}}}$

2. Evaluate the following without the use of a calculator:
   a) $\dfrac{25^{\frac{1}{2}}}{9^{-\frac{1}{2}}}$
   b) $\dfrac{4^{\frac{5}{2}}}{2^3}$
   c) $\dfrac{27^{\frac{4}{3}}}{3^3}$
   d) $25^{\frac{3}{2}} \times 5^2$
   e) $4^{\frac{6}{4}} \times 4^{-\frac{1}{2}}$
   f) $\dfrac{27^{\frac{2}{3}} \times 3^{-3}}{9^{-\frac{1}{2}}}$
   g) $\dfrac{(4^2)^{-\frac{1}{4}} \times 9^{\frac{3}{2}}}{\left(\frac{1}{4}\right)^{\frac{1}{2}}}$
   h) $\dfrac{(5^3)^{\frac{1}{2}} \times 5^{\frac{5}{6}}}{4^{-\frac{1}{2}}}$

3. Draw a pair of axes with x from -4 to 4 and y from 0 to 18.
   a) Plot a graph of $y = 4^{-\frac{x}{2}}$.
   b) Use your graph to estimate when $4^{-\frac{x}{2}} = 6$.

8 Money and finance

● **Currency conversions**

In 2012, 1 euro could be exchanged for 1.25 Australian dollars (A$).

Worked examples

a) How many Australian dollars can be bought for €400?
   €1 buys A$1.25.
   €400 buys 1.25 × 400 = A$500.

b) How much does it cost to buy A$940?
   A$1.25 costs €1.
   A$940 costs $\frac{1 \times 940}{1.25}$ = €752.

Exercise 8.1

The table shows the exchange rate for €1 into various currencies.

Australia	1.25 Australian dollars (A$)
India	70 rupees
Zimbabwe	470 Zimbabwe dollars (ZIM$)
South Africa	11 rand
Turkey	2.3 Turkish lira (L)
Japan	103 yen
Kuwait	0.4 dinar
USA	1.3 US dollars (US$)

1. Convert the following:
   a) €25 into Australian dollars
   b) €50 into rupees
   c) €20 into Zimbabwe dollars
   d) €300 into rand
   e) €130 into Turkish lira
   f) €40 into yen
   g) €400 into dinar
   h) €150 into US dollars

2. How many euro does it cost to buy the following:
   a) A$500
   b) 200 rupees
   c) ZIM$1000
   d) 500 rand
   e) 750 Turkish lira
   f) 1200 yen
   g) 50 dinar
   h) US$150

Number

● **Earnings**

Net pay is what is left after deductions such as tax, insurance and pension contributions are taken from **gross earnings**.
That is, Net pay = Gross pay − Deductions

A **bonus** is an extra payment sometimes added to an employee's basic pay.

In many companies there is a fixed number of hours that an employee is expected to work. Any work done in excess of this **basic week** is paid at a higher rate, referred to as **overtime**. Overtime may be 1.5 times basic pay, called **time and a half**, or twice basic pay, called **double time**.

Piece work is another method of payment. Employees are paid for the number of articles made, not for the time taken.

Exercise 8.2

1. Mr Ahmet's gross pay is $188.25. Deductions amount to $33.43. What is his net pay?

2. Miss Said's basic pay is $128. She earns $36 for overtime and receives a bonus of $18. What is her gross pay?

3. Mrs Hafar's gross pay is $203. She pays $54 in tax and $18 towards her pension. What is her net pay?

4. Mr Wong works 35 hours for an hourly rate of $8.30. What is his basic pay?

5. a) Miss Martinez works 38 hours for an hourly rate of $4.15. In addition she works 6 hours of overtime at time and a half. What is her total gross pay?
   b) Deductions amount to 32% of her total gross pay. What is her net pay?

6. Pepe is paid $5.50 for each basket of grapes he picks. One week he picks 25 baskets. How much is he paid?

7. Maria is paid €5 for every 12 plates that she makes. This is her record for one week.

Mon	240
Tues	360
Wed	288
Thurs	192
Fri	180

How much is she paid?

8. Neo works at home making clothes. The patterns and materials are provided by the company. The table shows the rates she is paid and the number of items she makes in one week:

Item	Rate	Number made
Jacket	25 rand	3
Trousers	11 rand	12
Shirt	13 rand	7
Dress	12 rand	0

a) What are her gross earnings?
b) Deductions amount to 15% of gross earnings. What is her net pay?

● **Profit and loss**

Foodstuffs and manufactured goods are produced at a cost, known as the **cost price**, and sold at the **selling price**. If the selling price is greater than the cost price, a profit is made.

Worked example A market trader buys oranges in boxes of 12 dozen for $14.40 per box. He buys three boxes and sells all the oranges for 12c each. What is his profit or loss?

Cost price: $3 \times \$14.40 = \$43.20$
Selling price: $3 \times 144 \times 12c = \$51.84$
In this case he makes a profit of $51.84 − $43.20
His profit is $8.64.

A second way of solving this problem would be:
$14.40 for a box of 144 oranges is 10c each.
So cost price of each orange is 10c, and selling price of each orange is 12c. The profit is 2c per orange.
So 3 boxes would give a profit of $3 \times 144 \times 2c$.
That is, $8.64.

Sometimes, particularly during sales or promotions, the selling price is reduced, this is known as a **discount**.

Worked example In a sale a skirt usually costing $35 is sold at a 15% discount. What is the discount?
15% of $35 = $0.15 \times \$35 = \$5.25$
The discount is $5.25.

Exercise 8.3 1. A market trader buys peaches in boxes of 120. He buys 4 boxes at a cost price of $13.20 per box. He sells 425 peaches at 12c each – the rest are ruined. How much profit or loss does he make?

2. A shopkeeper buys 72 bars of chocolate for $5.76. What is his profit if he sells them for 12c each?

3. A holiday company charters an aircraft to fly to Malta at a cost of $22 000. It then sells 150 seats at $185 each and a futher 35 seats at a 20% discount. Calculate the profit made per seat if the plane has 200 seats.

4. A car is priced at $7200. The car dealer allows a customer to pay a one-third deposit and 12 payments of $420 per month. How much extra does it cost the customer?

5. At an auction a company sells 150 television sets for an average of $65 each. The production cost was $10 000. How much loss did the company make?

● **Percentage profit and loss**

Most profits or losses are expressed as a percentage.
Profit or loss, divided by cost price, multiplied by 100
= % profit or loss.

Worked example A woman buys a car for $7500 and sells it two years later for $4500. Calculate her loss over two years as a percentage of the cost price.

cost price = $7500   selling price = $4500   loss = $3000

$$\text{Loss \%} = \frac{3000}{7500} \times 100 = 40$$

Her loss is 40%.

When something becomes worth less over a period of time, it is said to **depreciate**.

Exercise 8.4
1. Find the depreciation of the following cars as a percentage of the cost price. (C.P. = cost price, S.P. = selling price)
   a) VW   C.P. $4500   S.P. $4005
   b) Rover   C.P. $9200   S.P. $6900

2. A company manufactures electrical items for the kitchen. Find the percentage profit on each of the following:
   a) Fridge   C.P. $50   S.P. $65
   b) Freezer   C.P. $80   S.P. $96

3. A developer builds a number of different types of house. Which type gives the developer the largest percentage profit?
   Type A   C.P. $40 000   S.P. $52 000
   Type B   C.P. $65 000   S.P. $75 000
   Type C   C.P. $81 000   S.P. $108 000

4. Students in a school organise a disco. The disco company charges $350 hire charge. The students sell 280 tickets at $2.25. What is the percentage profit?

● Interest

Interest can be defined as money added by a bank to sums deposited by customers. The money deposited is called the **principal**. The **percentage interest** is the given rate and the money is left for a fixed period of time.

A formula can be obtained for **simple interest**:

$$SI = \frac{Ptr}{100}$$

where SI = simple interest, i.e. the interest paid
P = the principal
t = time in years
r = rate percent

Worked examples

a) Find the simple interest earned on $250 deposited for 6 years at 8% p.a.

$$SI = \frac{Ptr}{100}$$

$$SI = \frac{250 \times 6 \times 8}{100}$$

$$SI = 120$$

So the interest paid is $120.

b) How long will it take for a sum of $250 invested at 8% to earn interest of $80?

$$SI = \frac{Ptr}{100}$$

$$80 = \frac{250 \times t \times 8}{100}$$

$$80 = 20t$$

$$4 = t$$

It will take 4 years.

c) What rate per year must be paid for a principal of $750 to earn interest of $180 in 4 years?

$$SI = \frac{Ptr}{100}$$

$$180 = \frac{750 \times 4 \times r}{100}$$

$$180 = 30r$$

$$6 = r$$

The rate must be 6% per year.

d) Find the principal which will earn interest of $120 in 6 years at 4%.

$$SI = \frac{Ptr}{100}$$

$$120 = \frac{P \times 6 \times 4}{100}$$

$$120 = \frac{24P}{100}$$

$$12\,000 = 24P$$
$$500 = P$$

So the principal is $500.

Exercise 8.5 All rates of interest given here are annual rates.

1. Find the simple interest paid in the following cases:
   a) Principal $300   rate 6%   time 4 years
   b) Principal $750   rate 8%   time 7 years

2. Calculate how long it will take for the following amounts of interest to be earned at the given rate.
   a) $P = \$500$ $r = 6\%$ $SI = \$150$
   b) $P = \$400$ $r = 9\%$ $SI = \$252$

3. Calculate the rate of interest per year which will earn the given amount of interest:
   a) Principal $400   time 4 years   interest $112
   b) Principal $800   time 7 years   interest $224

4. Calculate the principal which will earn the interest below in the given number of years at the given rate:
   a) $SI = \$36$ time = 3 years   rate = 6%
   b) $SI = \$340$ time = 5 years   rate = 8%

5. What rate of interest is paid on a deposit of $2000 which earns $400 interest in 5 years?

6. How long will it take a principal of $350 to earn $56 interest at 8% per year?

7. A principal of $480 earns $108 interest in 5 years. What rate of interest was being paid?

8. A principal of $750 becomes a total of $1320 in 8 years. What rate of interest was being paid?

9. $1500 is invested for 6 years at 3.5% per year. What is the interest earned?

10. $500 is invested for 11 years and becomes $830 in total. What rate of interest was being paid?

Compound interest

Compound interest means interest is paid not only on the principal amount, but also on the interest itself: it is compounded (or added to).

This sounds complicated but the example below will make it clear.

e.g. A builder is going to build six houses on a plot of land in Spain. He borrows €500 000 at 10% interest and will pay off the loan in full after three years.

At the end of the first year he will owe:
€500 000 + 10% of €500 000 i.e. €500 000 × 1.10 = €550 000

At the end of the second year he will owe:
€550 000 + 10% of €550 000 i.e. €550 000 × 1.10 = €605 000

At the end of the third year he will owe:
€605 000 + 10% of €605 000 i.e. €605 000 × 1.10 = €665 500

The compound interest he has to pay is €665 500 − €500 000 i.e. €165 500

The time taken for a debt to grow at compound interest can be calculated as shown in the example below:

Worked example How long will it take for a debt to double at a compound interest rate of 27% p.a.?

An interest rate of 27% implies a multiplier of 1.27.

Time (years)	0	1	2	3
Debt	P	1.27P	$1.27^2P = 1.61P$	$1.27^3P = 2.05P$

× 1.27   × 1.27   × 1.27

The debt will have more than doubled after 3 years.

Using the example above of the builder's loan, if P represents the principal he borrows, then after 1 year his debt (D) will be given by the formula:

$D = P\left(1 + \dfrac{r}{100}\right)$ where r is the rate of interest.

After 2 years: $D = P\left(1 + \dfrac{r}{100}\right)\left(1 + \dfrac{r}{100}\right)$

After 3 years: $D = P\left(1 + \dfrac{r}{100}\right)\left(1 + \dfrac{r}{100}\right)\left(1 + \dfrac{r}{100}\right)$

After n years: $D = P\left(1 + \dfrac{r}{100}\right)^n$

Number

This formula for the debt includes the original loan. By subtracting P, the compound interest is calculated:

$$I = P\left(1 + \frac{r}{100}\right)^n - P$$

Compound interest is an example of a geometric sequence and therefore of exponential growth.

The interest is usually calculated annually, but there can be other time periods. Compound interest can be charged yearly, half-yearly, quarterly, monthly or daily. (In theory any time period can be chosen.)

Worked examples

a) Find the compound interest paid on a loan of $600 for 3 years at an annual percentage rate (APR) of 5%.

When the rate is 5%, $1 + \frac{5}{100} = 1.05$.

$D = 600 \times 1.05^3 = 694.58$ (to 2 d.p.)

The total payment is $694.58 so the interest due is $694.58 − $600 = $94.58.

b) Find the compound interest when $3000 is invested for 18 months at an APR of 8.5%. The interest is calculated every six months.

Note: The interest for each time period of 6 months is $\frac{8.5}{2}\% = 4.25\%$. There will therefore be 3 time periods of 6 months each.

When the rate is 4.25%, $1 + \frac{4.25}{100} = 1.0425$.

$D = 3000 \times 1.0425^3 = 3398.986 \ldots$

The final sum is $3399, so the interest is $3399 − $3000 = $399.

Exercise 8.6

1. A shipping company borrows $70 million at 5% p.a. compound interest to build a new cruise ship. If it repays the debt after 3 years, how much interest will the company pay?

2. A woman borrows $100 000 for home improvements. The compound interest rate is 15% p.a. and she repays it in full after 3 years. How much interest will she pay?

3. A man owes $5000 on his credit cards. The APR is 20%. If he doesn't repay any of the debt, how much will he owe after 4 years?

4. A school increases its intake by 10% each year. If it starts with 1000 students, how many will it have at the beginning of the fourth year of expansion?

5. 8 million tonnes of fish were caught in the North Sea in 2005. If the catch is reduced by 20% each year for 4 years, what weight is caught at the end of this time?

6. How many years will it take for a debt to double at 42% p.a. compound interest?

7. How many years will it take for a debt to double at 15% p.a. compound interest?

8. A car loses value at a rate of 27% each year. How long will it take for its value to halve?

Student assessment 1

1. A visitor from Hong Kong receives 12 Pakistan rupees for each Hong Kong dollar.
   a) How many Pakistan rupees would he get for HK$240?
   b) How many Hong Kong dollars does it cost for 1 thousand rupees?

2. Below is a currency conversion table showing the amount of foreign currency received for 1 euro.

New Zealand	1.6 dollars (NZ$)
Brazil	2.6 reals

   a) How many euro does it cost for NZ$1000?
   b) How many euro does it cost for 500 Brazilian reals?

3. A girl works in a shop on Saturdays for 8.5 hours. She is paid $3.60 per hour. What is her gross pay for 4 weeks' work?

4. A potter makes cups and saucers in a factory. He is paid $1.44 per batch of cups and $1.20 per batch of saucers. What is his gross pay if he makes 9 batches of cups and 11 batches of saucers in one day?

5. Calculate the missing numbers from the simple interest table below:

Principal ($)	Rate (%)	Time (years)	Interest ($)
300	6	4	(a)
250	(b)	3	60
480	5	(c)	96
650	(d)	8	390
(e)	3.75	4	187.50

6. A family house was bought for $48 000 twelve years ago. It is now valued at $120 000. What is the average annual increase in the value of the house?

7. An electrician bought five broken washing machines for $550. He repaired them and sold them for $143 each. What was his percentage profit?

Student assessment 2

1. Find the simple interest paid on the following principal sums P, deposited in a savings account for t years at a fixed rate of interest of $r\%$:
   a) $P = \$550$ $t = 5$ years $r = 3\%$
   b) $P = \$8000$ $t = 10$ years $r = 6\%$
   c) $P = \$12\,500$ $t = 7$ years $r = 2.5\%$

2. A sum of $25 000 is deposited in a bank. After 8 years, the simple interest gained was $7000. Calculate the annual rate of interest on the account assuming it remained constant over the 8 years.

3. A bank lends a business $250 000. The annual rate of interest is 8.4%. When paying back the loan, the business pays an amount of $105 000 in simple interest. Calculate the number of years the business took out the loan for.

4. Find the compound interest paid on the following principal sums P, deposited in a savings account for n years at a fixed rate of interest of $r\%$:
   a) $P = \$400$ $n = 2$ years $r = 3\%$
   b) $P = \$5000$ $n = 8$ years $r = 6\%$
   c) $P = \$18\,000$ $n = 10$ years $r = 4.5\%$

5. A car is bought for $12 500. Its value depreciates by 15% per year.
   a) Calculate its value after:
      i) 1 year   ii) 2 years
   b) After how many years will the car be worth less than $1000?

9 Time

Times may be given in terms of the 12-hour clock. We tend to say, 'I get up at seven o'clock in the morning, play football at half past two in the afternoon, and go to bed before eleven o'clock'.

These times can be written as 7 a.m., 2.30 p.m. and 11 p.m.

In order to save confusion, most timetables are written using the 24-hour clock.

    7 a.m. is written as 0700
    2.30 p.m. is written as 1430
    11.00 p.m. is written as 2300

Worked example A train covers the 480 km journey from Paris to Lyon at an average speed of 100 km/h. If the train leaves Paris at 0835, when does it arrive in Lyon?

Time taken = $\dfrac{\text{distance}}{\text{speed}}$

Paris to Lyon = $\dfrac{480}{100}$ hours, that is, 4.8 hours.

4.8 hours is 4 hours and (0.8 × 60 minutes), that is, 4 hours and 48 minutes.

Departure 0835; arrival 0835 + 0448

Arrival time is 1323.

Exercise 9.1

1. A journey to work takes a woman three quarters of an hour. If she catches the bus at 0755, when does she arrive?

2. The same woman catches a bus home each evening. The journey takes 55 minutes. If she catches the bus at 1750, when does she arrive?

3. A boy cycles to school each day. His journey takes 70 minutes. When will he arrive if he leaves home at 0715?

4. Find the time in hours and minutes for the following journeys of the given distance at the average speed stated:
   a) 230 km at 100 km/h   b) 70 km at 50 km/h

5. Grand Prix racing cars cover a 120 km race at the following average speeds. How long do the first five cars take to complete the race? Answer in minutes and seconds.

    First 240 km/h   Second 220 km/h   Third 210 km/h
    Fourth 205 km/h   Fifth 200 km/h

6. A train covers the 1500 km distance from Amsterdam to Barcelona at an average speed of 90 km/h. If the train leaves Amsterdam at 9.30 a.m. on Tuesday, when does it arrive in Barcelona?

7. A plane takes off at 16 25 for the 3200 km journey from Moscow to Athens. If the plane flies at an average speed of 600 km/h, when will it land in Athens?

8. A plane leaves London for Boston, a distance of 5200 km, at 09 45. The plane travels at an average speed of 800 km/h. If Boston time is five hours behind British time, what is the time in Boston when the aircraft lands?

Student assessment 1

1. A journey to school takes a girl 25 minutes. What time does she arrive if she leaves home at 08 38?

2. A car travels 295 km at 50 km/h. How long does the journey take? Give your answer in hours and minutes.

3. A bus leaves Deltaville at 11 32. It travels at an average speed of 42 km/h. If it arrives in Eastwich at 12 42, what is the distance between the two towns?

4. A plane leaves Betatown at 17 58 and arrives at Fleckley at 05 03 the following morning. How long does the journey take? Give your answer in hours and minutes.

Student assessment 2

1. A journey to school takes a boy 22 minutes. What is the latest time he can leave home if he must be at school at 08 40?

2. A plane travels 270 km at 120 km/h. How long does the journey take? Give your answer in hours and minutes.

3. A train leaves Alphaville at 13 27. It travels at an average speed of 56 km/h. If it arrives in Eastwich at 16 12, what is the distance between the two towns?

4. A car leaves Gramton at 16 39. It travels a distance of 315 km and arrives at Halfield at 20 09.

    a) How long does the journey take?
    b) What is the car's average speed?

10 Set notation and Venn diagrams

● **Sets**

A **set** is a well defined group of objects or symbols. The objects or symbols are called the **elements** of the set. If an element e belongs to a set S, this is represented as $e \in S$. If e does not belong to set S this is represented as $e \notin S$.

Worked examples

a) A particular set consists of the following elements:

{South Africa, Namibia, Egypt, Angola, ...}

i) Describe the set.

The elements of the set are countries of Africa.

ii) Add another two elements to the set.

e.g. Zimbabwe, Ghana

iii) Is the set finite or infinite?

Finite. There is a finite number of countries in Africa.

b) Consider the set $A = \{x: x \text{ is a natural number}\}$

i) Describe the set.

The elements of the set are the natural numbers.

ii) Write down two elements of the set.

e.g. 3 and 15

c) Consider the set $B = \{(x, y): y = 2x - 4\}$

i) Describe the set.

The elements of the set are the coordinates of points found on the straight line with equation $y = 2x - 4$.

ii) Write down two elements of the set.

e.g. $(0, -4)$ and $(10, 16)$

d) Consider the set $C = \{x: 2 \leq x \leq 8\}$

i) Describe the set.

The elements of the set include any number between 2 and 8 inclusive.

ii) Write down two elements of the set.

e.g. 5 and 6.3

Exercise 10.1

1. In the following questions:
   i) describe the set in words,
   ii) write down another two elements of the set.
      a) {Asia, Africa, Europe, ...}
      b) {2, 4, 6, 8, ...}
      c) {Sunday, Monday, Tuesday, ...}
      d) {January, March, July, ...}
      e) {1, 3, 6, 10, ...}
      f) {Mehmet, Michael, Mustapha, Matthew, ...}
      g) {11, 13, 17, 19, ...}
      h) {a, e, i, ...}
      i) {Earth, Mars, Venus, ...}
      j) $A = \{x : 3 \leqslant x \leqslant 12\}$
      k) $S = \{y : -5 \leqslant y \leqslant 5\}$

2. The number of elements in a set A is written as $n(A)$.
   Give the value of $n(A)$ for the finite sets in questions 1a–k above.

● Subsets

If all the elements of one set X are also elements of another set Y, then X is said to be a **subset** of Y.

This is written as $X \subseteq Y$.

If a set A is empty (i.e. it has no elements in it), then this is called the **empty set** and it is represented by the symbol ∅. Therefore $A = \emptyset$. The empty set is a subset of all sets.

   e.g. Three girls, Winnie, Natalie and Emma form a set A
   A = {Winnie, Natalie, Emma}
   All the possible subsets of A are given below:
   B = {Winnie, Natalie, Emma}
   C = {Winnie, Natalie}
   D = {Winnie, Emma}
   E = {Natalie, Emma}
   F = {Winnie}
   G = {Natalie}
   H = {Emma}
   I = ∅

Note that the sets B and I above are considered as subsets of A.

   i.e. $B \subseteq A$ and $I \subseteq A$

However, sets C, D, E, F, G and H are considered **proper subsets** of A. This distinction of subset is shown in the notation below.

   $C \subset A$ and $D \subset A$ etc.

Similarly $G \not\subseteq H$ implies that G is not a subset of H

   $G \not\subset H$ implies that G is not a proper subset of H

Set notation and Venn diagrams

Worked example $A = \{1, 2, 3, 4, 5, 6, 7, 8, 9, 10\}$
  i) List subset B {even numbers}.
  $B = \{2, 4, 6, 8, 10\}$

  ii) List subset C {prime numbers}.
  $C = \{2, 3, 5, 7\}$

Exercise 10.2

1. $P = $ {whole numbers less than 30}
   a) List the subset Q {even numbers}.
   b) List the subset R {odd numbers}.
   c) List the subset S {prime numbers}.
   d) List the subset T {square numbers}.
   e) List the subset U {triangle numbers}.

2. $A = $ {whole numbers between 50 and 70}
   a) List the subset B {multiples of 5}.
   b) List the subset C {multiples of 3}.
   c) List the subset D {square numbers}.

3. $J = \{p, q, r\}$
   a) List all the subsets of J.
   b) List all the proper subsets of J.

4. State whether each of the following statements is true or false:
   a) {Algeria, Mozambique} $\subseteq$ {countries in Africa}
   b) {mango, banana} $\subseteq$ {fruit}
   c) $\{1, 2, 3, 4\} \subseteq \{1, 2, 3, 4\}$
   d) $\{1, 2, 3, 4\} \subset \{1, 2, 3, 4\}$
   e) {volleyball, basketball} $\not\subseteq$ {team sport}
   f) $\{4, 6, 8, 10\} \not\subset \{4, 6, 8, 10\}$
   g) {potatoes, carrots} $\subseteq$ {vegetables}
   h) $\{12, 13, 14, 15\} \not\subseteq$ {whole numbers}

● **The universal set**

The **universal set** ($\mathscr{E}$) for any particular problem is the set which contains all the possible elements for that problem.

The **complement** of a set A is the set of elements which are in $\mathscr{E}$ but not in A. The complement of A is identified as A'. Notice that $\mathscr{E}' = \varnothing$ and $\varnothing' = \mathscr{E}$.

Worked examples a) If $\mathscr{E} = \{1, 2, 3, 4, 5, 6, 7, 8, 9, 10\}$ and $A = \{1, 2, 3, 4, 5\}$ what set is represented by A'?

  A' consists of those elements in $\mathscr{E}$ which are not in A.

  Therefore $A' = \{6, 7, 8, 9, 10\}$.

  b) If $\mathscr{E} = $ {all 3D shapes} and $P = $ {prisms} what set is represented by P'?

  $P' = $ {all 3D shapes except prisms}.

● Set notation and Venn diagrams

Venn diagrams are the principal way of showing sets diagrammatically. The method consists primarily of entering the elements of a set into a circle or circles.

Some examples of the uses of Venn diagrams are shown.

$A = \{2, 4, 6, 8, 10\}$ can be represented as:

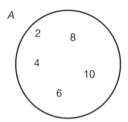

Elements which are in more than one set can also be represented using a Venn diagram.

$P = \{3, 6, 9, 12, 15, 18\}$ and $Q = \{2, 4, 6, 8, 10, 12\}$ can be represented as:

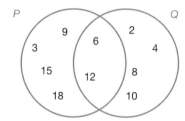

In the diagram above it can be seen that those elements which belong to both sets are placed in the region of overlap of the two circles.

When two sets P and Q overlap as they do above, the notation $P \cap Q$ is used to denote the set of elements in the **intersection**, i.e. $P \cap Q = \{6, 12\}$.

Note that $6 \in P \cap Q$; $8 \notin P \cap Q$.

$J = \{10, 20, 30, 40, 50, 60, 70, 80, 90, 100\}$ and
$K = \{60, 70, 80\}$; as discussed earlier, $K \subset J$ can be represented as shown below:

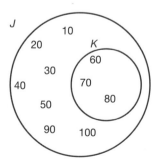

$X = \{1, 3, 6, 7, 14\}$ and $Y = \{3, 9, 13, 14, 18\}$ are represented as:

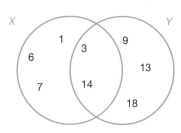

The **union** of two sets is everything which belongs to either or both sets and is represented by the symbol ∪.
Therefore in the example above $X \cup Y = \{1, 3, 6, 7, 9, 13, 14, 18\}$.

Exercise 10.3

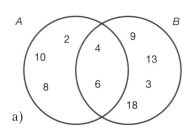

1. Using the Venn diagram (left), indicate whether the following statements are true or false. ∈ means 'is an element of' and ∉ means 'is not an element of'.
   a) $5 \in A$ b) $20 \in B$ c) $20 \notin A$
   d) $50 \in A$ e) $50 \notin B$ f) $A \cap B = \{10, 20\}$

2. Complete the statement $A \cap B = \{\ldots\}$ for each of the Venn diagrams below.

a) Venn diagram with A containing 2, 10, 8; intersection containing 4, 6; B containing 9, 13, 3, 18.

b) Venn diagram with A containing 1, 16; intersection containing 4, 9; B containing 5, 7, 6, 8.

c) Venn diagram with A containing Red, Orange, Blue, Indigo, Violet; intersection containing Yellow, Green; B containing Purple, Pink.

3. Complete the statement $A \cup B = \{\ldots\}$ for each of the Venn diagrams in question 2 above.

4.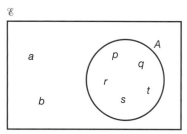

Copy and complete the following statements:
a) $\mathscr{E} = \{\ldots\}$ b) $A' = \{\ldots\}$

5.

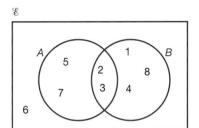

Copy and complete the following statements:
a) $\mathcal{E} = \{\ldots\}$ b) $A' = \{\ldots\}$ c) $A \cap B = \{\ldots\}$
d) $A \cup B = \{\ldots\}$ e) $(A \cap B)' = \{\ldots\}$ f) $A \cap B' = \{\ldots\}$

6. a) Describe in words the elements of:
   i) set A ii) set B iii) set C
   b) Copy and complete the following statements:
   i) $A \cap B = \{\ldots\}$ ii) $A \cap C = \{\ldots\}$
   iii) $B \cap C = \{\ldots\}$ iv) $A \cap B \cap C = \{\ldots\}$
   v) $A \cup B = \{\ldots\}$ vi) $C \cup B = \{\ldots\}$

7.

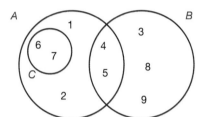

a) Copy and complete the following statements:
   i) $A = \{\ldots\}$ ii) $B = \{\ldots\}$
   iii) $C' = \{\ldots\}$ iv) $A \cap B = \{\ldots\}$
   v) $A \cup B = \{\ldots\}$ vi) $(A \cap B)' = \{\ldots\}$
b) State, using set notation, the relationship between C and A.

8.

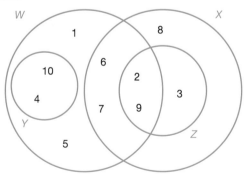

a) Copy and complete the following statements:
   i) $W = \{\ldots\}$ ii) $X = \{\ldots\}$
   iii) $Z' = \{\ldots\}$ iv) $W \cap Z = \{\ldots\}$
   v) $W \cap X = \{\ldots\}$ vi) $Y \cap Z = \{\ldots\}$
b) Which of the named sets is a subset of X?

Exercise 10.4

1. A = {Egypt, Libya, Morocco, Chad}
   B = {Iran, Iraq, Turkey, Egypt}
   a) Draw a Venn diagram to illustrate the above information.
   b) Copy and complete the following statements:
      i) $A \cap B$ = {...}   ii) $A \cup B$ = {...}

2. P = {2, 3, 5, 7, 11, 13, 17}
   Q = {11, 13, 15, 17, 19}
   a) Draw a Venn diagram to illustrate the above information.
   b) Copy and complete the following statements:
      i) $P \cap Q$ = {...}   ii) $P \cup Q$ = {...}

3. B = {2, 4, 6, 8, 10}
   $A \cup B$ = {1, 2, 3, 4, 6, 8, 10}
   $A \cap B$ = {2, 4}
   Represent the above information on a Venn diagram.

4. X = {a, c, d, e, f, g, l}
   Y = {b, c, d, e, h, i, k, l, m}
   Z = {c, f, i, j, m}
   Represent the above information on a Venn diagram.

5. P = {1, 4, 7, 9, 11, 15}
   Q = {5, 10, 15}
   R = {1, 4, 9}
   Represent the above information on a Venn diagram.

● Problems involving sets

Worked example In a class of 31 students, some study Physics and some study Chemistry. If 22 study Physics, 20 study Chemistry and 5 study neither, calculate the number of students who take both subjects.

The information given above can be entered in a Venn diagram in stages.

The students taking neither Physics nor Chemistry can be put in first (as shown left).

This leaves 26 students to be entered into the set circles.
If x students take both subjects then

$$n(P) = 22 - x + x$$
$$n(C) = 20 - x + x$$
$$P \cup C = 31 - 5 = 26$$

Therefore $22 - x + x + 20 - x = 26$
$$42 - x = 26$$
$$x = 16$$

Substituting the value of x into the Venn diagram gives:

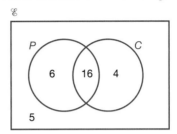

Therefore the number of students taking both Physics and Chemistry is 16.

Exercise 10.5

1. In a class of 35 students, 19 take Spanish, 18 take French and 3 take neither. Calculate how many take:
   a) both French and Spanish,
   b) just Spanish,
   c) just French.

2. In a year group of 108 students, 60 liked football, 53 liked tennis and 10 liked neither. Calculate the number of students who liked football but not tennis.

3. In a year group of 113 students, 60 liked hockey, 45 liked rugby and 18 liked neither. Calculate the number of students who:
   a) liked both hockey and rugby,
   b) liked only hockey.

4. One year 37 students sat an examination in Physics, 48 sat Chemistry and 45 sat Biology. 15 students sat Physics and Chemistry, 13 sat Chemistry and Biology, 7 sat Physics and Biology and 5 students sat all three.
   a) Draw a Venn diagram to represent this information.
   b) Calculate $n \, (P \cup C \cup B)$.

Student assessment 1

1. Describe the following sets in words:
   a) {2, 4, 6, 8}
   b) {2, 4, 6, 8, …}
   c) {1, 4, 9, 16, 25, …}
   d) {Arctic, Atlantic, Indian, Pacific}

2. Calculate the value of $n(A)$ for each of the sets shown below:
   a) $A = $ {days of the week}
   b) $A = $ {prime numbers between 50 and 60}
   c) $A = \{x : x$ is an integer and $5 \leqslant x \leqslant 10\}$
   d) $A = $ {days in a leap year}

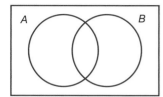

3. Copy out the Venn diagram (left) twice.
   a) On one copy shade and label the region which represents $A \cap B$.
   b) On the other copy shade and label the region which represents $A \cup B$.
4. If $A = \{a, b\}$ list all the subsets of A.
5. If $\mathscr{E} = \{m, a, t, h, s\}$ and $A = \{a, s\}$, what set is represented by A'?

Student assessment 2

1. Describe the following sets in words:
   a) $\{1, 3, 5, 7\}$
   b) $\{1, 3, 5, 7, \ldots\}$
   c) $\{1, 3, 6, 10, 15, \ldots\}$
   d) {Brazil, Chile, Argentina, Bolivia, ...}

2. Calculate the value of $n(A)$ for each of the sets shown below:
   a) $A = $ {months of the year}
   b) $A = $ {square numbers between 99 and 149}
   c) $A = \{x: x$ is an integer and $-9 \leqslant x \leqslant -3\}$
   d) $A = $ {students in your class}

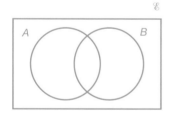

3. Copy out the Venn diagram (left) twice.
   a) On one copy shade and label the region which represents $\mathscr{E}$.
   b) On the other copy shade and label the region which represents $(A \cap B)'$.

4. If $A = \{w, o, r, k\}$ list all the subsets of A with at least three elements.

5. If $\mathscr{E} = \{1, 2, 3, 4, 5, 6, 7, 8\}$ and $P = \{2, 4, 6, 8\}$, what set is represented by P'?

Student assessment 3

1. If $A = \{2, 4, 6, 8\}$ write all the proper subsets of A with two or more elements.

2. $J = $ {London, Paris, Rome, Washington, Canberra, Ankara, Cairo}
   $K = $ {Cairo, Nairobi, Pretoria, Ankara}
   a) Draw a Venn diagram to represent the above information.
   b) Copy and complete the statement $J \cap K = \{\ldots\}$.
   c) Copy and complete the statement $J' \cap K = \{\ldots\}$.

3. $M = \{x: x \text{ is an integer and } 2 \leq x \leq 20\}$
   $N = \{\text{prime numbers less than } 30\}$
   a) Draw a Venn diagram to illustrate the information above.
   b) Copy and complete the statement $M \cap N = \{\ldots\}$.
   c) Copy and complete the statement $(M \cap N)' = \{\ldots\}$.

4. $\mathscr{E} = \{\text{natural numbers}\}$, $M = \{\text{even numbers}\}$ and $N = \{\text{multiples of } 5\}$.
   a) Draw a Venn diagram and place the numbers 1, 2, 3, 4, 5, 6, 7, 8, 9, 10 in the appropriate places in it.
   b) If $X = M \cap N$, describe set X in words.

5. In a region of mixed farming, farms keep goats, cattle or sheep. There are 77 farms altogether. 19 farms keep only goats, 8 keep only cattle and 13 keep only sheep. 13 keep both goats and cattle, 28 keep both cattle and sheep and 8 keep both goats and sheep.
   a) Draw a Venn diagram to show the above information.
   b) Calculate $n(G \cap C \cap S)$.

Student assessment 4

1. $M = \{a, e, i, o, u\}$
   a) How many subsets are there of M?
   b) List the subsets of M with four or more elements.

2. $X = \{\text{lion, tiger, cheetah, leopard, puma, jaguar, cat}\}$
   $Y = \{\text{elephant, lion, zebra, cheetah, gazelle}\}$
   $Z = \{\text{anaconda, jaguar, tarantula, mosquito}\}$
   a) Draw a Venn diagram to represent the above information.
   b) Copy and complete the statement $X \cap Y = \{\ldots\}$.
   c) Copy and complete the statement $Y \cap Z = \{\ldots\}$.
   d) Copy and complete the statement $X \cap Y \cap Z = \{\ldots\}$.

3. A group of 40 people were asked whether they like cricket (C) and football (F). The number liking both cricket and football was three times the number liking only cricket. Adding 3 to the number liking only cricket and doubling the answer equals the number of people liking only football. Four said they did not like sport at all.
   a) Draw a Venn diagram to represent this information.
   b) Calculate $n(C \cap F)$.
   c) Calculate $n(C \cap F')$.
   d) Calculate $n(C' \cap F)$.

4. The Venn diagram below shows the number of elements in three sets P, Q and R.

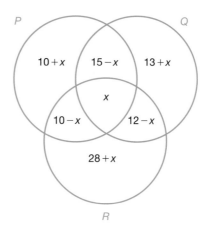

If $n(P \cup Q \cup R) = 93$ calculate:
a) x
b) $n(P)$
c) $n(Q)$
d) $n(R)$
e) $n(P \cap Q)$
f) $n(Q \cap R)$
g) $n(P \cap R)$
h) $n(R \cup Q)$
i) $n(P \cap Q)'$

Topic 1: Mathematical investigations and ICT

Investigations are an important part of mathematical learning. All mathematical discoveries stem from an idea that a mathematician has and then investigates.

Sometimes when faced with a mathematical investigation, it can seem difficult to know how to start. The structure and example below may help you.

1. Read the question carefully and start with simple cases.
2. Draw simple diagrams to help.
3. Put the results from simple cases in a table.
4. Look for a pattern in your results.
5. Try to find a general rule in words.
6. Express your rule algebraically.
7. Test the rule for a new example.
8. Check that the original question has been answered.

Worked example

A mystic rose is created by placing a number of points evenly spaced on the circumference of a circle. Straight lines are then drawn from each point to every other point. The diagram (left) shows a mystic rose with 20 points.

i) How many straight lines are there?
ii) How many straight lines would there be on a mystic rose with 100 points?

To answer these questions, you are not expected to draw either of the shapes and count the number of lines.

1/2. Try simple cases:
By drawing some simple cases and counting the lines, some results can be found:

Mystic rose with 2 points   Mystic rose with 3 points
Number of lines = 1   Number of lines = 3

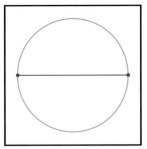
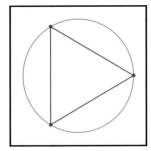

Mystic rose with 4 points Mystic rose with 5 points
Number of lines = 6 Number of lines = 10

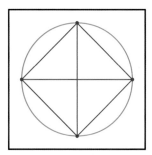 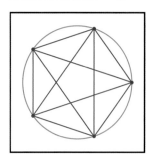

3. Enter the results in an ordered table:

Number of points	2	3	4	5
Number of lines	1	3	6	10

4/5. Look for a pattern in the results:
There are two patterns.
The first shows how the values change.

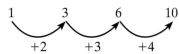

It can be seen that the difference between successive terms is increasing by one each time.

The problem with this pattern is that to find the 20th and 100th terms, it would be necessary to continue this pattern and find all the terms leading up to the 20th and 100th term.

The second is the relationship between the number of points and the number of lines.

Number of points	2	3	4	5
Number of lines	1	3	6	10

It is important to find a relationship that works for all values, for example subtracting 1 from the number of points gives the number of lines in the first example only, so is not useful. However, halving the number of points and multiplying this by 1 less than the number of points works each time,

i.e. Number of lines = (half the number of points) × (one less than the number of points).

6. Express the rule algebraically:
The rule expressed in words above can be written more elegantly using algebra. Let the number of lines be l and the number of points be p.

$$l = \tfrac{1}{2}p(p - 1)$$

Number

Note: Any letters can be used to represent the number of lines and the number of points, not just l and p.

7. **Test the rule:**
   The rule was derived from the original results. It can be tested by generating a further result.

   If the number of points $p = 6$, then the number of lines l is:
   $$l = \tfrac{1}{2} \times 6(6 - 1)$$
   $$= 3 \times 5$$
   $$= 15$$

   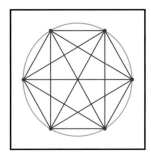

   From the diagram to the left, the number of lines can also be counted as 15.

8. **Check that the original questions have been answered:**
   Using the formula, the number of lines in a mystic rose with 20 points is:
   $$l = \tfrac{1}{2} \times 20(20 - 1)$$
   $$= 10 \times 19$$
   $$= 190$$

   The number of lines in a mystic rose with 100 points is:
   $$l = \tfrac{1}{2} \times 100(100 - 1)$$
   $$= 50 \times 99$$
   $$= 4950$$

● Primes and squares

13, 41 and 73 are prime numbers.
Two different square numbers can be added together to make these prime numbers, e.g. $3^2 + 8^2 = 73$.

1. Find the two square numbers that can be added to make 13 and 41.
2. List the prime numbers less than 100.
3. Which of the prime numbers less than 100 can be shown to be the sum of two different square numbers?
4. Is there a rule to the numbers in question 3?
5. Your rule is a predictive rule not a formula. Discuss the difference.

● Football leagues

There are 18 teams in a football league.

1. If each team plays the other teams twice, once at home and once away, then how many matches are played in a season?
2. If there are t teams in a league, how many matches are played in a season?

Topic 1   Mathematical investigations and ICT

● ICT activity 1

In this activity you will be using a spreadsheet to track the price of a company's shares over a period of time.

1. a) Using the internet or a newspaper as a resource, find the value of a particular company's shares.
    b) Over a period of a month (or week), record the value of the company's shares. This should be carried out on a daily basis.
2. When you have collected all the results, enter them into a spreadsheet similar to the one shown on the left.
3. In column C enter formulae that will calculate the value of the shares as a percentage of their value on day 1.
4. When the spreadsheet is complete, produce a graph showing how the percentage value of the share price changed over time.
5. Write a short report explaining the performance of the company's shares during that time.

	A	B	C
1		Company Name	
2	Day	Share Price	Percentage Value
3	1	3.26	100
4	2	3.29	
5	3	4.11	
6	4		
7	5		
8	↓	↓	
9			
10	↓	↓	
11	etc	etc	

● ICT activity 2

The following activity refers to the graphing package Autograph; however, a similar package may be used.

The velocity of a student at different parts of a 100 m sprint will be analysed.

A racecourse is set out as shown below:

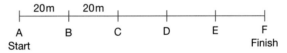

1. A student must stand at each of points A−F. The student at A runs the 100 m and is timed as he/she runs past each of the points B−F by the students at these points who each have a stopwatch.
2. In Autograph, plot a distance−time graph of the results by entering the data as pairs of coordinates, i.e. (time, distance).
3. Ensure that all the points are selected and draw a curve of best fit through them.
4. Select the curve and plot a coordinate of your choice on it. This point can now be moved along the curve using the cursor keys on the keyboard.
5. Draw a tangent to the curve through the point.
6. What does the gradient of the tangent represent?
7. At what point of the race was the student running fastest? How did you reach this answer?
8. Collect similar data for other students. Compare their graphs and running speeds.
9. Carefully analyse one of the graphs and write a brief report to the runner in which you should identify, giving reasons, the parts of the race he/she needs to improve on.

Topic 2
Algebra and graphs

○ Syllabus

E2.1

Use letters to express generalised numbers and express basic arithmetic processes algebraically.
Substitute numbers for words and letters in more complicated formulae.
Construct and transform more complicated formulae and equations.

E2.2

Manipulate directed numbers.
Use brackets and extract common factors.
Expand products of algebraic expressions.
Factorise where possible expressions of the form:
$ax + bx + kay + kby$
$a^2x^2 - b^2y^2$
$a^2 + 2ab + b^2$
$ax^2 + bx + c$

E2.3

Manipulate algebraic fractions.
Factorise and simplify rational expressions.

E2.4

Use and interpret positive, negative and zero indices.
Use and interpret fractional indices.
Use the rules of indices.

E2.5

Solve simple linear equations in one unknown.
Solve simultaneous linear equations in two unknowns.
Solve quadratic equations by factorisation, completing the square or by use of the formula.
Solve simple linear inequalities.

E2.6

Represent inequalities graphically and use this representation in the solution of simple linear programming problems.

E2.7

Continue a given number sequence.
Recognise patterns in sequences and relationships between different sequences.
Find the nth term of sequences.

E2.8

Express direct and inverse variation in algebraic terms and use this form of expression to find unknown quantities.

E2.9

Interpret and use graphs in practical situations including travel graphs and conversion graphs, draw graphs from given data.
Apply the idea of rate of change to easy kinematics involving distance–time and speed–time graphs, acceleration and deceleration.
Calculate distance travelled as area under a linear speed–time graph.

E2.10

Construct tables of values and draw graphs for functions of the form ax^n where a is a rational constant and $n = -2, -1, 0, 1, 2, 3$ and simple sums of not more than three of these and for functions of the form a^x where a is a positive integer.
Draw and interpret graphs representing exponential growth and decay problems.
Solve associated equations approximately by graphical methods.

E2.11
- Estimate gradients of curves by drawing tangents.

E2.12
Use function notation, e.g. $f(x) = 3x - 5$,
$f: x \mapsto 3x - 5$ to describe simple functions.
Find inverse functions $f^{-1}(x)$.
Form composite functions as defined by $gf(x) = g(f(x))$.

○ Contents

Chapter 11	Algebraic representation and manipulation (E2.1, E2.2, E2.3)
Chapter 12	Algebraic indices (E2.4)
Chapter 13	Equations and inequalities (E2.1, E2.5)
Chapter 14	Linear programming (E2.6)
Chapter 15	Sequences (E2.7)
Chapter 16	Variation (E2.8)
Chapter 17	Graphs in practical situations (E2.9)
Chapter 18	Graphs of functions (E2.10, E2.11)
Chapter 19	Functions (E2.12)

○ The Persians

Abu Ja'far Muhammad Ibn Musa al-Khwarizmi is called the 'father of algebra'. He was born in Baghdad in AD790. He wrote the book *Hisab al-jabr w'al-muqabala* in AD830 when Baghdad had the greatest university in the world and the greatest mathematicians studied there. He gave us the word 'algebra' and worked on quadratic equations. He also introduced the decimal system from India.

Muhammad al-Karaji was born in North Africa in what is now Morocco. He lived in the eleventh century and worked on the theory of indices. He also worked on an algebraic method of calculating square and cube roots. He may also have travelled to the University in Granada (then part of the Moorish Empire) where works of his can be found in the University library.

The poet Omar Khayyam is known for his long poem *The Rubaiyat*. He was also a fine mathematician working on the binomial theorem. He introduced the symbol 'shay', which became our 'x'.

Al-Khwarizmi (790–850)

11 Algebraic representation and manipulation

Expanding a bracket

When removing brackets, every term inside the bracket must be multiplied by whatever is outside the bracket.

Worked examples

a) $3(x + 4)$
   $= 3x + 12$

b) $5x(2y + 3)$
   $= 10xy + 15x$

c) $2a(3a + 2b - 3c)$
   $= 6a^2 + 4ab - 6ac$

d) $-4p(2p - q + r^2)$
   $= -8p^2 + 4pq - 4pr^2$

e) $-2x^2\left(x + 3y - \dfrac{1}{x}\right)$
   $= -2x^3 - 6x^2y + 2x$

f) $\dfrac{-2}{x}\left(-x + 4y + \dfrac{1}{x}\right)$
   $= 2 - \dfrac{8y}{x} - \dfrac{2}{x^2}$

Exercise 11.1 Expand the following:

1. a) $4(x - 3)$
    b) $5(2p - 4)$
    c) $-6(7x - 4y)$
    d) $3(2a - 3b - 4c)$
    e) $-7(2m - 3n)$
    f) $-2(8x - 3y)$

2. a) $3x(x - 3y)$
    b) $a(a + b + c)$
    c) $4m(2m - n)$
    d) $-5a(3a - 4b)$
    e) $-4x(-x + y)$
    f) $-8p(-3p + q)$

3. a) $-(2x^2 - 3y^2)$
    b) $-(-a + b)$
    c) $-(-7p + 2q)$
    d) $\frac{1}{2}(6x - 8y + 4z)$
    e) $\frac{3}{4}(4x - 2y)$
    f) $\frac{1}{5}x(10x - 15y)$

4. a) $3r(4r^2 - 5s + 2t)$
    b) $a^2(a + b + c)$
    c) $3a^2(2a - 3b)$
    d) $pq(p + q - pq)$
    e) $m^2(m - n + nm)$
    f) $a^3(a^3 + a^2b)$

Exercise 11.2 Expand and simplify the following:

1. a) $3a - 2(2a + 4)$
    b) $8x - 4(x + 5)$
    c) $3(p - 4) - 4$
    d) $7(3m - 2n) + 8n$
    e) $6x - 3(2x - 1)$
    f) $5p - 3p(p + 2)$

2. a) $7m(m + 4) + m^2 + 2$
    b) $3(x - 4) + 2(4 - x)$
    c) $6(p + 3) - 4(p - 1)$
    d) $5(m - 8) - 4(m - 7)$
    e) $3a(a + 2) - 2(a^2 - 1)$
    f) $7a(b - 2c) - c(2a - 3)$

3. a) $\frac{1}{2}(6x + 4) + \frac{1}{3}(3x + 6)$
    b) $\frac{1}{4}(2x + 6y) + \frac{3}{4}(6x - 4y)$
    c) $\frac{1}{8}(6x - 12y) + \frac{1}{2}(3x - 2y)$
    d) $\frac{1}{5}(15x + 10y) + \frac{3}{10}(5x - 5y)$
    e) $\frac{2}{3}(6x - 9y) + \frac{1}{3}(9x + 6y)$
    f) $\frac{x}{7}(14x - 21y) - \frac{x}{2}(4x - 6y)$

Algebraic representation and manipulation

● Expanding a pair of brackets

When multiplying together expressions in brackets, it is necessary to multiply all the terms in one bracket by all the terms in the other bracket.

Worked examples Expand the following:

a) $(x + 3)(x + 5)$

	x	+3
x	x^2	3x
+5	5x	15

$= x^2 + 5x + 3x + 15$
$= x^2 + 8x + 15$

b) $(x + 2)(x + 1)$

	x	+1
x	x^2	x
+2	2x	2

$= x^2 + x + 2x + 2$
$= x^2 + 3x + 2$

Exercise 11.3 Expand the following and simplify your answer:

1. a) $(x + 2)(x + 3)$ b) $(x + 3)(x + 4)$
   c) $(x + 5)(x + 2)$ d) $(x + 6)(x + 1)$
   e) $(x - 2)(x + 3)$ f) $(x + 8)(x - 3)$

2. a) $(x - 4)(x + 6)$ b) $(x - 7)(x + 4)$
   c) $(x + 5)(x - 7)$ d) $(x + 3)(x - 5)$
   e) $(x + 1)(x - 3)$ f) $(x - 7)(x + 9)$

3. a) $(x - 2)(x - 3)$ b) $(x - 5)(x - 2)$
   c) $(x - 4)(x - 8)$ d) $(x + 3)(x + 3)$
   e) $(x - 3)(x - 3)$ f) $(x - 7)(x - 5)$

4. a) $(x + 3)(x - 3)$ b) $(x + 7)(x - 7)$
   c) $(x - 8)(x + 8)$ d) $(x + y)(x - y)$
   e) $(a + b)(a - b)$ f) $(p - q)(p + q)$

● Simple factorising

When factorising, the largest possible factor is removed from each of the terms and placed outside the brackets.

Worked examples Factorise the following expressions:

a) $10x + 15$
   $= 5(2x + 3)$

b) $8p - 6q + 10r$
   $= 2(4p - 3q + 5r)$

c) $-2q - 6p + 12$
   $= 2(-q - 3p + 6)$

d) $2a^2 + 3ab - 5ac$
   $= a(2a + 3b - 5c)$

e) $6ax - 12ay - 18a^2$
   $= 6a(x - 2y - 3a)$

f) $3b + 9ba - 6bd$
   $= 3b(1 + 3a - 2d)$

Algebra and graphs

Exercise 11.4 Factorise the following:

1. a) $4x - 6$ b) $18 - 12p$
    c) $6y - 3$ d) $4a + 6b$
    e) $3p - 3q$ f) $8m + 12n + 16r$

2. a) $3ab + 4ac - 5ad$ b) $8pq + 6pr - 4ps$
    c) $a^2 - ab$ d) $4x^2 - 6xy$
    e) $abc + abd + fab$ f) $3m^2 + 9m$

3. a) $3pqr - 9pqs$ b) $5m^2 - 10mn$
    c) $8x^2y - 4xy^2$ d) $2a^2b^2 - 3b^2c^2$
    e) $12p - 36$ f) $42x - 54$

4. a) $18 + 12y$ b) $14a - 21b$
    c) $11x + 11xy$ d) $4s - 16t + 20r$
    e) $5pq - 10qr + 15qs$ f) $4xy + 8y^2$

5. a) $m^2 + mn$ b) $3p^2 - 6pq$
    c) $pqr + qrs$ d) $ab + a^2b + ab^2$
    e) $3p^3 - 4p^4$ f) $7b^3c + b^2c^2$

6. a) $m^3 - m^2n + mn^2$ b) $4r^3 - 6r^2 + 8r^2s$
    c) $56x^2y - 28xy^2$ d) $72m^2n + 36mn^2 - 18m^2n^2$

● Substitution

Worked examples Evaluate the expressions below if $a = 3$, $b = 4$, $c = -5$:

a) $2a + 3b - c$ b) $3a - 4b + 2c$
    $= 6 + 12 + 5$ $= 9 - 16 - 10$
    $= 23$ $= -17$

c) $-2a + 2b - 3c$ d) $a^2 + b^2 + c^2$
    $= -6 + 8 + 15$ $= 9 + 16 + 25$
    $= 17$ $= 50$

e) $3a(2b - 3c)$ f) $-2c(-a + 2b)$
    $= 9(8 + 15)$ $= 10(-3 + 8)$
    $= 9 \times 23$ $= 10 \times 5$
    $= 207$ $= 50$

Exercise 11.5 Evaluate the following expressions if $p = 4$, $q = -2$, $r = 3$ and $s = -5$:

1. a) $2p + 4q$ b) $5r - 3s$
    c) $3q - 4s$ d) $6p - 8q + 4s$
    e) $3r - 3p + 5q$ f) $-p - q + r + s$

2. a) $2p - 3q - 4r + s$ b) $3s - 4p + r + q$
    c) $p^2 + q^2$ d) $r^2 - s^2$
    e) $p(q - r + s)$ f) $r(2p - 3q)$

3. a) $2s(3p - 2q)$ b) $pq + rs$
    c) $2pr - 3rq$ d) $q^3 - r^2$
    e) $s^3 - p^3$ f) $r^4 - q^5$

4. a) $-2pqr$ b) $-2p(q + r)$
    c) $-2rq + r$ d) $(p + q)(r - s)$
    e) $(p + s)(r - q)$ f) $(r + q)(p - s)$

5. a) $(2p + 3q)(p - q)$ b) $(q + r)(q - r)$
    c) $q^2 - r^2$ d) $p^2 - r^2$
    e) $(p + r)(p - r)$ f) $(-s + p)q^2$

● **Transformation of formulae**

In the formula $a = 2b + c$, 'a' is the subject. In order to make either b or c the subject, the formula has to be rearranged.

Worked examples   Rearrange the following formulae to make the **bold** letter the subject:

a) $a = 2b + \mathbf{c}$
$a - 2b = \mathbf{c}$

b) $2r + \mathbf{p} = q$
$\mathbf{p} = q - 2r$

c) $ab = cd$
$\dfrac{ab}{d} = \mathbf{c}$

d) $\dfrac{a}{b} = \dfrac{c}{\mathbf{d}}$
$a\mathbf{d} = cb$
$\mathbf{d} = \dfrac{cb}{a}$

Exercise 11.6   In the following questions, make the letter in **bold** the subject of the formula:

1. a) $m + \mathbf{n} = r$ b) $\mathbf{m} + n = p$ c) $2m + \mathbf{n} = 3p$
    d) $3x = 2p + \mathbf{q}$ e) $ab = c\mathbf{d}$ f) $ab = c\mathbf{d}$

2. a) $3xy = 4m$ b) $7pq = 5r$ c) $3\mathbf{x} = c$
    d) $3\mathbf{x} + 7 = y$ e) $5y - 9 = 3r$ f) $5y - 9 = 3\mathbf{x}$

3. a) $6\mathbf{b} = 2a - 5$ b) $6b = 2\mathbf{a} - 5$ c) $3x - 7y = 4\mathbf{z}$
    d) $3\mathbf{x} - 7y = 4z$ e) $3x - 7\mathbf{y} = 4z$ f) $2pr - \mathbf{q} = 8$

4. a) $\dfrac{\mathbf{p}}{4} = r$ b) $\dfrac{4}{\mathbf{p}} = 3r$ c) $\dfrac{1}{5}n = 2\mathbf{p}$
    d) $\dfrac{1}{5}\mathbf{n} = 2p$ e) $\mathbf{p}(q + r) = 2t$ f) $p(\mathbf{q} + r) = 2t$

5. a) $3\mathbf{m} - n = rt(p + q)$ b) $3m - n = rt(p + q)$
    c) $3\mathbf{m} - n = rt(p + q)$ d) $3m - \mathbf{n} = rt(p + q)$
    e) $3m - n = rt(\mathbf{p} + q)$ f) $3m - n = rt(p + \mathbf{q})$

6. a) $\dfrac{ab}{c} = de$ b) $\dfrac{a\mathbf{b}}{c} = de$ c) $\dfrac{ab}{\mathbf{c}} = de$
    d) $\dfrac{a + b}{c} = \mathbf{d}$ e) $\dfrac{a}{c} + \mathbf{b} = d$ f) $\dfrac{\mathbf{a}}{c} + b = d$

Algebra and graphs

● Further expansion

You will have seen earlier in this chapter how to expand a pair of brackets of the form $(x - 3)(x + 4)$. A similar method can be used to expand a pair of brackets of the form $(2x - 3)(3x - 6)$.

Worked example Expand $(2x - 3)(3x - 6)$.

	$2x$	-3
$3x$	$6x^2$	$-9x$
-6	$-12x$	18

$= 6x^2 - 9x - 12x + 18$
$= 6x^2 - 21x + 18$

Exercise 11.7

1. a) $(y + 2)(2y + 3)$ b) $(y + 7)(3y + 4)$
   c) $(2y + 1)(y + 8)$ d) $(2y + 1)(2y + 2)$
   e) $(3y + 4)(2y + 5)$ f) $(6y + 3)(3y + 1)$

2. a) $(2p - 3)(p + 8)$ b) $(4p - 5)(p + 7)$
   c) $(3p - 4)(2p + 3)$ d) $(4p - 5)(3p + 7)$
   e) $(6p + 2)(3p - 1)$ f) $(7p - 3)(4p + 8)$

3. a) $(2x - 1)(2x - 1)$ b) $(3x + 1)^2$
   c) $(4x - 2)^2$ d) $(5x - 4)^2$
   e) $(2x + 6)^2$ f) $(2x + 3)(2x - 3)$

4. a) $(3 + 2x)(3 - 2x)$ b) $(4x - 3)(4x + 3)$
   c) $(3 + 4x)(3 - 4x)$ d) $(7 - 5y)(7 + 5y)$
   e) $(3 + 2y)(4y - 6)$ f) $(7 - 5y)^2$

● Further factorisation

Factorisation by grouping

Worked examples Factorise the following expressions:

a) $6x + 3 + 2xy + y$
    $= 3(2x + 1) + y(2x + 1)$
    $= (3 + y)(2x + 1)$
    Note that $(2x + 1)$ was a common factor of both terms.

b) $ax + ay - bx - by$
    $= a(x + y) - b(x + y)$
    $= (a - b)(x + y)$

c) $2x^2 - 3x + 2xy - 3y$
    $= x(2x - 3) + y(2x - 3)$
    $= (x + y)(2x - 3)$

Algebraic representation and manipulation

Exercise 11.8 Factorise the following by grouping:

1. a) $ax + bx + ay + by$
   b) $ax + bx - ay - by$
   c) $3m + 3n + mx + nx$
   d) $4m + mx + 4n + nx$
   e) $3m + mx - 3n - nx$
   f) $6x + xy + 6z + zy$

2. a) $pr - ps + qr - qs$
   b) $pq - 4p + 3q - 12$
   c) $pq + 3q - 4p - 12$
   d) $rs + rt + 2ts + 2t^2$
   e) $rs - 2ts + rt - 2t^2$
   f) $ab - 4cb + ac - 4c^2$

3. a) $xy + 4y + x^2 + 4x$
   b) $x^2 - xy - 2x + 2y$
   c) $ab + 3a - 7b - 21$
   d) $ab - b - a + 1$
   e) $pq - 4p - 4q + 16$
   f) $mn - 5m - 5n + 25$

4. a) $mn - 2m - 3n + 6$
   b) $mn - 2mr - 3rn + 6r^2$
   c) $pr - 4p - 4qr + 16q$
   d) $ab - a - bc + c$
   e) $x^2 - 2xz - 2xy + 4yz$
   f) $2a^2 + 2ab + b^2 + ab$

Difference of two squares

On expanding $(x + y)(x - y)$
$= x^2 - xy + xy - y^2$
$= x^2 - y^2$

The reverse is that $x^2 - y^2$ factorises to $(x + y)(x - y)$.
x^2 and y^2 are both square and therefore $x^2 - y^2$ is known as the **difference of two squares**.

Worked examples

a) $p^2 - q^2$
$= (p + q)(p - q)$

b) $4a^2 - 9b^2$
$= (2a)^2 - (3b)^2$
$= (2a + 3b)(2a - 3b)$

c) $(mn)^2 - 25k^2$
$= (mn)^2 - (5k)^2$
$= (mn + 5k)(mn - 5k)$

d) $4x^2 - (9y)^2$
$= (2x)^2 - (9y)^2$
$= (2x + 9y)(2x - 9y)$

Exercise 11.9 Factorise the following:

1. a) $a^2 - b^2$
   b) $m^2 - n^2$
   c) $x^2 - 25$
   d) $m^2 - 49$
   e) $81 - x^2$
   f) $100 - y^2$

2. a) $144 - y^2$
   b) $q^2 - 169$
   c) $m^2 - 1$
   d) $1 - t^2$
   e) $4x^2 - y^2$
   f) $25p^2 - 64q^2$

3. a) $9x^2 - 4y^2$
   b) $16p^2 - 36q^2$
   c) $64x^2 - y^2$
   d) $x^2 - 100y^2$
   e) $(qr)^2 - 4p^2$
   f) $(ab)^2 - (cd)^2$

4. a) $m^2n^2 - 9y^2$
   b) $\frac{1}{4}x^2 - \frac{1}{9}y^2$
   c) $(2x)^2 - (3y)^4$
   d) $p^4 - q^4$
   e) $4m^4 - 36y^4$
   f) $16x^4 - 81y^4$

Algebra and graphs

● Evaluation

Once factorised, numerical expressions can be evaluated.

Worked examples Evaluate the following expressions:

a) $13^2 - 7^2$
$= (13 + 7)(13 - 7)$
$= 20 \times 6$
$= 120$

b) $6.25^2 - 3.75^2$
$= (6.25 + 3.75)(6.25 - 3.75)$
$= 10 \times 2.5$
$= 25$

Exercise 11.10 By factorising, evaluate the following:

1. a) $8^2 - 2^2$ b) $16^2 - 4^2$ c) $49^2 - 1$
   d) $17^2 - 3^2$ e) $88^2 - 12^2$ f) $96^2 - 4^2$

2. a) $45^2 - 25$ b) $99^2 - 1$ c) $27^2 - 23^2$
   d) $66^2 - 34^2$ e) $999^2 - 1$ f) $225 - 8^2$

3. a) $8.4^2 - 1.6^2$ b) $9.3^2 - 0.7^2$ c) $42.8^2 - 7.2^2$
   d) $(8\frac{1}{2})^2 - (1\frac{1}{2})^2$ e) $(7\frac{3}{4})^2 - (2\frac{1}{4})^2$ f) $5.25^2 - 4.75^2$

4. a) $8.62^2 - 1.38^2$ b) $0.9^2 - 0.1^2$ c) $3^4 - 2^4$
   d) $2^4 - 1$ e) $1111^2 - 111^2$ f) $2^8 - 25$

● Factorising quadratic expressions

$x^2 + 5x + 6$ is known as a quadratic expression as the highest power of any of its terms is squared − in this case x^2.

It can be factorised by writing it as a product of two brackets.

Worked examples a) Factorise $x^2 + 5x + 6$.

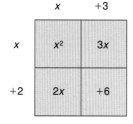

On setting up a 2×2 grid, some of the information can immediately be entered.

As there is only one term in x^2, this can be entered, as can the constant $+6$. The only two values which multiply to give x^2 are x and x. These too can be entered.

We now need to find two values which multiply to give $+6$ and which add to give $+5x$.

The only two values which satisfy both these conditions are $+3$ and $+2$.

Therefore $x^2 + 5x + 6 = (x + 3)(x + 2)$

b) Factorise $x^2 + 2x - 24$.

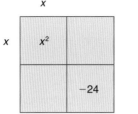 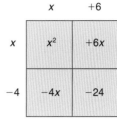

Therefore $x^2 + 2x - 24 = (x + 6)(x - 4)$

c) Factorise $2x^2 + 11x + 12$.

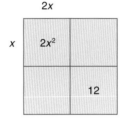 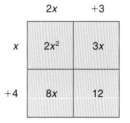

Therefore $2x^2 + 11x + 12 = (2x + 3)(x + 4)$

d) Factorise $3x^2 + 7x - 6$.

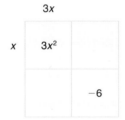 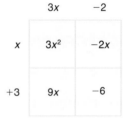

Therefore $3x^2 + 7x - 6 = (3x - 2)(x + 3)$

Exercise 11.11 Factorise the following quadratic expressions:

1. a) $x^2 + 7x + 12$ b) $x^2 + 8x + 12$ c) $x^2 + 13x + 12$
   d) $x^2 - 7x + 12$ e) $x^2 - 8x + 12$ f) $x^2 - 13x + 12$

2. a) $x^2 + 6x + 5$ b) $x^2 + 6x + 8$ c) $x^2 + 6x + 9$
   d) $x^2 + 10x + 25$ e) $x^2 + 22x + 121$ f) $x^2 - 13x + 42$

3. a) $x^2 + 14x + 24$ b) $x^2 + 11x + 24$ c) $x^2 - 10x + 24$
   d) $x^2 + 15x + 36$ e) $x^2 + 20x + 36$ f) $x^2 - 12x + 36$

4. a) $x^2 + 2x - 15$ b) $x^2 - 2x - 15$ c) $x^2 + x - 12$
   d) $x^2 - x - 12$ e) $x^2 + 4x - 12$ f) $x^2 - 15x + 36$

5. a) $x^2 - 2x - 8$ b) $x^2 - x - 20$ c) $x^2 + x - 30$
   d) $x^2 - x - 42$ e) $x^2 - 2x - 63$ f) $x^2 + 3x - 54$

6. a) $2x^2 + 3x + 1$ b) $2x^2 + 7x + 6$ c) $2x^2 + x - 6$
   d) $2x^2 - 7x + 6$ e) $3x^2 + 8x + 4$ f) $3x^2 + 11x - 4$
   g) $4x^2 + 12x + 9$ h) $9x^2 - 6x + 1$ i) $6x^2 - x - 1$

Algebra and graphs

Transformation of complex formulae

Worked examples Make the letters in **bold** the subject of each formula:

a) $C = 2\pi \mathbf{r}$

$$\frac{C}{2\pi} = \mathbf{r}$$

b) $A = \pi \mathbf{r}^2$

$$\frac{A}{\pi} = r^2$$

$$\pm\sqrt{\frac{A}{\pi}} = \mathbf{r}$$

c) $x^2 + \mathbf{y}^2 = h^2$

$$y^2 = h^2 - x^2$$

$$\mathbf{y} = \pm\sqrt{h^2 - x^2}$$

Note: not $y = h - x$

d) $f = \sqrt{\frac{\mathbf{x}}{k}}$

$$f^2 = \frac{\mathbf{x}}{k}$$

$$f^2 k = \mathbf{x}$$

Square both sides

e) $m = 3a\sqrt{\frac{p}{\mathbf{x}}}$

$$m^2 = \frac{9a^2 p}{\mathbf{x}}$$

$$m^2 \mathbf{x} = 9a^2 p$$

$$\mathbf{x} = \frac{9a^2 p}{m^2}$$

f) $A = \frac{y+x}{p+\mathbf{q}^2}$

$$A(p+\mathbf{q}^2) = y+x$$

$$p+\mathbf{q}^2 = \frac{y+x}{A}$$

$$\mathbf{q}^2 = \frac{y+x}{A} - p$$

$$\mathbf{q} = \pm\sqrt{\frac{y+x}{A} - p}$$

g) $\frac{\mathbf{x}}{4} = \frac{a-b}{3\mathbf{x}}$

$$3\mathbf{x}^2 = 4(a-b)$$

$$\mathbf{x}^2 = \frac{a-b}{3}$$

$$\mathbf{x} = \pm\sqrt{\frac{a-b}{3}}$$

h) $\frac{a}{b\mathbf{x}+1} = \frac{b}{\mathbf{x}}$

$$a\mathbf{x} = b(b\mathbf{x}+1)$$

$$a\mathbf{x} = b^2 \mathbf{x} + b$$

$$a\mathbf{x} - b^2 \mathbf{x} = b$$

$$\mathbf{x}(a-b^2) = b$$

$$\mathbf{x} = \frac{b}{a-b^2}$$

Exercise 11.12 In the formulae below, make x the subject:

1. a) $P = 2mx$
   b) $T = 3x^2$
   c) $mx^2 = y^2$
   d) $x^2 + y^2 = p^2 - q^2$
   e) $m^2 + x^2 = y^2 - n^2$
   f) $p^2 - q^2 = 4x^2 - y^2$

2. a) $\frac{P}{Q} = rx$
   b) $\frac{P}{Q} = rx^2$
   c) $\frac{P}{Q} = \frac{x^2}{r}$
   d) $\frac{m}{n} = \frac{1}{x^2}$
   e) $\frac{r}{st} = \frac{w}{x^2}$
   f) $\frac{p+q}{r} = \frac{w}{x^2}$

3. a) $\sqrt{x} = rp$ b) $\dfrac{mn}{p} = \sqrt{x}$ c) $g = \sqrt{\dfrac{k}{x}}$

    d) $r = 2\pi\sqrt{\dfrac{x}{g}}$ e) $p^2 = \dfrac{4m^2 r}{x}$ f) $p = 2m\sqrt{\dfrac{r}{x}}$

Exercise 11.13 In the following questions, make the letter in **bold** the subject of the formula:

1. a) $v = u + a\mathbf{t}$ b) $v^2 = u^2 + 2a\mathbf{s}$ c) $v^2 = \mathbf{u}^2 + 2as$
    d) $s = u\mathbf{t} + \tfrac{1}{2}at^2$ e) $s = ut + \tfrac{1}{2}\mathbf{a}t^2$ f) $s = ut + \tfrac{1}{2}at^2$ (for $\mathbf{t}$)

2. a) $A = \pi r\sqrt{s^2 + \mathbf{t}^2}$ b) $A = \pi r\sqrt{\mathbf{h}^2 + r^2}$

    c) $\dfrac{1}{f} = \dfrac{1}{\mathbf{u}} + \dfrac{1}{v}$ d) $\dfrac{1}{f} = \dfrac{1}{u} + \dfrac{1}{\mathbf{v}}$

    e) $t = 2\pi\sqrt{\dfrac{l}{\mathbf{g}}}$ f) $t = 2\pi\sqrt{\dfrac{\mathbf{l}}{g}}$

3. a) $\dfrac{\mathbf{x}t}{7} = \dfrac{p+2}{3x}$ b) $\sqrt{\mathbf{a}+2} = \dfrac{b-3}{\sqrt{a-2}}$

Exercise 11.14

1. The volume of a cylinder is given by the formula $V = \pi r^2 h$, where h is the height of the cylinder and r is the radius.
    a) Find the volume of a cylindrical post of length 7.5 m and a diameter of 30 cm.
    b) Make r the subject of the formula.
    c) A cylinder of height 75 cm has a volume of 6000 cm³, find its radius correct to 3 s.f.

2. The formula $C = \tfrac{5}{9}(F - 32)$ can be used to convert temperatures in degrees Fahrenheit (°F) into degrees Celsius (°C).
    a) What temperature in °C is equivalent to 150 °F?
    b) What temperature in °C is equivalent to 12 °F?
    c) Make F the subject of the formula.
    d) Use your rearranged formula to find what temperature in °F is equivalent to 160 °C.

3. The height of Mount Kilamanjaro is given as 5900 m. The formula for the time taken, T hours, to climb to a height H metres is:
    $$T = \dfrac{H}{1200} + k$$
    where k is a constant.
    a) Calculate the time taken, to the nearest hour, to climb to the top of the mountain if $k = 9.8$.
    b) Make H the subject of the formula.
    c) How far up the mountain, to the nearest 100 m, could you expect to be after 14 hours?

Algebra and graphs

4. The formula for the volume V of a sphere is given as $V = \frac{4}{3}\pi r^3$.
   a) Find V if $r = 5$ cm.
   b) Make r the subject of the formula.
   c) Find the radius of a sphere of volume 2500 m³.

5. The cost $\$x$ of printing n newspapers is given by the formula $x = 1.50 + 0.05n$.
   a) Calculate the cost of printing 5000 newspapers.
   b) Make n the subject of the formula.
   c) How many newspapers can be printed for $25?

● Algebraic fractions

Simplifying algebraic fractions

The rules for fractions involving algebraic terms are the same as those for numeric fractions. However the actual calculations are often easier when using algebra.

Worked examples

a) $\dfrac{3}{4} \times \dfrac{5}{7} = \dfrac{15}{28}$

b) $\dfrac{a}{c} \times \dfrac{b}{d} = \dfrac{ab}{cd}$

c) $\dfrac{\cancel{3}}{4} \times \dfrac{5}{\cancel{6}_2} = \dfrac{5}{8}$

d) $\dfrac{\cancel{a}}{c} \times \dfrac{b}{2\cancel{a}} = \dfrac{b}{2c}$

e) $\dfrac{\cancel{a}b}{e\cancel{c}} \times \dfrac{\cancel{c}d}{f\cancel{a}} = \dfrac{bd}{ef}$

f) $\dfrac{x^5}{x^3} = \dfrac{x \times x \times x \times x \times x}{x \times x \times x} = x^2$

g) $\dfrac{2b}{5} \div \dfrac{b}{7} = \dfrac{2b}{5} \times \dfrac{7}{b} = \dfrac{14\cancel{b}}{5\cancel{b}} = \dfrac{14}{5} = 2.8$

Exercise 11.15 Simplify the following algebraic fractions:

1. a) $\dfrac{x}{y} \times \dfrac{p}{q}$ b) $\dfrac{x}{y} \times \dfrac{q}{x}$ c) $\dfrac{p}{q} \times \dfrac{q}{r}$

   d) $\dfrac{ab}{c} \times \dfrac{d}{ab}$ e) $\dfrac{ab}{c} \times \dfrac{d}{ac}$ f) $\dfrac{p^2}{q^2} \times \dfrac{q^2}{p}$

2. a) $\dfrac{m^3}{m}$ b) $\dfrac{r^7}{r^2}$ c) $\dfrac{x^9}{x^3}$

   d) $\dfrac{x^2 y^4}{xy^2}$ e) $\dfrac{a^2 b^3 c^4}{ab^2 c}$ f) $\dfrac{pq^2 r^4}{p^2 q^3 r}$

3. a) $\dfrac{4ax}{2ay}$ b) $\dfrac{12pq^2}{3p}$ c) $\dfrac{15mn^2}{3mn}$

   d) $\dfrac{24x^5 y^3}{8x^2 y^2}$ e) $\dfrac{36p^2 qr}{12pqr}$ f) $\dfrac{16m^2 n}{24m^3 n^2}$

4. a) $\dfrac{2}{b} \times \dfrac{a}{3}$ b) $\dfrac{4}{x} \times \dfrac{y}{2}$ c) $\dfrac{8}{x} \times \dfrac{x}{4}$

   d) $\dfrac{9y}{2} \times \dfrac{2x}{3}$ e) $\dfrac{12x}{7} \times \dfrac{7}{4x}$ f) $\dfrac{4x^3}{3y} \times \dfrac{9y^2}{2x^2}$

Algebraic representation and manipulation

5. a) $\dfrac{2ax}{3bx} \times \dfrac{4by}{a}$ b) $\dfrac{3p^2}{2q} \times \dfrac{5q}{3p}$

   c) $\dfrac{p^2q}{rs} \times \dfrac{pr}{q}$ d) $\dfrac{a^2b}{fc^2} \times \dfrac{cd}{bd} \times \dfrac{ef^2}{ca^2}$

6. a) $\dfrac{8x^2}{3} \div \dfrac{2x}{5}$ b) $\dfrac{3b^3}{2} \div \dfrac{4b^2}{3}$

Addition and subtraction of fractions

In arithmetic it is easy to add or subtract fractions with the same denominator. It is the same process when dealing with algebraic fractions.

Worked examples

a) $\dfrac{4}{11} + \dfrac{3}{11}$ b) $\dfrac{a}{11} + \dfrac{b}{11}$ c) $\dfrac{4}{x} + \dfrac{3}{x}$

$= \dfrac{7}{11}$ $= \dfrac{a+b}{11}$ $= \dfrac{7}{x}$

If the denominators are different, the fractions need to be changed to form fractions with the same denominator.

Worked examples

a) $\dfrac{2}{9} + \dfrac{1}{3}$ b) $\dfrac{a}{9} + \dfrac{b}{3}$ c) $\dfrac{4}{5a} + \dfrac{7}{10a}$

$= \dfrac{2}{9} + \dfrac{3}{9}$ $= \dfrac{a}{9} + \dfrac{3b}{9}$ $= \dfrac{8}{10a} + \dfrac{7}{10a}$

$= \dfrac{5}{9}$ $= \dfrac{a+3b}{9}$ $= \dfrac{15}{10a}$

$= \dfrac{3}{2a}$

Similarly, with subtraction, the denominators need to be the same.

Worked examples

a) $\dfrac{7}{a} - \dfrac{1}{2a}$ b) $\dfrac{p}{3} - \dfrac{q}{15}$ c) $\dfrac{5}{3b} - \dfrac{8}{9b}$

$= \dfrac{14}{2a} - \dfrac{1}{2a}$ $= \dfrac{5p}{15} - \dfrac{q}{15}$ $= \dfrac{15}{9b} - \dfrac{8}{9b}$

$= \dfrac{13}{2a}$ $= \dfrac{5p-q}{15}$ $= \dfrac{7}{9b}$

Exercise 11.16 Simplify the following fractions:

1. a) $\dfrac{1}{7} + \dfrac{3}{7}$ b) $\dfrac{a}{7} + \dfrac{b}{7}$ c) $\dfrac{5}{13} + \dfrac{6}{13}$

   d) $\dfrac{c}{13} + \dfrac{d}{13}$ e) $\dfrac{x}{3} + \dfrac{y}{3} + \dfrac{z}{3}$ f) $\dfrac{p^2}{5} + \dfrac{q^2}{5}$

2. a) $\dfrac{5}{11} - \dfrac{2}{11}$ b) $\dfrac{c}{11} - \dfrac{d}{11}$ c) $\dfrac{6}{a} - \dfrac{2}{a}$

   d) $\dfrac{2a}{3} - \dfrac{5b}{3}$ e) $\dfrac{2x}{7} - \dfrac{3y}{7}$ f) $\dfrac{3}{4x} - \dfrac{5}{4x}$

Algebra and graphs

3. a) $\dfrac{5}{6} - \dfrac{1}{3}$ b) $\dfrac{5}{2a} - \dfrac{1}{a}$ c) $\dfrac{2}{3c} + \dfrac{1}{c}$

   d) $\dfrac{2}{x} + \dfrac{3}{2x}$ e) $\dfrac{5}{2p} - \dfrac{1}{p}$ f) $\dfrac{1}{w} - \dfrac{3}{2w}$

4. a) $\dfrac{p}{4} - \dfrac{q}{12}$ b) $\dfrac{x}{4} - \dfrac{y}{2}$ c) $\dfrac{m}{3} - \dfrac{n}{9}$

   d) $\dfrac{x}{12} - \dfrac{y}{6}$ e) $\dfrac{r}{2} + \dfrac{m}{10}$ f) $\dfrac{s}{3} - \dfrac{t}{15}$

5. a) $\dfrac{3x}{4} - \dfrac{2x}{12}$ b) $\dfrac{3x}{5} - \dfrac{2y}{15}$ c) $\dfrac{3m}{7} + \dfrac{m}{14}$

   d) $\dfrac{4m}{3p} - \dfrac{3m}{9p}$ e) $\dfrac{4x}{3y} - \dfrac{5x}{6y}$ f) $\dfrac{3r}{7s} + \dfrac{2r}{14s}$

Often one denominator is not a multiple of the other. In these cases the **lowest common multiple** of both denominators has to be found.

Worked examples   a) $\dfrac{1}{4} + \dfrac{1}{3}$

$= \dfrac{3}{12} + \dfrac{4}{12}$

$= \dfrac{7}{12}$

b) $\dfrac{1}{5} + \dfrac{2}{3}$

$= \dfrac{3}{15} + \dfrac{10}{15}$

$= \dfrac{13}{15}$

c) $\dfrac{a}{3} + \dfrac{b}{4}$

$= \dfrac{4a}{12} + \dfrac{3b}{12}$

$= \dfrac{4a + 3b}{12}$

d) $\dfrac{2a}{3} + \dfrac{3b}{5}$

$= \dfrac{10a}{15} + \dfrac{9b}{15}$

$= \dfrac{10a + 9b}{15}$

Exercise 11.17

Simplify the following fractions:

1. a) $\dfrac{a}{2} + \dfrac{b}{3}$ b) $\dfrac{a}{3} + \dfrac{b}{5}$ c) $\dfrac{p}{4} + \dfrac{q}{7}$

   d) $\dfrac{2a}{5} + \dfrac{b}{3}$ e) $\dfrac{x}{4} + \dfrac{5y}{9}$ f) $\dfrac{2x}{7} + \dfrac{2y}{5}$

2. a) $\dfrac{a}{2} - \dfrac{a}{3}$ b) $\dfrac{a}{3} - \dfrac{a}{5}$ c) $\dfrac{p}{4} + \dfrac{p}{7}$

   d) $\dfrac{2a}{5} + \dfrac{a}{3}$ e) $\dfrac{x}{4} + \dfrac{5x}{9}$ f) $\dfrac{2x}{7} + \dfrac{2x}{5}$

3. a) $\dfrac{3m}{5} - \dfrac{m}{2}$ b) $\dfrac{3r}{5} - \dfrac{r}{2}$ c) $\dfrac{5x}{4} - \dfrac{3x}{2}$

   d) $\dfrac{2x}{7} + \dfrac{3x}{4}$ e) $\dfrac{11x}{2} - \dfrac{5x}{3}$ f) $\dfrac{2p}{3} - \dfrac{p}{2}$

4. a) $p - \frac{p}{2}$ b) $c - \frac{c}{3}$ c) $x - \frac{x}{5}$

   d) $m - \frac{2m}{3}$ e) $q - \frac{4q}{5}$ f) $w - \frac{3w}{4}$

5. a) $2m - \frac{m}{2}$ b) $3m - \frac{2m}{3}$ c) $2m - \frac{5m}{2}$

   d) $4m - \frac{3m}{2}$ e) $2p - \frac{5p}{3}$ f) $6q - \frac{6q}{7}$

6. a) $p - \frac{p}{r}$ b) $\frac{x}{y} + x$ c) $m + \frac{m}{n}$

   d) $\frac{a}{b} + a$ e) $2x - \frac{x}{y}$ f) $2p - \frac{3p}{q}$

7. a) $\frac{a}{3} + \frac{a+4}{2}$ b) $\frac{2b}{5} + \frac{b-4}{3}$

   c) $\frac{c+2}{4} - \frac{2-c}{2}$ d) $\frac{2(d-3)}{7} - \frac{3(2-d)}{2}$

● **Simplifying complex algebraic fractions**

With more complex algebraic fractions, the method of getting a common denominator is still required.

Worked examples a) $\frac{2}{x+1} + \frac{3}{x+2}$

$= \frac{2(x+2)}{(x+1)(x+2)} + \frac{3(x+1)}{(x+1)(x+2)}$

$= \frac{2(x+2) + 3(x+1)}{(x+1)(x+2)}$

$= \frac{2x+4+3x+3}{(x+1)(x+2)}$

$= \frac{5x+7}{(x+1)(x+2)}$

b) $\frac{5}{p+3} - \frac{3}{p-5}$

$= \frac{5(p-5)}{(p+3)(p-5)} - \frac{3(p+3)}{(p+3)(p-5)}$

$= \frac{5(p-5) - 3(p+3)}{(p+3)(p-5)}$

$= \frac{5p - 25 - 3p - 9}{(p+3)(p-5)}$

$= \frac{2p - 34}{(p+3)(p-5)}$

Algebra and graphs

c) $\dfrac{x^2-2x}{x^2+x-6}$

$=\dfrac{x(x-2)}{(x+3)(x-2)}$

$=\dfrac{x}{x+3}$

d) $\dfrac{x^2-3x}{x^2+2x-15}$

$=\dfrac{x(x-3)}{(x-3)(x+5)}$

$=\dfrac{x}{x+5}$

Exercise 11.18 Simplify the following algebraic fractions:

1. a) $\dfrac{1}{x+1}+\dfrac{2}{x+2}$ b) $\dfrac{3}{m+2}-\dfrac{2}{m-1}$

   c) $\dfrac{2}{p-3}+\dfrac{1}{p-2}$ d) $\dfrac{3}{w-1}-\dfrac{2}{w+3}$

   e) $\dfrac{4}{y+4}-\dfrac{4}{y+1}$ f) $\dfrac{2}{m-2}-\dfrac{3}{m+3}$

2. a) $\dfrac{x(x-4)}{(x-4)(x+2)}$ b) $\dfrac{y(y-3)}{(y+3)(y-3)}$

   c) $\dfrac{(m+2)(m-2)}{(m-2)(m-3)}$ d) $\dfrac{p(p+5)}{(p-5)(p+5)}$

   e) $\dfrac{m(2m+3)}{(m+4)(2m+3)}$ f) $\dfrac{(m+1)(m-1)}{(m+2)(m-1)}$

3. a) $\dfrac{x^2-5x}{(x+3)(x-5)}$ b) $\dfrac{x^2-3x}{(x+4)(x-3)}$

   c) $\dfrac{y^2-7y}{(y-7)(y-3)}$ d) $\dfrac{x(x-1)}{x^2+2x-3}$

   e) $\dfrac{x(x+2)}{x^2+4x+4}$ f) $\dfrac{x(x+4)}{x^2+5x+4}$

4. a) $\dfrac{x^2-x}{x^2-1}$ b) $\dfrac{x^2+2x}{x^2+5x+6}$

   c) $\dfrac{x^2+4x}{x^2+x-12}$ d) $\dfrac{x^2-5x}{x^2-3x-10}$

   e) $\dfrac{x^2+3x}{x^2-9}$ f) $\dfrac{x^2-7x}{x^2-49}$

Algebraic representation and manipulation

Student assessment 1

1. Expand the following and simplify where possible:
   a) $5(2a - 6b + 3c)$
   b) $3x(5x - 9)$
   c) $-5y(3xy + y^2)$
   d) $3x^2(5xy + 3y^2 - x^3)$
   e) $5p - 3(2p - 4)$
   f) $4m(2m - 3) + 2(3m^2 - m)$
   g) $\frac{1}{3}(6x - 9) + \frac{1}{4}(8x + 24)$
   h) $\frac{m}{4}(6m - 8) + \frac{m}{2}(10m - 2)$

2. Factorise the following:
   a) $12a - 4b$
   b) $x^2 - 4xy$
   c) $8p^3 - 4p^2q$
   d) $24xy - 16x^2y + 8xy^2$

3. If $x = 2$, $y = -3$ and $z = 4$, evaluate the following:
   a) $2x + 3y - 4z$
   b) $10x + 2y^2 - 3z$
   c) $z^2 - y^3$
   d) $(x + y)(y - z)$
   e) $z^2 - x^2$
   f) $(z + x)(z - x)$

4. Rearrange the following formulae to make the **bold** letter the subject:
   a) $x = 3p + \mathbf{q}$
   b) $3m - 5\mathbf{n} = 8r$
   c) $2m = \frac{3\mathbf{y}}{t}$
   d) $x(\mathbf{w} + y) = 2y$
   e) $\frac{xy}{2\mathbf{p}} = \frac{rs}{t}$
   f) $\frac{x+y}{\mathbf{w}} = m + n$

Student assessment 2

1. Expand the following and simplify where possible:
   a) $3(2x - 3y + 5z)$
   b) $4p(2m - 7)$
   c) $-4m(2mn - n^2)$
   d) $4p^2(5pq - 2q^2 - 2p)$
   e) $4x - 2(3x + 1)$
   f) $4x(3x - 2) + 2(5x^2 - 3x)$
   g) $\frac{1}{5}(15x - 10) - \frac{1}{3}(9x - 12)$
   h) $\frac{x}{2}(4x - 6) + \frac{x}{4}(2x + 8)$

2. Factorise the following:
   a) $16p - 8q$
   b) $p^2 - 6pq$
   c) $5p^2q - 10pq^2$
   d) $9pq - 6p^2q + 12q^2p$

3. If $a = 4$, $b = 3$ and $c = -2$, evaluate the following:
   a) $3a - 2b + 3c$
   b) $5a - 3b^2$
   c) $a^2 + b^2 + c^2$
   d) $(a + b)(a - b)$
   e) $a^2 - b^2$
   f) $b^3 - c^3$

4. Rearrange the following formulae to make the **bold** letter the subject:
   a) $p = 4m + \mathbf{n}$
   b) $4x - 3\mathbf{y} = 5z$
   c) $2x = \frac{3\mathbf{y}}{5p}$
   d) $m(\mathbf{x} + y) = 3w$
   e) $\frac{pq}{4\mathbf{r}} = \frac{mn}{t}$
   f) $\frac{p+q}{r} = m - n$

Algebra and graphs

Student assessment 3

1. Factorise the following fully:
   a) $mx - 5m - 5nx + 25n$
   b) $4x^2 - 81y^2$
   c) $88^2 - 12^2$
   d) $x^4 - y^4$

2. Expand the following and simplify where possible:
   a) $(x + 3)(x + 5)$
   b) $(x - 7)(x - 7)$
   c) $(x + 5)^2$
   d) $(x - 7)(x + 2)$
   e) $(2x - 1)(3x + 8)$
   f) $(7 - 5y)^2$

3. Factorise the following:
   a) $x^2 - 18x + 32$
   b) $x^2 - 2x - 24$
   c) $x^2 - 9x + 18$
   d) $x^2 - 2x + 1$
   e) $2x^2 + 5x - 3$
   f) $9x^2 - 12x + 4$

4. Make the letter in **bold** the subject of the formula:
   a) $v^2 = u^2 + 2a\mathbf{s}$
   b) $r^2 + \mathbf{h}^2 = p^2$
   c) $\dfrac{m}{\mathbf{n}} = \dfrac{r}{s^2}$
   d) $t = 2\pi\sqrt{\dfrac{\mathbf{l}}{g}}$
   e) $3\mathbf{x} - 2y = 5\mathbf{x} - 7$
   f) $\dfrac{w - x}{2} = x - 2w$

5. Simplify the following algebraic fractions:
   a) $\dfrac{ab}{c} \times \dfrac{bc}{a}$
   b) $\dfrac{(x^3)^2}{x^2}$
   c) $\dfrac{12mn^2}{3m^2}$
   d) $\dfrac{p^2q^2}{r^4} \times \dfrac{pr^5}{qr^3}$
   e) $\dfrac{5t}{4} \div \dfrac{3}{2t}$
   f) $\dfrac{rs^3}{2} \div \dfrac{3r^2}{4}$

6. Simplify the following algebraic fractions:
   a) $\dfrac{3m}{4} + \dfrac{5n}{16}$
   b) $\dfrac{5m}{4y} - \dfrac{3m}{6y}$
   c) $\dfrac{2r}{3x} + \dfrac{5r}{4x} - \dfrac{3r}{2x}$
   d) $\dfrac{2x}{3} - \dfrac{3y}{4}$
   e) $r - \dfrac{r}{7}$
   f) $\dfrac{-(x - 2)}{6} - \dfrac{3x + 2}{2}$

7. Simplify the following:
   a) $\dfrac{3}{m + 2} + \dfrac{2}{m + 3}$
   b) $\dfrac{(y + 3)(y - 3)}{(y - 3)^2}$
   c) $\dfrac{x^2 - 3x}{x^2 + 4x - 21}$

Student assessment 4

1. Factorise the following fully:
   a) $pq - 3rq + pr - 3r^2$
   b) $1 - t^4$
   c) $875^2 - 125^2$
   d) $7.5^2 - 2.5^2$

2. Expand the following and simplify where possible:
   a) $(x - 4)(x + 2)$
   b) $(x - 8)^2$
   c) $(x + y)^2$
   d) $(x - 11)(x + 11)$
   e) $(3x - 2)(2x - 3)$
   f) $(5 - 3x)^2$

3. Factorise the following:
   a) $x^2 - 4x - 77$
   b) $x^2 - 6x + 9$
   c) $x^2 - 144$
   d) $3x^2 + 3x - 18$
   e) $2x^2 + 5x - 12$
   f) $4x^2 - 20x + 25$

4. Make the letter in **bold** the subject of the formula:
   a) $m\mathbf{f}^2 = p$
   b) $\mathbf{m} = 5t^2$
   c) $A = \pi r\sqrt{\mathbf{p}+q}$
   d) $\dfrac{1}{x} + \dfrac{1}{y} = \dfrac{1}{\mathbf{t}}$
   e) $\dfrac{1}{\mathbf{p}-3q} = \dfrac{5}{p+q}$
   f) $r(s-t) = 2t + r$

5. Simplify the following algebraic fractions:
   a) $\dfrac{x^7}{x^3}$
   b) $\dfrac{mn}{p} \times \dfrac{pq}{m}$
   c) $\dfrac{(y^3)^3}{(y^2)^3}$
   d) $\dfrac{28pq^2}{7pq^3}$
   e) $\dfrac{m^2 n}{2} \div \dfrac{m^2}{n^2}$
   f) $\dfrac{7b^3}{c} \div \dfrac{4b^2}{3c^3}$

6. Simplify the following algebraic fractions:
   a) $\dfrac{m}{11} + \dfrac{3m}{11} - \dfrac{2m}{11}$
   b) $\dfrac{3p}{8} - \dfrac{9p}{16}$
   c) $\dfrac{4x}{3y} - \dfrac{7x}{12y}$
   d) $\dfrac{3m}{15p} + \dfrac{4n}{5p} - \dfrac{11n}{30p}$
   e) $\dfrac{2(y+4)}{3} - (y - 2)$
   f) $3(y+2) - \dfrac{2y+3}{2}$

7. Simplify the following:
   a) $\dfrac{4}{(x-5)} + \dfrac{3}{(x-2)}$
   b) $\dfrac{a^2 - b^2}{(a+b)^2}$
   c) $\dfrac{x-2}{x^2+x-6}$

Algebra and graphs

Student assessment 5

1. The volume V of a cylinder is given by the formula $V = \pi r^2 h$, where h is the height of the cylinder and r is the radius.
   a) Find the volume of a cylindrical post 6.5 m long and with a diameter of 20 cm.
   b) Make r the subject of the formula.
   c) A cylinder of height 60 cm has a volume of 5500 cm³: find its radius correct to 3 s.f.

2. The formula for the surface area of a closed cylinder is $A = 2\pi r(r + h)$, where r is the radius of the cylinder and h is its height.
   a) Find the surface area of a cylinder of radius 12 cm and height 20 cm, giving your answer to 3 s.f.
   b) Rearrange the formula to make h the subject.
   c) What is the height of a cylinder of surface area 500 cm² and radius 5 cm? Give your answer to 3 s.f.

3. The formula for finding the length d of the body diagonal of a cuboid whose dimensions are x, y and z is:

$$d = \sqrt{x^2 + y^2 + z^2}$$

   a) Find d when $x = 2$, $y = 3$ and $z = 4$.
   b) How long is the body diagonal of a block of concrete in the shape of a rectangular prism of dimensions 2 m, 3 m and 75 cm?
   c) Rearrange the formula to make x the subject.
   d) Find x when $d = 0.86$, $y = 0.25$ and $z = 0.41$.

4. A pendulum of length l metres takes T seconds to complete one full oscillation. The formula for T is:

$$T = 2\pi\sqrt{\frac{l}{g}}$$

   where g m/s² is the acceleration due to gravity.
   a) Find T if $l = 5$ and $g = 10$.
   b) Rearrange the formula to make l the subject of the formula.
   c) How long is a pendulum which takes 3 seconds for one oscillation, if $g = 10$?

12 Algebraic indices

In Chapter 7 you saw how numbers can be expressed using indices. For example, $5 \times 5 \times 5 = 125$, therefore $125 = 5^3$. The 3 is called the **index**. **Indices** is the plural of index.
Three laws of indices were introduced:

(1) $a^m \times a^n = a^{m+n}$
(2) $a^m \div a^n$ or $\dfrac{a^m}{a^n} = a^{m-n}$
(3) $(a^m)^n = a^{mn}$

● Positive indices

Worked examples **a)** Simplify $d^3 \times d^4$.

$d^3 \times d^4 = d^{3+4}$
$= d^7$

b) Simplify $\dfrac{(p^2)^4}{p^2 \times p^4}$.

$\dfrac{(p^2)^4}{p^2 \times p^4} = \dfrac{p^{2 \times 4}}{p^{2+4}}$
$= \dfrac{p^8}{p^6}$
$= p^{8-6}$
$= p^2$

Exercise 12.1

1. Simplify the following:
a) $c^5 \times c^3$
b) $m^4 \div m^2$
c) $(b^3)^5 \div b^6$
d) $\dfrac{m^4 n^9}{mn^3}$
e) $\dfrac{6a^6 b^4}{3a^2 b^3}$
f) $\dfrac{12x^5 y^7}{4x^2 y^5}$
g) $\dfrac{4u^3 v^6}{8u^2 v^3}$
h) $\dfrac{3x^6 y^5 z^3}{9x^4 y^2 z}$

2. Simplify the following:
a) $4a^2 \times 3a^3$
b) $2a^2 b \times 4a^3 b^2$
c) $(2p^2)^3$
d) $(4m^2 n^3)^2$
e) $(5p^2)^2 \times (2p^3)^3$
f) $(4m^2 n^2) \times (2mn^3)^3$
g) $\dfrac{(6x^2 y^4)^2 \times (2xy)^3}{12x^6 y^8}$
h) $(ab)^d \times (ab)^e$

● The zero index

As shown in Chapter 7, the zero index indicates that a number or algebraic term is raised to the power of zero. A term raised to the power of zero is always equal to 1. This is shown below.

Algebra and graphs

$$a^m \div a^n = a^{m-n} \quad \text{therefore} \quad \frac{a^m}{a^m} = a^{m-m}$$
$$= a^0$$

However, $\quad \dfrac{a^m}{a^m} = 1$

therefore $a^0 = 1$

● Negative indices

A negative index indicates that a number or an algebraic term is being raised to a negative power e.g. a^{-4}.

As shown in Chapter 7, one law of indices states that $a^{-m} = \dfrac{1}{a^m}$. This is proved as follows.

$$a^{-m} = a^{0-m}$$
$$= \frac{a^0}{a^m} \quad \text{(from the second law of indices)}$$
$$= \frac{1}{a^m}$$

therefore $a^{-m} = \dfrac{1}{a^m}$

Exercise 12.2

1. Simplify the following:
   a) $c^3 \times c^0$
   b) $g^{-2} \times g^3 \div g^0$
   c) $(p^0)^3 (q^2)^{-1}$
   d) $(m^3)^3 (m^{-2})^5$

2. Simplify the following:
   a) $\dfrac{a^{-3} \times a^5}{(a^2)^0}$
   b) $\dfrac{(r^3)^{-2}}{(p^{-2})^3}$
   c) $(t^3 \div t^{-5})^2$
   d) $\dfrac{m^0 \div m^{-6}}{(m^{-1})^3}$

● Fractional indices

It was shown in Chapter 7 that $16^{\frac{1}{2}} = \sqrt{16}$ and that $27^{\frac{1}{3}} = \sqrt[3]{27}$.

This can be applied to algebraic indices too.

In general:

$$a^{\frac{1}{n}} = \sqrt[n]{a}$$

$$a^{\frac{m}{n}} = \sqrt[n]{a^m} \quad \text{or} \quad (\sqrt[n]{a})^m$$

The last rule can be proved as shown below:

Using the laws of indices:

$a^{\frac{m}{n}}$ can be written as $(a^m)^{\frac{1}{n}}$ which in turn can be written as $\sqrt[n]{a^m}$.

Similarly:

$a^{\frac{m}{n}}$ can be written as $(a^{\frac{1}{n}})^m$ which in turn can be written as $(\sqrt[n]{a})^m$.

Worked examples

a) Express $(\sqrt[3]{a})^4$ in the form $a^{\frac{m}{n}}$.

$(\sqrt[3]{a}) = a^{\frac{1}{3}}$

Therefore $(\sqrt[3]{a})^4 = (a^{\frac{1}{3}})^4 = a^{\frac{4}{3}}$

b) Express $b^{\frac{2}{5}}$ in the form $(\sqrt[n]{b})^m$.

$b^{\frac{2}{5}}$ can be expressed as $(b^{\frac{1}{5}})^2$

$b^{\frac{1}{5}} = \sqrt[5]{b}$

Therefore $b^{\frac{2}{5}} = (b^{\frac{1}{5}})^2 = (\sqrt[5]{b})^2$

c) Simplify $\dfrac{p^{\frac{1}{2}} \times p^{\frac{1}{3}}}{p}$.

Using the laws of indices, the numerator $p^{\frac{1}{2}} \times p^{\frac{1}{3}}$ can be simplified to $p^{(\frac{1}{2}+\frac{1}{3})} = p^{\frac{5}{6}}$.

Therefore $\dfrac{p^{\frac{5}{6}}}{p}$ can now be written as $p^{\frac{5}{6}} \times p^{-1}$.

Using the laws of indices again, this can be simplified as $p^{(\frac{5}{6}-1)} = p^{-\frac{1}{6}}$.

Therefore $\dfrac{p^{\frac{1}{2}} \times p^{\frac{1}{3}}}{p} = p^{-\frac{1}{6}}$

Other possible simplifications are $(\sqrt[6]{p})^{-1}$ or $\dfrac{1}{(\sqrt[6]{p})}$.

Exercise 12.3

1. Rewrite the following in the form $a^{\frac{m}{n}}$:

   a) $(\sqrt[5]{a})^3$ b) $(\sqrt[6]{a})^2$ c) $(\sqrt[4]{a})^4$ d) $(\sqrt[7]{a})^3$

2. Rewrite the following in the form $(\sqrt[n]{b})^m$:

   a) $b^{\frac{2}{7}}$ b) $b^{\frac{8}{3}}$ c) $b^{-\frac{2}{5}}$ d) $b^{-\frac{4}{3}}$

3. Simplify the following algebraic expressions, giving your answer in the form $a^{\frac{m}{n}}$:

   a) $a^{\frac{1}{2}} \times a^{\frac{1}{4}}$
   b) $a^{\frac{2}{5}} \times a^{-\frac{1}{4}}$
   c) $\dfrac{\sqrt{a}}{a^{-2}}$
   d) $\dfrac{\sqrt[3]{a}}{\sqrt{a}}$

4. Simplify the following algebraic expressions, giving your answer in the form $(\sqrt[n]{b})^m$:

   a) $\dfrac{\sqrt{b} \times b^{\frac{1}{4}}}{b^{-\frac{1}{5}}}$
   b) $\dfrac{b^{-\frac{1}{3}} \times \sqrt[3]{b}}{b^{\frac{2}{5}} \times b}$
   c) $\dfrac{b^3 \times b^{-\frac{1}{3}}}{b^{-2}}$
   d) $\dfrac{b^{-2} \times \sqrt[3]{b}}{\sqrt{b} \times (\sqrt[3]{b})^{-1}}$

Algebra and graphs

5. Simplify the following:

   a) $\dfrac{1}{3}x^{\frac{1}{2}} \div 4x^{-2}$

   b) $\dfrac{2}{5}y^{\frac{1}{3}} \times 5y^{-\frac{2}{3}}$

   c) $\left(2p^{-\frac{1}{4}}\right)^2 \div \dfrac{1}{2}p^2$

   d) $3x^{-\frac{2}{3}} \div \dfrac{2}{3}x^{-\frac{1}{3}}$

Student assessment 1

1. Simplify the following using indices:
   a) $a \times a \times a \times b \times b$
   b) $d \times d \times e \times e \times e \times e \times e$

2. Write the following out in full:
   a) m^3
   b) r^4

3. Simplify the following using indices:

   a) $a^4 \times a^3$

   b) $p^3 \times p^2 \times q^4 \times q^5$

   c) $\dfrac{b^7}{b^4}$

   d) $\dfrac{(e^4)^5}{e^{14}}$

4. Simplify the following:

   a) $r^4 \times t^0$

   b) $\dfrac{(a^3)^0}{b^2}$

   c) $\dfrac{(m^0)^5}{n^{-3}}$

5. Simplify the following:

   a) $\dfrac{(p^2 \times p^{-5})^2}{p^3}$

   b) $\dfrac{(h^{-2} \times h^{-5})^{-1}}{h^0}$

Student assessment 2

1. Rewrite the following in the form $a^{\frac{m}{n}}$:

   a) $(\sqrt[8]{a})^7$

   b) $(\sqrt[5]{a})^{-2}$

2. Rewrite the following in the form $(\sqrt[n]{b^m})$:

   a) $b^{\frac{4}{9}}$

   b) $b^{-\frac{2}{3}}$

3. Simplify the following algebraic expressions, giving your answer in the form $a^{\frac{m}{n}}$:

   a) $a^{\frac{1}{3}} \times a^{\frac{3}{2}}$

   b) $\dfrac{\sqrt[3]{a}}{a^{-\frac{5}{6}}} \times a^2$

4. Simplify the following algebraic expressions, giving your answer in the form $(\sqrt[n]{t})^m$:

   a) $\dfrac{\sqrt{t} \times t^{\frac{2}{3}}}{t^{-\frac{1}{3}}}$

   b) $\dfrac{\sqrt[3]{t}}{t^2 \times t^{-\frac{2}{5}}}$

13 Equations and inequalities

An equation is formed when the value of an unknown quantity is needed.

Simple linear equations

Worked examples Solve the following linear equations:

a) $3x + 8 = 14$
    $3x = 6$
    $x = 2$

b) $12 = 20 + 2x$
    $-8 = 2x$
    $-4 = x$

c) $3(p + 4) = 21$
    $3p + 12 = 21$
    $3p = 9$
    $p = 3$

d) $4(x - 5) = 7(2x - 5)$
    $4x - 20 = 14x - 35$
    $4x + 15 = 14x$
    $15 = 10x$
    $1.5 = x$

Exercise 13.1 Solve the following linear equations:

1. a) $3x = 2x - 4$
   b) $5y = 3y + 10$
   c) $2y - 5 = 3y$
   d) $p - 8 = 3p$
   e) $3y - 8 = 2y$
   f) $7x + 11 = 5x$

2. a) $3x - 9 = 4$
   b) $4 = 3x - 11$
   c) $6x - 15 = 3x + 3$
   d) $4y + 5 = 3y - 3$
   e) $8y - 31 = 13 - 3y$
   f) $4m + 2 = 5m - 8$

3. a) $7m - 1 = 5m + 1$
   b) $5p - 3 = 3 + 3p$
   c) $12 - 2k = 16 + 2k$
   d) $6x + 9 = 3x - 54$
   e) $8 - 3x = 18 - 8x$
   f) $2 - y = y - 4$

4. a) $\frac{x}{2} = 3$
   b) $\frac{1}{2}y = 7$
   c) $\frac{x}{4} = 1$
   d) $\frac{1}{4}m = 3$
   e) $7 = \frac{x}{5}$
   f) $4 = \frac{1}{5}p$

5. a) $\frac{x}{3} - 1 = 4$
   b) $\frac{x}{5} + 2 = 1$
   c) $\frac{2}{3}x = 5$
   d) $\frac{3}{4}x = 6$
   e) $\frac{1}{5}x = \frac{1}{2}$
   f) $\frac{2x}{5} = 4$

Algebra and graphs

6. a) $\dfrac{x+1}{2} = 3$ b) $4 = \dfrac{x-2}{3}$

 c) $\dfrac{x-10}{3} = 4$ d) $8 = \dfrac{5x-1}{3}$

 e) $\dfrac{2(x-5)}{3} = 2$ f) $\dfrac{3(x-2)}{4} = 4x - 8$

7. a) $6 = \dfrac{2(y-1)}{3}$ b) $2(x + 1) = 3(x - 5)$

 c) $5(x - 4) = 3(x + 2)$ d) $\dfrac{3+y}{2} = \dfrac{y+1}{4}$

 e) $\dfrac{7+2x}{3} = \dfrac{9x-1}{7}$ f) $\dfrac{2x+3}{4} = \dfrac{4x-2}{6}$

● Constructing equations

NB: All diagrams are not to scale.

In many cases, when dealing with the practical applications of mathematics, equations need to be constructed first before they can be solved. Often the information is either given within the context of a problem or in a diagram.

Worked examples

a) Find the size of each of the angles in the triangle (left) by constructing an equation and solving it to find the value of x.

The sum of the angles of a triangle is 180°.

$(x + 30) + (x - 30) + 90 = 180$
$2x + 90 = 180$
$2x = 90$
$x = 45$

The three angles are therefore: 90°, $x + 30 = 75°$, $x - 30 = 15°$.

Check: $90° + 75° + 15° = 180°$.

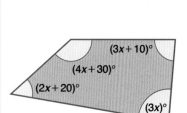

b) Find the size of each of the angles in the quadrilateral (left) by constructing an equation and solving it to find the value of x.

The sum of the angles of a quadrilateral is 360°.

$4x + 30 + 3x + 10 + 3x + 2x + 20 = 360$
$12x + 60 = 360$
$12x = 300$
$x = 25$

The angles are:

$4x + 30 = (4 \times 25) + 30 = 130°$
$3x + 10 = (3 \times 25) + 10 = 85°$
$3x \quad\quad = 3 \times 25 \quad\quad = 75°$
$2x + 20 = (2 \times 25) + 20 = 70°$
$\quad\quad\quad\quad\quad\quad\text{Total} = 360°$

13   Equations and inequalities

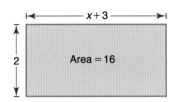

c) Construct an equation and solve it to find the value of x in the diagram (left).

Area of rectangle = base × height
$$2(x + 3) = 16$$
$$2x + 6 = 16$$
$$2x = 10$$
$$x = 5$$

Exercise 13.2

In questions 1–3:
i) construct an equation in terms of x,
ii) solve the equation,
iii) calculate the size of each of the angles,
iv) check your answers.

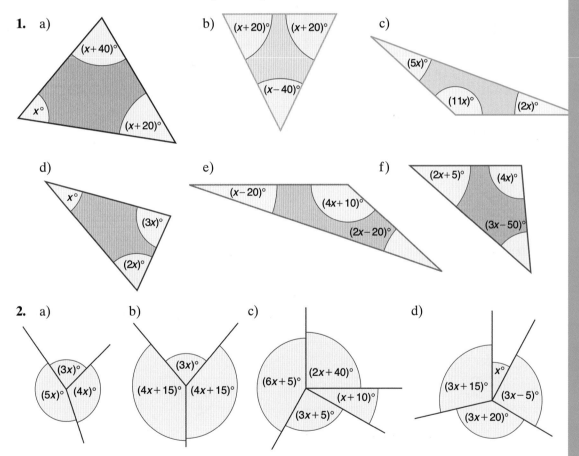

Algebra and graphs

3. a) Quadrilateral with angles $x°$, $(4x)°$, $(3x-40)°$, $(3x-40)°$

    b) Quadrilateral with angles $(3x)°$, $(3x+40)°$, $(x+15)°$, $(3x+5)°$

    c) Quadrilateral with angles $(5x-10)°$, $(5x+10)°$, $(2x)°$, $(4x+8)°$

    d) Quadrilateral with angles $x°$, $(4x+10)°$, $(3x+15)°$, $(x+20)°$

    e) Parallelogram with angles $(x+23)°$ and $(5x-5)°$

4. By constructing an equation and solving it, find the value of x in each of these isosceles triangles:

    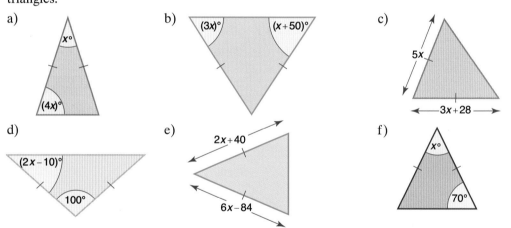

    a) Isosceles triangle with apex $x°$ and base angle $(4x)°$

    b) Isosceles triangle with angles $(3x)°$ and $(x+50)°$

    c) Isosceles triangle with side $5x$ and base $3x+28$

    d) Isosceles triangle with angle $(2x-10)°$ and $100°$

    e) Isosceles triangle with sides $2x+40$ and $6x-84$

    f) Isosceles triangle with apex $x°$ and base angle $70°$

5. Using angle properties, calculate the value of x in each of these questions:

    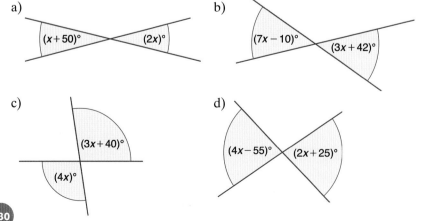

    a) Vertically opposite angles $(x+50)°$ and $(2x)°$

    b) Angles on a line/vertex $(7x-10)°$ and $(3x+42)°$

    c) Angles $(3x+40)°$ and $(4x)°$

    d) Vertically opposite angles $(4x-55)°$ and $(2x+25)°$

Equations and inequalities

6. Calculate the value of x:

a)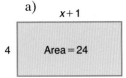

b)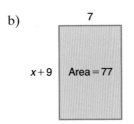

c)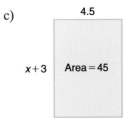

d)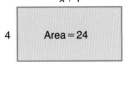

e)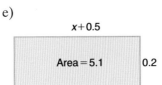

f)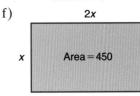

Simultaneous equations

When the values of two unknowns are needed, two equations need to be formed and solved. The process of solving two equations and finding a common solution is known as solving equations simultaneously.

The two most common ways of solving simultaneous equations algebraically are by **elimination** and by **substitution**.

By elimination

The aim of this method is to eliminate one of the unknowns by either adding or subtracting the two equations.

Worked examples   Solve the following simultaneous equations by finding the values of x and y which satisfy both equations.

a) $3x + y = 9$ (1)
$5x - y = 7$ (2)

By adding equations (1) + (2) we eliminate the variable y:

$8x = 16$
$x = 2$

To find the value of y we substitute $x = 2$ into either equation (1) or (2).
Substituting $x = 2$ into equation (1):

$3x + y = 9$
$6 + y = 9$
$y = 3$

To check that the solution is correct, the values of x and y are substituted into equation (2). If it is correct then the left-hand side of the equation will equal the right-hand side.

$5x - y = 7$
$\text{LHS} = 10 - 3 = 7$
$= \text{RHS} \checkmark$

b) $4x + y = 23$ (1)
    $x + y = 8$ (2)

By subtracting the equations, i.e. (1) − (2), we eliminate the variable y:

$$3x = 15$$
$$x = 5$$

By substituting $x = 5$ into equation (2), y can be calculated:

$$x + y = 8$$
$$5 + y = 8$$
$$y = 3$$

Check by substituting both values into equation (1):

$$4x + y = 23$$
$$20 + 3 = 23$$
$$23 = 23$$

By substitution

The same equations can also be solved by the method known as **substitution**.

Worked examples   **a)** $3x + y = 9$ (1)
                            $5x - y = 7$ (2)

Equation (2) can be rearranged to give: $y = 5x - 7$
This can now be substituted into equation (1):

$$3x + (5x - 7) = 9$$
$$3x + 5x - 7 = 9$$
$$8x - 7 = 9$$
$$8x = 16$$
$$x = 2$$

To find the value of y, $x = 2$ is substituted into either equation (1) or (2) as before giving $y = 3$.

b) $4x + y = 23$ (1)
    $x + y = 8$ (2)

Equation (2) can be rearranged to give $y = 8 - x$.
This can be substituted into equation (1):

$$4x + (8 - x) = 23$$
$$4x + 8 - x = 23$$
$$3x + 8 = 23$$
$$3x = 15$$
$$x = 5$$

y can be found as before, giving a result of $y = 3$.

Equations and inequalities

Exercise 13.3 Solve the following simultaneous equations either by elimination or by substitution:

1. a) $x + y = 6$
   $x - y = 2$
   b) $x + y = 11$
   $x - y - 1 = 0$
   c) $x + y = 5$
   $x - y = 7$
   d) $2x + y = 12$
   $2x - y = 8$
   e) $3x + y = 17$
   $3x - y = 13$
   f) $5x + y = 29$
   $5x - y = 11$

2. a) $3x + 2y = 13$
   $4x = 2y + 8$
   b) $6x + 5y = 62$
   $4x - 5y = 8$
   c) $x + 2y = 3$
   $8x - 2y = 6$
   d) $9x + 3y = 24$
   $x - 3y = -14$
   e) $7x - y = -3$
   $4x + y = 14$
   f) $3x = 5y + 14$
   $6x + 5y = 58$

3. a) $2x + y = 14$
   $x + y = 9$
   b) $5x + 3y = 29$
   $x + 3y = 13$
   c) $4x + 2y = 50$
   $x + 2y = 20$
   d) $x + y = 10$
   $3x = -y + 22$
   e) $2x + 5y = 28$
   $4x + 5y = 36$
   f) $x + 6y = -2$
   $3x + 6y = 18$

4. a) $x - y = 1$
   $2x - y = 6$
   b) $3x - 2y = 8$
   $2x - 2y = 4$
   c) $7x - 3y = 26$
   $2x - 3y = 1$
   d) $x = y + 7$
   $3x - y = 17$
   e) $8x - 2y = -2$
   $3x - 2y = -7$
   f) $4x - y = -9$
   $7x - y = -18$

5. a) $x + y = -7$
   $x - y = -3$
   b) $2x + 3y = -18$
   $2x = 3y + 6$
   c) $5x - 3y = 9$
   $2x + 3y = 19$
   d) $7x + 4y = 42$
   $9x - 4y = -10$
   e) $4x - 4y = 0$
   $8x + 4y = 12$
   f) $x - 3y = -25$
   $5x - 3y = -17$

6. a) $2x + 3y = 13$
   $2x - 4y + 8 = 0$
   b) $2x + 4y = 50$
   $2x + y = 20$
   c) $x + y = 10$
   $3y = 22 - x$
   d) $5x + 2y = 28$
   $5x + 4y = 36$
   e) $2x - 8y = 2$
   $2x - 3y = 7$
   f) $x - 4y = 9$
   $x - 7y = 18$

7. a) $-4x = 4y$
   $4x - 8y = 12$
   b) $3x = 19 + 2y$
   $-3x + 5y = 5$
   c) $3x + 2y = 12$
   $-3x + 9y = -12$
   d) $3x + 5y = 29$
   $3x + y = 13$
   e) $-5x + 3y = 14$
   $5x + 6y = 58$
   f) $-2x + 8y = 6$
   $2x = 3 - y$

Algebra and graphs

Further simultaneous equations

If neither x nor y can be eliminated by simply adding or subtracting the two equations then it is necessary to multiply one or both of the equations. The equations are multiplied by a number in order to make the coefficients of x (or y) numerically equal.

Worked examples

a) $3x + 2y = 22$ (1)
$x + y = 9$ (2)

To eliminate y, equation (2) is multiplied by 2:

$3x + 2y = 22$ (1)
$2x + 2y = 18$ (3)

By subtracting (3) from (1), the variable y is eliminated:

$x = 4$

Substituting $x = 4$ into equation (2), we have:

$x + y = 9$
$4 + y = 9$
$y = 5$

Check by substituting both values into equation (1):

$3x + 2y = 22$
LHS $= 12 + 10 = 22$
$=$ RHS ✓

b) $5x - 3y = 1$ (1)
$3x + 4y = 18$ (2)

To eliminate the variable y, equation (1) is multiplied by 4, and equation (2) is multiplied by 3.

$20x - 12y = 4$ (3)
$9x + 12y = 54$ (4)

By adding equations (3) and (4) the variable y is eliminated:

$29x = 58$
$x = 2$

Substituting $x = 2$ into equation (2) gives:

$3x + 4y = 18$
$6 + 4y = 18$
$4y = 12$
$y = 3$

Check by substituting both values into equation (1):

$5x - 3y = 1$
LHS $= 10 - 9 = 1$
$=$ RHS ✓

Equations and inequalities

Exercise 13.4 Solve the following:

1. a) $2x + y = 7$ b) $5x + 4y = 21$ c) $x + y = 7$
   $3x + 2y = 12$ $x + 2y = 9$ $3x + 4y = 23$
   d) $2x - 3y = -3$ e) $4x = 4y + 8$ f) $x + 5y = 11$
   $3x + 2y = 15$ $x + 3y = 10$ $2x - 2y = 10$

2. a) $x + y = 5$ b) $2x - 2y = 6$ c) $2x + 3y = 15$
   $3x - 2y + 5 = 0$ $x - 5y = -5$ $2y = 15 - 3x$
   d) $x - 6y = 0$ e) $2x - 5y = -11$ f) $x + y = 5$
   $3x - 3y = 15$ $3x + 4y = 18$ $2x - 2y = -2$

3. a) $3y = 9 + 2x$ b) $x + 4y = 13$ c) $2x = 3y - 19$
   $3x + 2y = 6$ $3x - 3y = 9$ $3x + 2y = 17$
   d) $2x - 5y = -8$ e) $5x - 2y = 0$ f) $8y = 3 - x$
   $-3x - 2y = -26$ $2x + 5y = 29$ $3x - 2y = 9$

4. a) $4x + 2y = 5$ b) $4x + y = 14$ c) $10x - y = -2$
   $3x + 6y = 6$ $6x - 3y = 3$ $-15x + 3y = 9$
   d) $-2y = 0.5 - 2x$ e) $x + 3y = 6$ f) $5x - 3y = -0.5$
   $6x + 3y = 6$ $2x - 9y = 7$ $3x + 2y = 3.5$

Exercise 13.5

1. The sum of two numbers is 17 and their difference is 3. Find the two numbers by forming two equations and solving them simultaneously.

2. The difference between two numbers is 7. If their sum is 25, find the two numbers by forming two equations and solving them simultaneously.

3. Find the values of x and y.

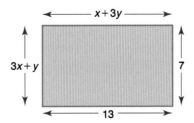

4. Find the values of x and y.

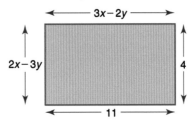

Algebra and graphs

5. A man's age is three times his son's age. Ten years ago he was five times his son's age. By forming two equations and solving them simultaneously, find both of their ages.

6. A grandfather is ten times older than his granddaughter. He is also 54 years older than her. How old is each of them?

● Constructing further equations

Earlier in this chapter we looked at some simple examples of constructing and solving equations when we were given geometrical diagrams. This section extends this work with more complicated formulae and equations.

Worked examples   Construct and solve the equations below.

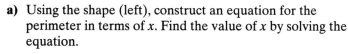

a) Using the shape (left), construct an equation for the perimeter in terms of x. Find the value of x by solving the equation.

$$x + 3 + x + x - 5 + 8 + 8 + x + 8 = 54$$
$$4x + 22 = 54$$
$$4x = 32$$
$$x = 8$$

b) A number is doubled, 5 is subtracted from it, and the total is 17. Find the number.

Let x be the unknown number.

$$2x - 5 = 17$$
$$2x = 22$$
$$x = 11$$

c) 3 is added to a number. The result is multiplied by 8. If the answer is 64 calculate the value of the original number.

Let x be the unknown number.

$$8(x + 3) = 64$$
$$8x + 24 = 64$$
$$8x = 40$$
$$x = 5$$

or
$$8(x + 3) = 64$$
$$x + 3 = 8$$
$$x = 5$$

The original number $= 5$

Exercise 13.6

1. Calculate the value of x:

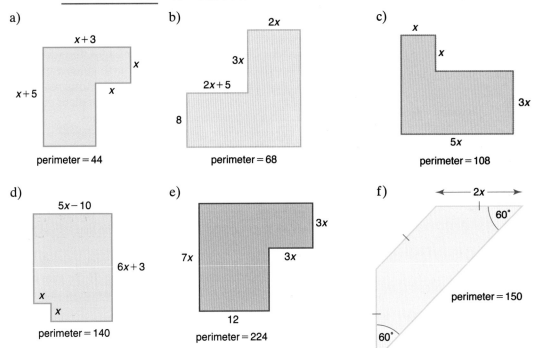

2. a) A number is trebled and then 7 is added to it. If the total is 28, find the number.
  b) Multiply a number by 4 and then add 5 to it. If the total is 29, find the number.
  c) If 31 is the result of adding 1 to 5 times a number, find the number.
  d) Double a number and then subtract 9. If the answer is 11, what is the number?
  e) If 9 is the result of subtracting 12 from 7 times a number, find the number.

3. a) Add 3 to a number and then double the result. If the total is 22, find the number.
  b) 27 is the answer when you add 4 to a number and then treble it. What is the number?
  c) Subtract 1 from a number and multiply the result by 5. If the answer is 35, what is the number?
  d) Add 3 to a number. If the result of multiplying this total by 7 is 63, find the number.
  e) Add 3 to a number. Quadruple the result. If the answer is 36, what is the number?

Algebra and graphs

4. a) Gabriella is x years old. Her brother is 8 years older and her sister is 12 years younger than she is. If their total age is 50 years, how old are they?
   b) A series of mathematics textbooks consists of four volumes. The first volume has x pages, the second 54 pages more. The third and fourth volume each have 32 pages more than the second. If the total number of pages in all four volumes is 866, calculate the number of pages in each of the volumes.
   c) The five interior angles (in °) of a pentagon are x, $x + 30$, $2x$, $2x + 40$ and $3x + 20$. The sum of the interior angles of a pentagon is 540°. Calculate the size of each of the angles.
   d) A hexagon consists of three interior angles of equal size and a further three which are double this size. The sum of all six angles is 720°. Calculate the size of each of the angles.
   e) Four of the exterior angles of an octagon are the same size. The other four are twice as big. If the sum of the exterior angles is 360°, calculate the size of the interior angles.

● Solving quadratic equations by factorising

Students will need to be familiar with the work covered in Chapter 11 on the factorising of quadratics.

$x^2 - 3x - 10 = 0$ is a quadratic equation, which when factorised can be written as $(x - 5)(x + 2) = 0$.

Therefore either $x - 5 = 0$ or $x + 2 = 0$ since, if two things multiply to make zero, then one or both of them must be zero.

$$x - 5 = 0 \quad \text{or} \quad x + 2 = 0$$
$$x = 5 \quad \text{or} \quad x = -2$$

Worked examples Solve the following equations to give two solutions for x:

a) $x^2 - x - 12 = 0$
$(x - 4)(x + 3) = 0$
so either $x - 4 = 0$ or $x + 3 = 0$
$x = 4$ or $x = -3$

b) $x^2 + 2x = 24$
This becomes $x^2 + 2x - 24 = 0$
$(x + 6)(x - 4) = 0$
so either $x + 6 = 0$ or $x - 4 = 0$
$x = -6$ or $x = 4$

Equations and inequalities

c) $x^2 - 6x = 0$
$x(x - 6) = 0$
so either $x = 0$ or $x - 6 = 0$
or $x = 6$

d) $x^2 - 4 = 0$
$(x - 2)(x + 2) = 0$
so either $x - 2 = 0$ or $x + 2 = 0$
$x = 2$ or $x = -2$

Exercise 13.7 Solve the following quadratic equations by factorising:

1. a) $x^2 + 7x + 12 = 0$ b) $x^2 + 8x + 12 = 0$
   c) $x^2 + 13x + 12 = 0$ d) $x^2 - 7x + 10 = 0$
   e) $x^2 - 5x + 6 = 0$ f) $x^2 - 6x + 8 = 0$

2. a) $x^2 + 3x - 10 = 0$ b) $x^2 - 3x - 10 = 0$
   c) $x^2 + 5x - 14 = 0$ d) $x^2 - 5x - 14 = 0$
   e) $x^2 + 2x - 15 = 0$ f) $x^2 - 2x - 15 = 0$

3. a) $x^2 + 5x = -6$ b) $x^2 + 6x = -9$
   c) $x^2 + 11x = -24$ d) $x^2 - 10x = -24$
   e) $x^2 + x = 12$ f) $x^2 - 4x = 12$

4. a) $x^2 - 2x = 8$ b) $x^2 - x = 20$
   c) $x^2 + x = 30$ d) $x^2 - x = 42$
   e) $x^2 - 2x = 63$ f) $x^2 + 3x = 54$

Exercise 13.8 Solve the following quadratic equations:

1. a) $x^2 - 9 = 0$ b) $x^2 - 16 = 0$
   c) $x^2 = 25$ d) $x^2 = 121$
   e) $x^2 - 144 = 0$ f) $x^2 - 220 = 5$

2. a) $4x^2 - 25 = 0$ b) $9x^2 - 36 = 0$
   c) $25x^2 = 64$ d) $x^2 = \frac{1}{4}$
   e) $x^2 - \frac{1}{9} = 0$ f) $16x^2 - \frac{1}{25} = 0$

3. a) $x^2 + 5x + 4 = 0$ b) $x^2 + 7x + 10 = 0$
   c) $x^2 + 6x + 8 = 0$ d) $x^2 - 6x + 8 = 0$
   e) $x^2 - 7x + 10 = 0$ f) $x^2 + 2x - 8 = 0$

4. a) $x^2 - 3x - 10 = 0$ b) $x^2 + 3x - 10 = 0$
   c) $x^2 - 3x - 18 = 0$ d) $x^2 + 3x - 18 = 0$
   e) $x^2 - 2x - 24 = 0$ f) $x^2 - 2x - 48 = 0$

5. a) $x^2 + x = 12$ b) $x^2 + 8x = -12$
   c) $x^2 + 5x = 36$ d) $x^2 + 2x = -1$
   e) $x^2 + 4x = -4$ f) $x^2 + 17x = -72$

6. a) $x^2 - 8x = 0$ b) $x^2 - 7x = 0$
   c) $x^2 + 3x = 0$ d) $x^2 + 4x = 0$
   e) $x^2 - 9x = 0$ f) $4x^2 - 16x = 0$

Algebra and graphs

7. a) $2x^2 + 5x + 3 = 0$ b) $2x^2 - 3x - 5 = 0$
    c) $3x^2 + 2x - 1 = 0$ d) $2x^2 + 11x + 5 = 0$
    e) $2x^2 - 13x + 15 = 0$ f) $12x^2 + 10x - 8 = 0$

8. a) $x^2 + 12x = 0$ b) $x^2 + 12x + 27 = 0$
    c) $x^2 + 4x = 32$ d) $x^2 + 5x = 14$
    e) $2x^2 = 72$ f) $3x^2 - 12 = 288$

Exercise 13.9 In the following questions construct equations from the information given and then solve to find the unknown.

1. When a number x is added to its square, the total is 12. Find two possible values for x.

2. A number x is equal to its own square minus 42. Find two possible values for x.

3. If the area of the rectangle (below) is 10 cm², calculate the only possible value for x.

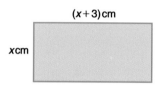

4. If the area of the rectangle (below) is 52 cm², calculate the only possible value for x.

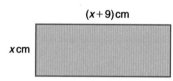

5. A triangle has a base length of $2x$ cm and a height of $(x - 3)$ cm. If its area is 18 cm², calculate its height and base length.

6. A triangle has a base length of $(x - 8)$ cm and a height of $2x$ cm. If its area is 20 cm² calculate its height and base length.

7. A right-angled triangle has a base length of x cm and a height of $(x - 1)$ cm. If its area is 15 cm² calculate the base length and height.

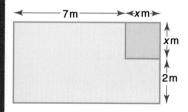

8. A rectangular garden has a square flower bed of side length x m in one of its corners. The remainder of the garden consists of lawn and has dimensions as shown (left). If the total area of the lawn is 50 m²:
   a) form an equation in terms of x
   b) solve the equation
   c) calculate the length and width of the whole garden.

The quadratic formula

In general a quadratic equation takes the form $ax^2 + bx + c = 0$ where a, b and c are integers. Quadratic equations can be solved by the use of the quadratic formula which states that:

$$x = \frac{-b \pm \sqrt{b^2 - 4ac}}{2a}$$

Worked examples

a) Solve the quadratic equation $x^2 + 7x + 3 = 0$.

$a = 1$, $b = 7$ and $c = 3$.

Substituting these values into the quadratic formula gives:

$$x = \frac{-7 \pm \sqrt{7^2 - 4 \times 1 \times 3}}{2 \times 1}$$

$$x = \frac{-7 \pm \sqrt{49 - 12}}{2}$$

$$x = \frac{-7 \pm \sqrt{37}}{2}$$

Therefore $x = \dfrac{-7 + 6.083}{2}$ or $x = \dfrac{-7 - 6.083}{2}$

$x = -0.459$ (3 s.f.) or $x = -6.54$ (3 s.f.)

b) Solve the quadratic equation $x^2 - 4x - 2 = 0$.

$a = 1$, $b = -4$ and $c = -2$.

Substituting these values into the quadratic formula gives:

$$x = \frac{-(-4) \pm \sqrt{(-4)^2 - (4 \times 1 \times -2)}}{2 \times 1}$$

$$x = \frac{4 \pm \sqrt{16 + 8}}{2}$$

$$x = \frac{4 \pm \sqrt{24}}{2}$$

Therefore $x = \dfrac{4 + 4.899}{2}$ or $x = \dfrac{4 - 4.899}{2}$

$x = 4.45$ (3 s.f.) or $x = -0.449$ (3 s.f.)

Algebra and graphs

● Completing the square

Quadratics can also be solved by expressing them in terms of a perfect square. We look once again at the quadratic $x^2 - 4x - 2 = 0$.

The perfect square $(x - 2)^2$ can be expanded to give $x^2 - 4x + 4$. Notice that the x^2 and x terms are the same as those in the original quadratic.

Therefore $(x - 2)^2 - 6 = x^2 - 4x - 2$ and can be used to solve the quadratic.

$$(x - 2)^2 - 6 = 0$$
$$(x - 2)^2 = 6$$
$$x - 2 = \pm\sqrt{6}$$
$$x = 2 \pm \sqrt{6}$$
$$x = 4.45 \text{ (3 s.f.)} \quad \text{or} \quad x = -0.449 \text{ (3 s.f.)}$$

Exercise 13.10 Solve the following quadratic equations using either the quadratic formula or by completing the square. Give your answers to 2 d.p.

1. a) $x^2 - x - 13 = 0$ b) $x^2 + 4x - 11 = 0$
   c) $x^2 + 5x - 7 = 0$ d) $x^2 + 6x + 6 = 0$
   e) $x^2 + 5x - 13 = 0$ f) $x^2 - 9x + 19 = 0$

2. a) $x^2 + 7x + 9 = 0$ b) $x^2 - 35 = 0$
   c) $x^2 + 3x - 3 = 0$ d) $x^2 - 5x - 7 = 0$
   e) $x^2 + x - 18 = 0$ f) $x^2 - 8 = 0$

3. a) $x^2 - 2x - 2 = 0$ b) $x^2 - 4x - 11 = 0$
   c) $x^2 - x - 5 = 0$ d) $x^2 + 2x - 7 = 0$
   e) $x^2 - 3x + 1 = 0$ f) $x^2 - 8x + 3 = 0$

4. a) $2x^2 - 3x - 4 = 0$ b) $4x^2 + 2x - 5 = 0$
   c) $5x^2 - 8x + 1 = 0$ d) $-2x^2 - 5x - 2 = 0$
   e) $3x^2 - 4x - 2 = 0$ f) $-7x^2 - x + 15 = 0$

Linear inequalities

The statement

   6 is less than 8

can be written as:

   $6 < 8$

This inequality can be manipulated in the following ways:

adding 2 to each side:	$8 < 10$	this inequality is still true
subtracting 2 from each side:	$4 < 6$	this inequality is still true
multiplying both sides by 2:	$12 < 16$	this inequality is still true
dividing both sides by 2:	$3 < 4$	this inequality is still true
multiplying both sides by -2:	$-12 < -16$	this inequality is not true
dividing both sides by -2:	$-3 < -4$	this inequality is not true

As can be seen, when both sides of an inequality are either multiplied or divided by a negative number, the inequality is no longer true. For it to be true the inequality sign needs to be changed around.

   i.e. $-12 > -16$ and $-3 > -4$

When solving linear inequalities, the procedure is very similar to that of solving linear equations.

Worked examples   Remember:

 implies that the number is not included in the solution. It is associated with $>$ and $<$.

implies that the number is included in the solution. It is associated with $\geqslant$ and $\leqslant$.

Solve the following inequalities and represent the solution on a number line:

a) $15 + 3x < 6$
  $3x < -9$
  $x < -3$

b) $17 \leqslant 7x + 3$
  $14 \leqslant 7x$
  $2 \leqslant x$ that is $x \geqslant 2$

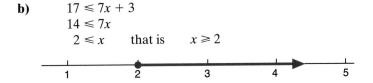

Algebra and graphs

c) $9 - 4x \geqslant 17$
$-4x \geqslant 8$
$x \leqslant -2$ Note the inequality sign has changed direction.

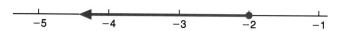

Exercise 13.11 Solve the following inequalities and illustrate your solution on a number line:

1. a) $x + 3 < 7$ b) $5 + x > 6$
   c) $4 + 2x \leqslant 10$ d) $8 \leqslant x + 1$
   e) $5 > 3 + x$ f) $7 < 3 + 2x$

2. a) $x - 3 < 4$ b) $x - 6 \geqslant -8$
   c) $8 + 3x > -1$ d) $5 \geqslant -x - 7$
   e) $12 > -x - 12$ f) $4 \leqslant 2x + 10$

3. a) $\frac{x}{2} < 1$ b) $4 \geqslant \frac{x}{3}$
   c) $1 \leqslant \frac{x}{2}$ d) $9x \geqslant -18$
   e) $-4x + 1 < 3$ f) $1 \geqslant -3x + 7$

Worked example Find the range of values for which $7 < 3x + 1 \leqslant 13$ and illustrate the solutions on a number line.

This is in fact two inequalities which can therefore be solved separately.

$$7 < 3x + 1 \qquad \text{and} \qquad 3x + 1 \leqslant 13$$
$(-1) \to \quad 6 < 3x \qquad\qquad\qquad (-1) \to \quad 3x \quad \leqslant 12$
$(\div 3) \to \quad 2 < x \text{ that is } x > 2 \qquad (\div 3) \to \quad x \quad \leqslant 4$

Exercise 13.12 Find the range of values for which the following inequalities are satisfied. Illustrate each solution on a number line:

1. a) $4 < 2x \leqslant 8$ b) $3 \leqslant 3x < 15$
   c) $7 \leqslant 2x < 10$ d) $10 \leqslant 5x < 21$

2. a) $5 < 3x + 2 \leqslant 17$ b) $3 \leqslant 2x + 5 < 7$
   c) $12 < 8x - 4 < 20$ d) $15 \leqslant 3(x - 2) < 9$

Equations and inequalities

Student assessment 1

Solve the following equations:
1. a) $x + 7 = 16$
   b) $2x - 9 = 13$
   c) $8 - 4x = 24$
   d) $5 - 3x = -13$

2. a) $7 - m = 4 + m$
   b) $5m - 3 = 3m + 11$
   c) $6m - 1 = 9m - 13$
   d) $18 - 3p = 6 + p$

3. a) $\dfrac{x}{-5} = 2$
   b) $4 = \dfrac{1}{3}x$
   c) $\dfrac{x+2}{3} = 4$
   d) $\dfrac{2x-5}{7} = \dfrac{5}{2}$

4. a) $\dfrac{2}{3}(x - 4) = 8$
   b) $4(x - 3) = 7(x + 2)$
   c) $4 = \dfrac{2}{7}(3x + 8)$
   d) $\dfrac{3}{4}(x - 1) = \dfrac{5}{8}(2x - 4)$

Solve the following simultaneous equations:
5. a) $2x + 3y = 16$
      $2x - 3y = 4$
   b) $4x + 2y = 22$
      $-2x + 2y = 2$
   c) $x + y = 9$
      $2x + 4y = 26$
   d) $2x - 3y = 7$
      $-3x + 4y = -11$

Student assessment 2

Solve the following equations:
1. a) $y + 9 = 3$
   b) $3x - 5 = 13$
   c) $12 - 5p = -8$
   d) $2.5y + 1.5 = 7.5$

2. a) $5 - p = 4 + p$
   b) $8m - 9 = 5m + 3$
   c) $11p - 4 = 9p + 15$
   d) $27 - 5r = r - 3$

3. a) $\dfrac{p}{-2} = -3$
   b) $6 = \dfrac{2}{5}x$
   c) $\dfrac{m-7}{5} = 3$
   d) $\dfrac{4t-3}{3} = 7$

4. a) $\dfrac{2}{5}(t - 1) = 3$
   b) $5(3 - m) = 4(m - 6)$
   c) $5 = \dfrac{2}{3}(x - 1)$
   d) $\dfrac{4}{5}(t - 2) = \dfrac{1}{4}(2t + 8)$

Solve the following simultaneous equations:
5. a) $x + y = 11$
      $x - y = 3$
   b) $5p - 3q = -1$
      $-2p - 3q = -8$
   c) $3x + 5y = 26$
      $x - y = 6$
   d) $2m - 3n = -9$
      $3m + 2n = 19$

Algebra and graphs

Student assessment 3

1. The angles of a triangle are x, $2x$ and $(x + 40)$ degrees.
   a) Construct an equation in terms of x.
   b) Solve the equation.
   c) Calculate the size of each of the three angles.

2. Seven is added to three times a number. The result is then doubled. If the answer is 68, calculate the number.

3. A decagon has six equal exterior angles. Each of the remaining four is fifteen degrees larger than these six angles. Construct an equation and then solve it to find the sizes of the angles.

4. Solve the following quadratic equation by factorisation:
   $$x^2 + 6x = -5$$

5. Solve the following quadratic equation by using the quadratic formula:
   $$x^2 + 6 = 8x$$

6. Solve the inequality below and illustrate your answer on a number line:
   $$12 \leqslant 3(x - 2) < 15$$

7. For what values of p is the following inequality true?
   $$\frac{1}{p^2} \leqslant -1$$

Student assessment 4

1. The angles of a quadrilateral are x, $3x$, $(2x - 40)$ and $(3x - 50)$ degrees.
   a) Construct an equation in terms of x.
   b) Solve the equation.
   c) Calculate the size of the four angles.

2. Three is subtracted from seven times a number. The result is multiplied by 5. If the answer is 55, calculate the value of the number by constructing an equation and solving it.

3. The interior angles of a pentagon are $9x$, $5x + 10$, $6x + 5$, $8x - 25$ and $10x - 20$ degrees. If the sum of the interior angles of a pentagon is 540°, find the size of each of the angles.

4. Solve the following quadratic equation by factorisation:
   $$x^2 - x = 20$$

5. Solve the following quadratic equation by using the quadratic formula:
$$2x^2 - 7 = 3x$$

6. Solve the inequality below and illustrate your answer on a number line
$$6 < 2x \leq 10$$

7. For what values of m is the following inequality true?
$$\frac{1}{m^2} > 0$$

Student assessment 5

1. A rectangle is x cm long. The length is 3 cm more than the width. The perimeter of the rectangle is 54 cm.
   a) Draw a diagram to illustrate the above information.
   b) Construct an equation in terms of x.
   c) Solve the equation and hence calculate the length and width of the rectangle.

2. At the end of a football season the leading goal scorer in a league has scored eight more goals than the second leading goal scorer. The second has scored fifteen more than the third. The total number of goals scored by all three players is 134.
   a) Write an expression for each of the three scores.
   b) Form an equation and then solve it to find the number of goals scored by each player.

3. a) Show that $x = 1 + \dfrac{7}{x-5}$ can be written as $x^2 - 6x - 2 = 0$.
   b) Use the quadratic formula to solve the equation
   $$x = 1 + \frac{7}{x-5}.$$

4. The angles of a quadrilateral are $x°$, $y°$, $70°$ and $40°$. The difference between the two unknowns is $18°$.
   a) Write two equations from the information given above.
   b) Solve the equations to find x and y.

5. A right-angled triangle has sides of length x, $x - 1$ and $x - 8$.
   a) Illustrate this information on a diagram.
   b) Show from the information given that $x^2 - 18x + 65 = 0$.
   c) Solve the quadratic equation and calculate the length of each of the three sides.

Algebra and graphs

Student assessment 6

1. The angles of a triangle are $x°$, $y°$ and $40°$. The difference between the two unknown angles is $30°$.
   a) Write down two equations from the information given above.
   b) What is the size of the two unknown angles?

2. The interior angles of a pentagon increase by $10°$ as you progress clockwise.
   a) Illustrate this information in a diagram.
   b) Write an expression for the sum of the interior angles.
   c) The sum of the interior angles of a pentagon is $540°$. Use this to calculate the largest **exterior** angle of the pentagon.
   d) Illustrate on your diagram the size of each of the five exterior angles.
   e) Show that the sum of the exterior angles is $360°$.

3. A flat sheet of card measures 12 cm by 10 cm. It is made into an open box by cutting a square of side x cm from each corner and then folding up the sides.
   a) Illustrate the box and its dimensions on a simple 3D sketch.
   b) Write an expression for the surface area of the outside of the box.
   c) If the surface area is 56 cm², form and solve a quadratic equation to find the value of x.

4. a) Show that $x - 2 = \dfrac{4}{x-3}$ can be written as $x^2 - 5x + 2 = 0$.
   b) Use the quadratic formula to solve $x - 2 = \dfrac{4}{x-3}$.

5. A right-angled triangle ABC has side lengths as follows: $AB = x$ cm, AC is 2 cm shorter than AB, and BC is 2 cm shorter than AC.
   a) Illustrate this information on a diagram.
   b) Using this information show that $x^2 - 12x + 20 = 0$.
   c) Solve the above quadratic and hence find the length of each of the three sides of the triangle.

14 Linear programming

● Revision

An understanding of the following symbols is necessary:

> means 'is greater than'
⩾ means 'is greater than or equal to'
< means 'is less than'
⩽ means 'is less than or equal to'

Exercise 14.1

1. Solve each of the following inequalities:
   a) $15 + 3x < 21$
   b) $18 \leqslant 7y + 4$
   c) $19 - 4x \geqslant 27$
   d) $2 \geqslant \frac{y}{3}$
   e) $-4t + 1 < 1$
   f) $1 \geqslant 3p + 10$

2. Solve each of the following inequalities:
   a) $7 < 3y + 1 \leqslant 13$
   b) $3 \leqslant 3p < 15$
   c) $9 \leqslant 3(m - 2) < 15$
   d) $20 < 8x - 4 < 28$

● Graphing an inequality

The solution to an inequality can also be illustrated on a graph.

Worked examples a) On a pair of axes, shade the region which satisfies the inequality $x \geqslant 3$.

To do this the line $x = 3$ is drawn.
The region to the right of $x = 3$ represents the inequality $x \geqslant 3$ and therefore is shaded as shown below.

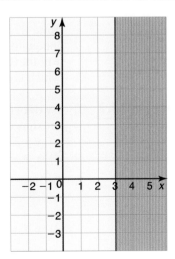

Algebra and graphs

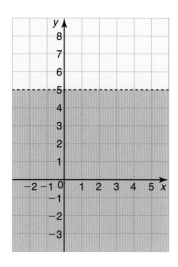

b) On a pair of axes, shade the region which satisfies the inequality $y < 5$.

The line $y = 5$ is drawn first (in this case it is drawn as a broken line).
The region below the line $y = 5$ represents the inequality $y < 5$ and therefore is shaded as shown (left).

Note that a broken (dashed) line shows $<$ or $>$ whilst a solid line shows $\leqslant$ or $\geqslant$.

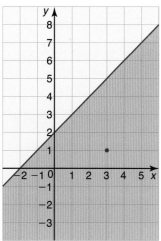

c) On a pair of axes, shade the region which satisfies the inequality $y \leqslant x + 2$.

The line $y = x + 2$ is drawn first (since it is included, this line is solid).
To know which region satisfies the inequality, and hence to know which side of the line to shade, the following steps are taken.

- Choose a point at random which does not lie on the line e.g. (3, 1).
- Substitute those values of x and y into the inequality i.e. $1 \leqslant 3 + 2$
- If the inequality holds true, then the region in which the point lies satisfies the inequality and can therefore be shaded.

Note that in some questions the region which satisfies the inequality is left **unshaded** whilst in others it is **shaded**. You will therefore need to read the question carefully to see which is required.

Exercise 14.2

1. By drawing appropriate axes, shade the region which satisfies each of the following inequalities:
   a) $y > 2$
   b) $x < 3$
   c) $y \leqslant 4$
   d) $x \geqslant -1$
   e) $y > 2x + 1$
   f) $y \leqslant x - 3$

2. By drawing appropriate axes, leave unshaded the region which satisfies each of the following inequalities:
   a) $y \geqslant -x$
   b) $y \leqslant 2 - x$
   c) $x \leqslant y - 3$
   d) $x + y \geqslant 4$
   e) $2x - y \geqslant 3$
   f) $2y - x < 4$

Graphing more than one inequality

Several inequalities can be graphed on the same set of axes. If the regions which satisfy each inequality are left unshaded then a solution can be found which satisfies all the inequalities, i.e. the region left unshaded by all the inequalities.

Worked example On the same pair of axes leave unshaded the regions which satisfy the following inequalities simultaneously:

$$x \leq 2 \quad y > -1 \quad y \leq 3 \quad y \leq x + 2$$

Hence find the region which satisfies all four inequalities.

If the four inequalities are graphed on separate axes the solutions are as shown below:

$x \leq 2$

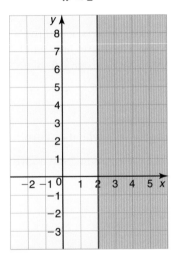

$y > -1$

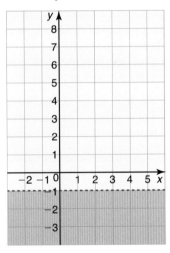

$y \leq 3$

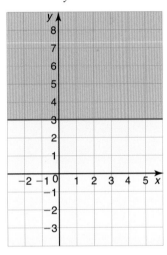

$y \leq x + 2$

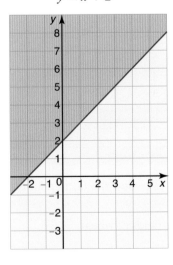

Combining all four on one pair of axes gives this diagram.

The unshaded region therefore gives a solution which satisfies all four inequalities.

Algebra and graphs

Exercise 14.3 On the same pair of axes plot the following inequalities and leave unshaded the region which satisfies all of them simultaneously.

1. $y \leq x$ $y > 1$ $x \leq 5$
2. $x + y \leq 6$ $y < x$ $y \geq 1$
3. $y \geq 3x$ $y \leq 5$ $x + y > 4$
4. $2y \geq x + 4$ $y \leq 2x + 2$ $y < 4$ $x \leq 3$

● Linear programming

Linear programming is a way of finding a number of possible solutions to a problem given a number of constraints. But it is more than this – it is also a method for minimising a linear function in two (or more) variables.

Worked example The number of fields a farmer plants with wheat is w and the number of fields he plants with corn is c. There are, however, certain restrictions which govern how many fields he can plant of each. These are as follows.

- There must be at least two fields of corn.
- There must be at least two fields of wheat.
- Not more than 10 fields are to be sown with wheat or corn.

i) Construct three inequalities from the information given above.

$$c \geq 2 \quad w \geq 2 \quad c + w \leq 10$$

ii) On one pair of axes, graph the three inequalities and leave unshaded the region which satisfies all three simultaneously.

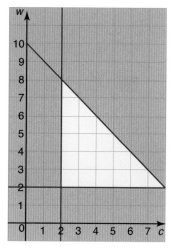

iii) Give one possible arrangement as to how the farmer should plant his fields.

Four fields of corn and four fields of wheat.

Exercise 14.4 In the following questions draw both axes numbered from 0 to 12. For each question:
a) write an inequality for each statement,
b) graph the inequalities, leaving the region which satisfies the inequalities unshaded,
c) using your graph, state one solution which satisfies all the inequalities simultaneously.

1. A taxi firm has at its disposal one morning a car and a minibus for hire. During the morning it makes x car trips and y minibus trips.
   - It makes at least five car trips.
   - It makes between two and eight minibus trips.
   - The total number of car and minibus trips does not exceed 12.

2. A woman is baking bread and cakes. She makes p loaves and q cakes. She bakes at least five loaves and at least two cakes but no more than ten loaves and cakes altogether.

3. A couple are buying curtains for their house. They buy m long curtains and n short curtains. They buy at least two long curtains. They also buy at least twice as many short curtains as long curtains. A maximum of 11 curtains are bought altogether.

4. A shop sells large and small oranges. A girl buys L large oranges and S small oranges. She buys at least three but fewer than nine large oranges. She also buys fewer than six small oranges. The maximum number of oranges she needs to buy is 10.

Student assessment 1

1. Solve the following inequalities:
   a) $17 + 5x \leq 42$
   b) $3 \geq \frac{y}{3} + 2$

2. Find the range of values for which:
   a) $7 < 4y - 1 \leq 15$
   b) $18 < 3(p + 2) \leq 30$

3. A garage stocks two kinds of engine oil. They have r cases of regular oil and s cases of super oil.
   They have fewer than ten cases of regular oil in stock and between three and nine cases of super oil in stock. They also have fewer than 12 cases in stock altogether.
   a) Express the three constraints described above as inequalities.
   b) Draw an appropriate pair of axes and identify the region which satisfies all the inequalities by shading the unwanted regions.
   c) State two possible solutions for the number of each case in stock.

Algebra and graphs

4. Students from Argentina and students from England are meeting on a cultural exchange.
   In total there will be between 12 and 20 students. There are fewer than 10 students from Argentina, whilst the number from England cannot be more than three greater than the number from Argentina.
   a) Write inequalities for the number (A) of students from Argentina and the number (E) of students from England.
   b) On an appropriate pair of axes, graph the inequalities, leaving unshaded the region which satisfies all of them.
   c) State two of the possible combinations of A and E which satisfy the given conditions.

Student assessment 2

1. Solve the following inequalities:
   a) $5 + 6x \leqslant 47$
   b) $4 \geqslant \dfrac{y+3}{3}$

2. Find the range of values for which:
   a) $3 \leqslant 3p < 12$
   b) $24 < 8(x-1) \leqslant 48$

3. A breeder of horses has x stallions and y mares. She has fewer than four stallions and more than two mares. She has enough room for a maximum of eight animals in total.
   a) Express the three conditions above as inequalities.
   b) Draw an appropriate pair of axes and leave unshaded the region which satisfies all the inequalities.
   c) State two of the possible combinations of stallions and mares which she can have.

4. Antonio is employed by a company to do two jobs. He mends cars and also repairs electrical goods. His terms of employment are listed below.

   - He is employed for a maximum of 40 hours.
   - He must spend at least 16 hours mending cars.
   - He must spend at least 5 hours repairing electrical goods.
   - He must spend more than twice as much time mending cars as repairing electrical goods.

   a) Express the conditions above as inequalities, using c to represent the number of hours spent mending cars, and e to represent the number of hours spent mending electrical goods.
   b) On an appropriate pair of axes, graph the inequalities, leaving unshaded the region which satisfies all four.
   c) State two of the possible combinations which satisfy the above conditions.

15 Sequences

A **sequence** is a collection of terms arranged in a specific order, where each term is obtained according to a rule. Examples of some simple sequences are given below:

$$2, 4, 6, 8, 10 \qquad 1, 4, 9, 16, 25 \qquad 1, 2, 4, 8, 16$$
$$1, 1, 2, 3, 5, 8 \qquad 1, 8, 27, 64, 125 \qquad 10, 5, \tfrac{5}{2}, \tfrac{5}{4}, \tfrac{5}{8}$$

You could discuss with another student the rules involved in producing the sequences above.

The terms of a sequence can be expressed as $u_1, u_2, u_3, \ldots, u_n$ where:

u_1 is the first term
u_2 is the second term
u_n is the nth term

Therefore in the sequence 2, 4, 6, 8, 10, $u_1 = 2$, $u_2 = 4$, etc.

● Arithmetic sequences

In an **arithmetic sequence** there is a **common difference** (d) between successive terms. Examples of some arithmetic sequences are given below:

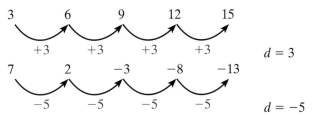

Formulae for the terms of an arithmetic sequence
There are two main ways of describing a sequence.

1. A term-to-term rule
   In the following sequence,

   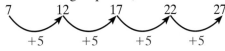

   the term-to-term rule is $+5$, i.e. $u_2 = u_1 + 5$, $u_3 = u_2 + 5$, etc. The general form is therefore written as $u_{n+1} = u_n + 5$, $u_1 = 7$, where u_n is the nth term and u_{n+1} the term after the nth term.

   Note: It is important to give one of the terms, e.g. u_1, so that the exact sequence can be generated.

2. **A formula for the nth term of a sequence**
   This type of rule links each term to its position in the sequence, e.g.

Position	1	2	3	4	5	n
Term	7	12	17	22	27	

   We can deduce from the figures above that each term can be calculated by multiplying its position number by 5 and adding 2. Algebraically this can be written as the formula for the nth term:

   $u_n = 5n + 2$

This textbook focuses on the generation and use of the rule for the nth term.

With an arithmetic sequence, the rule for the nth term can easily be deduced by looking at the common difference, e.g.

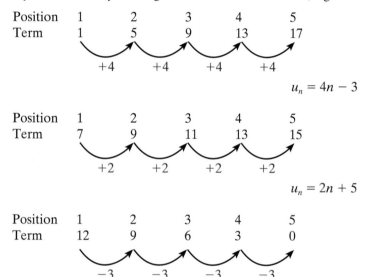

The common difference is the coefficient of n (i.e. the number by which n is multiplied). The constant is then worked out by calculating the number needed to make the term.

Worked example Find the rule for the nth term of the sequence
12, 7, 2, −3, −8, …

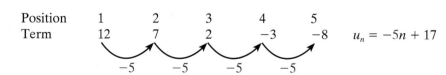

15 Sequences

Exercise 15.1

1. For each of the following sequences:
   i) deduce the formula for the nth term
   ii) calculate the 10th term.

   a) 5, 8, 11, 14, 17
   b) 0, 4, 8, 12, 16
   c) $\frac{1}{2}, 1\frac{1}{2}, 2\frac{1}{2}, 3\frac{1}{2}, 4\frac{1}{2}$
   d) 6, 3, 0, −3, −6
   e) −7, −4, −1, 2, 5
   f) −9, −13, −17, −21, −25

2. Copy and complete each of the following tables of arithmetic sequences:

   a)
Position	1	2	5		50	n
Term			45			$4n - 3$

   b)
Position	1	2	5			n
Term			59		449	$6n - 1$

   c)
Position	1				100	n
Term		0	−5	−47		$-n + 3$

   d)
Position	1	2	3			n
Term	3	0	−3	−24	−294	

   e)
Position		5	7			n
Term	1	10	16	25	145	

   f)
Position	1	2	5		50	n
Term	−5.5	−7		−34		

3. For each of the following arithmetic sequences:
   i) deduce the common difference d
   ii) give the formula for the nth term
   iii) calculate the 50th term.

   a) 5, 9, 13, 17, 21
   b) 0, ..., 2, ..., 4
   c) −10, ..., ..., ..., 2
   d) $u_1 = 6, u_9 = 10$
   e) $u_3 = -50, u_{20} = 18$
   f) $u_5 = 60, u_{12} = 39$

● Sequences with quadratic and cubic rules

So far all the sequences we have looked at have been arithmetic, i.e. the rule for the nth term is linear and takes the form $u_n = an + b$. The rule for the nth term can be found algebraically using the method of differences and this method is particularly useful for more complex sequences.

Algebra and graphs

Worked examples **a)** Deduce the rule for the nth term for the sequence
4, 7, 10, 13, 16, … .

Firstly, produce a table of the terms and their positions in the sequence:

Position	1	2	3	4	5
Term	4	7	10	13	16

Extend the table to look at the differences:

Position	1	2	3	4	5
Term	4	7	10	13	16
1st difference		3	3	3	3

As the row of 1st differences is constant, the rule for the nth term is linear and takes the form $u_n = an + b$.

By substituting the values of n into the rule, each term can be expressed in terms of a and b:

Position	1	2	3	4	5
Term	$a + b$	$2a + b$	$3a + b$	$4a + b$	$5a + b$
1st difference		a	a	a	a

Compare the two tables in order to deduce the values of a and b:
$a = 3$
$a + b = 4$ therefore $b = 1$

The rule for the nth term $u_n = an + b$ can be written as $u_n = 3n + 1$.

For a linear rule, this method is perhaps overcomplicated. However it is very efficient for quadratic and cubic rules.

b) Deduce the rule for the nth term for the sequence
0, 7, 18, 33, 52, … .

Entering the sequence in a table gives:

Position	1	2	3	4	5
Term	0	7	18	33	52

Extending the table to look at the differences gives:

Position	1	2	3	4	5
Term	0	7	18	33	52
1st difference		7	11	15	19

The row of 1st differences is not constant, and so the rule for the nth term is not linear. Extend the table again to look at the row of 2nd differences:

Position	1		2		3		4		5
Term	0		7		18		33		52
1st difference		7		11		15		19	
2nd difference			4		4		4		

The row of 2nd differences is constant, and so the rule for the nth term is therefore a quadratic which takes the form $u_n = an^2 + bn + c$.

By substituting the values of n into the rule, each term can be expressed in terms of a, b and c as shown:

Position	1		2		3		4		5
Term	$a + b + c$		$4a + 2b + c$		$9a + 3b + c$		$16a + 4b + c$		$25a + 5b + c$
1st difference		$3a + b$		$5a + b$		$7a + b$		$9a + b$	
2nd difference			$2a$		$2a$		$2a$		

Comparing the two tables, the values of a, b and c can be deduced:

$2a = 4$ therefore $a = 2$
$3a + b = 7$ therefore $6 + b = 7$ giving $b = 1$
$a + b + c = 0$ therefore $2 + 1 + c = 0$ giving $c = -3$

The rule for the nth term $u_n = an^2 + bn + c$ can be written as $u_n = 2n^2 + n - 3$.

c) Deduce the rule for the nth term for the sequence $-6, -8, -6, 6, 34, \ldots$.

Entering the sequence in a table gives:

Position	1	2	3	4	5
Term	-6	-8	-6	6	34

Extending the table to look at the differences:

Position	1		2		3		4		5
Term	-6		-8		-6		6		34
1st difference		-2		2		12		28	

Algebra and graphs

The row of 1st differences is not constant, and so the rule for the nth term is not linear. Extend the table again to look at the row of 2nd differences:

Position	1	2	3	4	5
Term	−6	−8	−6	6	34
1st difference		−2	2	12	28
2nd difference			4	10	16

The row of 2nd differences is not constant either, and so the rule for the nth term is not quadratic. Extend the table by a further row to look at the row of 3rd differences:

Position	1	2	3	4	5
Term	−6	−8	−6	6	34
1st difference		−2	2	12	28
2nd difference			4	10	16
3rd difference				6	6

The row of 3rd differences is constant, and so the rule for the nth term is therefore a cubic which takes the form $u_n = an^3 + bn^2 + cn + d$.

By substituting the values of n into the rule, each term can be expressed in terms of a, b, c, and d as shown:

Position	1	2	3	4	5
Term	$a + b + c + d$	$8a + 4b + 2c + d$	$27a + 9b + 3c + d$	$64a + 16b + 4c + d$	$125a + 25b + 5c + d$
1st difference		$7a + 3b + c$	$19a + 5b + c$	$37a + 7b + c$	$61a + 9b + c$
2nd difference			$12a + 2b$	$18a + 2b$	$24a + 2b$
3rd difference				$6a$	$6a$

By comparing the two tables, equations can be formed and the values of a, b, c, and d can be found:

$6a = 6$
    therefore $a = 1$
$12a + 2b = 4$
    therefore $12 + 2b = 4$ giving $b = -4$
$7a + 3b + c = -2$
    therefore $7 - 12 + c = -2$ giving $c = 3$
$a + b + c + d = -6$
    therefore $1 - 4 + 3 + d = -6$ giving $d = -6$

Therefore the equation for the nth term is
$u_n = n^3 - 4n^2 + 3n - 6$.

Exercise 15.2

By using a table if necessary, find the formula for the nth term of each of the following sequences:

1. 2, 5, 10, 17, 26
2. 0, 3, 8, 15, 24
3. 6, 9, 14, 21, 30
4. 9, 12, 17, 24, 33
5. −2, 1, 6, 13, 22
6. 4, 10, 20, 34, 52
7. 0, 6, 16, 30, 48
8. 5, 14, 29, 50, 77
9. 0, 12, 32, 60, 96
10. 1, 16, 41, 76, 121

Exercise 15.3

Use a table to find the formula for the nth term of the following sequences:

1. 11, 18, 37, 74, 135
2. 0, 6, 24, 60, 120
3. −4, 3, 22, 59, 120
4. 2, 12, 36, 80, 150
5. 7, 22, 51, 100, 175
6. 7, 28, 67, 130, 223
7. 1, 10, 33, 76, 145
8. 13, 25, 49, 91, 157

Algebra and graphs

● Geometric sequences

So far we have looked at sequences where there is a common difference between successive terms. There are, however, other types of sequences, e.g. 2, 4, 8, 16, 32. There is clearly a pattern to the way the numbers are generated as each term is double the previous term, but there is no common difference.

A sequence where there is a **common ratio** (r) between successive terms is known as a **geometric sequence**.
e.g.

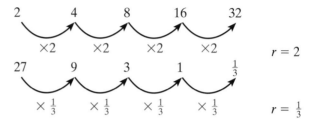

As with an arithmetic sequence, there are two main ways of describing a geometric sequence.

1. The term-to-term rule

    For example, for the following sequence,

    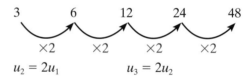

    $u_2 = 2u_1$ $u_3 = 2u_2$

    the general rule is $u_{n+1} = 2u_n$; $u_1 = 3$.

2. The formula for the nth term of a geometric sequence

    As with an arithmetic sequence, this rule links each term to its position in the sequence,

    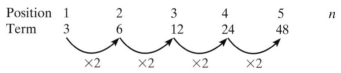

    to reach the second term the calculation is 3×2 or 3×2^1
    to reach the third term, the calculation is $3 \times 2 \times 2$ or 3×2^2
    to reach the fourth term, the calculation is $3 \times 2 \times 2 \times 2$ or 3×2^3

    In general therefore

    $$u_n = ar^{n-1}$$

    where a is the first term and r is the common ratio.

Applications of geometric sequences

In Chapter 8 simple and compound interest were shown as different ways that interest could be earned on money left in a bank account for a period of time. Here we look at compound interest as an example of a geometric sequence.

Compound interest

e.g. $100 is deposited in a bank account and left untouched. After 1 year the amount has increased to $110 as a result of interest payments. To work out the interest rate, calculate the multiplier from $100 → $110:

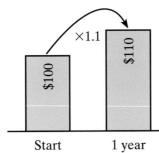

$$\frac{110}{100} = 1.1$$

The multiplier is 1.1. This corresponds to a 10% increase. Therefore the simple interest rate is 10% in the first year.

Assume the money is left in the account and that the interest rate remains unchanged. Calculate the amount in the account after 5 years.

This is an example of a geometric sequence.

Number of years	0	1	2	3	4	5
Amount	100.00	110.00	121.00	133.10	146.41	161.05

×1.1   ×1.1   ×1.1   ×1.1   ×1.1

Alternatively the amount after 5 years can be calculated using a variation of $u_n = ar^{n-1}$, i.e. $u_5 = 100 \times 1.1^5 = 161.05$. Note: As the number of years starts at 0, ×1.1 is applied 5 times to get to the fifth year.

This is an example of compound interest as the previous year's interest is added to the total and included in the following year's calculation.

Algebra and graphs

Worked examples **a)** Alex deposits $1500 in his savings account. The interest rate offered by the savings account is 6% each year for a 10-year period. Assuming Alex leaves the money in the account, calculate how much interest he has gained after the 10 years.

An interest rate of 6% implies a common ratio of 1.06

Therefore $u_{10} = 1500 \times 1.06^{10} = 2686.27$

The amount of interest gained is $2686.27 - 1500 = \$1186.27$

b) Adrienne deposits $2000 in her savings account. The interest rate offered by the bank for this account is 8% compound interest per year. Calculate the number of years Adrienne needs to leave the money in her account for it to double in value.

An interest rate of 8% implies a common ratio of 1.08

The amount each year can be found using the term-to-term rule $u_{n+1} = 1.08 \times u_n$

$u_1 = 2000 \times 1.08 = 2160$

$u_2 = 2160 \times 1.08 = 2332.80$

$u_3 = 2332.80 \times 1.08 = 2519.42$

...

$u_9 = 3998.01$

$u_{10} = 4317.85$

Adrienne needs to leave the money in the account for 10 years in order for it to double in value.

Exercise 15.4

1. Identify which of the following are geometric sequences and which are not.
   a) 2, 6, 18, 54
   b) 25, 5, 1, $\frac{1}{5}$
   c) 1, 4, 9, 16,
   d) −3, 9, −27, 81
   e) $\frac{1}{2}, \frac{2}{3}, \frac{3}{4}, \frac{4}{5}$
   f) $\frac{1}{2}, \frac{2}{4}, \frac{4}{8}, \frac{8}{16}$

2. For the sequences in question 1 that are geometric, calculate:
   i) the common ratio r
   ii) the next two terms
   iii) a formula for the nth term.

3. The nth term of a geometric sequence is given by the formula $u_n = -6 \times 2^{n-1}$.
   a) Calculate u_1, u_2 and u_3.
   b) What is the value of n, if $u_n = -768$?

4. Part of a geometric sequence is given below:

..., −1, ..., ..., 64, ...   where $u_2 = -1$ and $u_5 = 64$.

Calculate:
a) the common ratio r
b) the value of u_1
c) the value of u_{10}.

5. A homebuyer takes out a loan with a mortgage company for $200 000. The interest rate is 6% per year. If she is unable to repay any of the loan during the first 3 years, calculate the extra amount she will have to pay by the end of the third year, due to interest.

6. A car is bought for $10 000. It loses value at a rate of 20% each year.
a) Explain why the car is not worthless after 5 years.
b) Calculate its value after 5 years.
c) Explain why a depreciation of 20% per year means, in theory, that the car will never be worthless.

Student assessment I

1. For each of the sequences given below:
   i) calculate the next two terms,
   ii) explain the pattern in words.
   a) 9, 18, 27, 36, ...   b) 54, 48, 42, 36, ...
   c) 18, 9, 4.5, ...   d) 12, 6, 0, −6, ...
   e) 216, 125, 64, ...   f) 1, 3, 9, 27, ...

2. For each of the sequences shown below give an expression for the nth term:
   a) 6, 10, 14, 18, 22, ...   b) 13, 19, 25, 31, ...
   c) 3, 9, 15, 21, 27, ...   d) 4, 7, 12, 19, 28, ...
   e) 0, 10, 20, 30, 40, ...   f) 0, 7, 26, 63, 124, ...

3. For each of the following arithmetic sequences:
   i) write down a formula for the nth term
   ii) calculate the 10th term.
   a) 1, 5, 9, 13, ...   b) 1, −2, −5, −8, ...

4. For both of the following, calculate u_5 and u_{100}:
   a) $u_n = 6n - 3$ b) $u_n = -\frac{1}{2}n + 4$

Algebra and graphs

5. Copy and complete both of the following tables of arithmetic sequences:

a)
Position	1	2	3	10		n
Term	17	14			−55	

b)
Position	2	6	10		n
Term	−4	−2		35	

Student assessment 2

1. A girl deposits $300 in a bank account. The bank offers 7% interest per year.
   Assuming the girl does not take any money out of the account calculate:
   a) the amount of money in the account after 8 years
   b) the minimum number of years the money must be left in the account, for the amount to be greater than $350.

2. A computer loses 35% of its value each year. If the computer cost $600 new, calculate:
   a) its value after 2 years
   b) its value after 10 years.

3. Part of a geometric sequence is given below:
   $$\ldots, \ldots, 27, \ldots, \ldots, -1$$
   where $u_3 = 27$ and $u_6 = -1$.
   Calculate:
   a) the common ratio r
   b) the value u_1
   c) the value of n if $u_n = -\frac{1}{81}$.

4. Using a table of differences if necessary, calculate the rule for the nth term of the sequence 8, 24, 58, 116, 204, … .

5. Using a table of differences, calculate the rule for the nth term of the sequence 10, 23, 50, 97, 170, … .

16 Variation

● Direct variation
Consider the tables below:

x	0	1	2	3	5	10
y	0	2	4	6	10	20

$y = 2x$

x	0	1	2	3	5	10
y	0	3	6	9	15	30

$y = 3x$

x	0	1	2	3	5	10
y	0	2.5	5	7.5	12.5	25

$y = 2.5x$

In each case y is directly proportional to x. This is written $y \propto x$. If any of these three tables is shown on a graph, the graph will be a straight line passing through the origin.

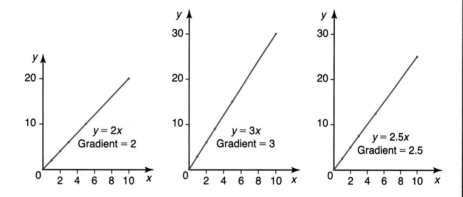

For any statement where $y \propto x$,

$y = kx$

where k is a constant equal to the gradient of the graph and is called the **constant of proportionality** or constant of variation.

Consider the tables below:

x	1	2	3	4	5
y	2	7	18	32	50

$y = 2x^2$

x	1	2	3	4	5
y	$\frac{1}{2}$	4	$13\frac{1}{2}$	32	$62\frac{1}{2}$

$y = \frac{1}{2}x^3$

x	1	2	3	4	5
y	1	$\sqrt{2}$	$\sqrt{3}$	2	$\sqrt{5}$

$y = \sqrt{x} = x^{\frac{1}{2}}$

In the cases above, y is directly proportional to x^n, where $n > 0$. This can be written as $y \propto x^n$.

The graphs of each of the three equations are shown below:

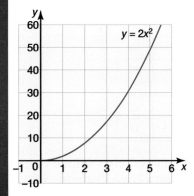
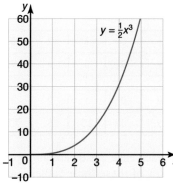
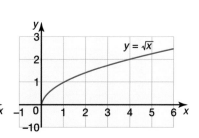

The graphs above, with (x, y) plotted, are not linear. However if the graph of $y = 2x^2$ is plotted as (x^2, y), then the graph is linear and passes through the origin demonstrating that $y \propto x^2$ as shown in the graph below.

x	1	2	3	4	5
x^2	1	4	9	16	25
y	2	8	18	32	50

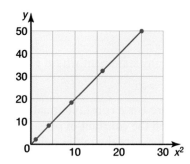

Similarly, the graph of $y = \frac{1}{2}x^3$ is curved when plotted as (x, y), but is linear and passes through the origin if it is plotted as (x^3, y) as shown:

x	1	2	3	4	5
x^3	1	8	27	64	125
y	$\frac{1}{2}$	4	$13\frac{1}{2}$	32	$62\frac{1}{2}$

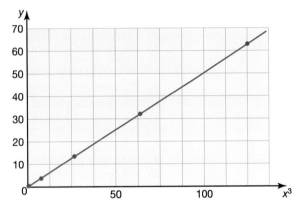

The graph of $y = \sqrt{x}$ is also linear if plotted as $(\sqrt{x}, y)$.

● Inverse variation

If y is inversely proportional to x, then $y \propto \frac{1}{x}$ and $y = \frac{k}{x}$.

If a graph of y against $\frac{1}{x}$ is plotted, this too will be a straight line passing through the origin.

Worked examples   **a)** $y \propto x$. If $y = 7$ when $x = 2$, find y when $x = 5$.

$y = kx$
$7 = k \times 2$ so $k = 3.5$

When $x = 5$,
$y = 3.5 \times 5 = 17.5$

b) $y \propto \frac{1}{x}$. If $y = 5$ when $x = 3$, find y when $x = 30$.

$y = \frac{k}{x}$

$5 = \frac{k}{3}$ so $k = 15$

When $x = 30$,
$y = \frac{15}{30} = 0.5$

Algebra and graphs

Exercise 16.1

1. y is directly proportional to x. If $y = 6$ when $x = 2$, find:
   a) the constant of proportionality
   b) the value of y when $x = 7$
   c) the value of y when $x = 9$
   d) the value of x when $y = 9$
   e) the value of x when $y = 30$.

2. y is directly proportional to x^2. If $y = 18$ when $x = 6$, find:
   a) the constant of proportionality
   b) the value of y when $x = 4$
   c) the value of y when $x = 7$
   d) the value of x when $y = 32$
   e) the value of x when $y = 128$.

3. y is inversely proportional to x^3. If $y = 3$ when $x = 2$, find:
   a) the constant of proportionality
   b) the value of y when $x = 4$
   c) the value of y when $x = 6$
   d) the value of x when $y = 24$.

4. y is inversely proportional to x^2. If $y = 1$ when $x = 0.5$, find:
   a) the constant of proportionality
   b) the value of y when $x = 0.1$
   c) the value of y when $x = 0.25$
   d) the value of x when $y = 64$.

Exercise 16.2

1. Write each of the following in the form:
   i) $y \propto x$
   ii) $y = kx$.

   a) y is directly proportional to x^3
   b) y is inversely proportional to x^3
   c) t is directly proportional to P
   d) s is inversely proportional to t
   e) A is directly proportional to r^2
   f) T is inversely proportional to the square root of g

2. If $y \propto x$ and $y = 6$ when $x = 2$, find y when $x = 3.5$.

3. If $y \propto \dfrac{1}{x}$ and $y = 4$ when $x = 2.5$ find:
   a) y when $x = 20$
   b) x when $y = 5$.

4. If $p \propto r^2$ and $p = 2$ when $r = 2$, find p when $r = 8$.

5. If $m \propto \dfrac{1}{r^3}$ and $m = 1$ when $r = 2$, find:
   a) m when $r = 4$
   b) r when $m = 125$.

6. If $y \propto x^2$ and $y = 12$ when $x = 2$, find y when $x = 5$.

Variation

Exercise 16.3

1. If a stone is dropped off the edge of a cliff, the height (h metres) of the cliff is proportional to the square of the time (t seconds) taken for the stone to reach the ground.
   A stone takes 5 seconds to reach the ground when dropped off a cliff 125 m high.
   a) Write down a relationship between h and t, using k as the constant of variation.
   b) Calculate the constant of variation.
   c) Find the height of a cliff if a stone takes 3 seconds to reach the ground.
   d) Find the time taken for a stone to fall from a cliff 180 m high.

2. The velocity (v metres per second) of a body is known to be proportional to the square root of its kinetic energy (e joules). When the velocity of a body is 120 m/s, its kinetic energy is 1600 J.
   a) Write down a relationship between v and e, using k as the constant of variation.
   b) Calculate the value of k.
   c) If $v = 21$, calculate the kinetic energy of the body in joules.

3. The length (l cm) of an edge of a cube is proportional to the cube root of its mass (m grams). It is known that if $l = 15$, then $m = 125$. Let k be the constant of variation.
   a) Write down the relationship between l, m and k.
   b) Calculate the value of k.
   c) Calculate the value of l when $m = 8$.

4. The power (P) generated in an electrical circuit is proportional to the square of the current (I amps).
   When the power is 108 watts, the current is 6 amps.
   a) Write down a relationship between P, I and the constant of variation, k.
   b) Calculate the value of I when $P = 75$ watts.

Algebra and graphs

Student assessment 1

1. $y = kx$. When $y = 12, x = 8$.
   a) Calculate the value of k.
   b) Calculate y when $x = 10$.
   c) Calculate y when $x = 2$.
   d) Calculate x when $y = 18$.

2. $y = \dfrac{k}{x}$. When $y = 2, x = 5$.
   a) Calculate the value of k.
   b) Calculate y when $x = 4$.
   c) Calculate x when $y = 10$.
   d) Calculate x when $y = 0.5$.

3. $p = kq^3$. When $p = 9, q = 3$.
   a) Calculate the value of k.
   b) Calculate p when $q = 6$.
   c) Calculate p when $q = 1$.
   d) Calculate q when $p = 576$.

4. $m = \dfrac{k}{\sqrt{n}}$. When $m = 1, n = 25$.
   a) Calculate the value of k.
   b) Calculate m when $n = 16$.
   c) Calculate m when $n = 100$.
   d) Calculate n when $m = 5$.

5. $y = \dfrac{k}{x^2}$. When $y = 3, x = \frac{1}{3}$.
   a) Calculate the value of k.
   b) Calculate y when $x = 0.5$.
   c) Calculate both values of x when $y = \frac{1}{12}$.
   d) Calculate both values of x when $y = \frac{1}{3}$.

Student assessment 2

1. y is inversely proportional to x.
   a) Copy and complete the table below:

x	1	2	4	8	16	32
y				4		

   b) What is the value of x when $y = 20$?

2. Copy and complete the tables below:
   a) $y \propto x$

x	1	2	4	5	10
y		10			

   b) $y \propto \dfrac{1}{x}$

x	1	2	4	5	10
y	20				

   c) $y \propto \sqrt{x}$

x	4	16	25	36	64
y	4				

3. The pressure (P) of a given mass of gas is inversely proportional to its volume (V) at a constant temperature. If $P = 4$ when $V = 6$, calculate:
   a) P when $V = 30$
   b) V when $P = 30$.

4. The gravitational force (F) between two masses is inversely proportional to the square of the distance (d) between them. If $F = 4$ when $d = 5$, calculate:
   a) F when $d = 8$
   b) d when $F = 25$.

17 Graphs in practical situations

● **Conversion graphs**

A straight-line graph can be used to convert one set of units to another. Examples include converting from one currency to another, converting distance in miles to kilometres and converting temperature from degrees Celsius to degrees Fahrenheit.

Worked example The graph below converts South African rand into euros based on an exchange rate of €1 = 8.80 rand.

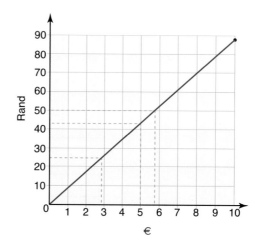

i) Using the graph estimate the number of rand equivalent to €5.
A line is drawn up from €5 until it reaches the plotted line, then across to the vertical axis.
From the graph it can be seen that €5 ≈ 44 rand.
(≈ is the symbol for 'is approximately equal to')

ii) Using the graph, what would be the cost in euros of a drink costing 25 rand?
A line is drawn across from 25 rand until it reaches the plotted line, then down to the horizontal axis.
From the graph it can be seen that the cost of the drink ≈ €2.80.

iii) If a meal costs 200 rand, use the graph to estimate its cost in euros.
The graph does not go up to 200 rand, therefore a factor of 200 needs to be used e.g. 50 rand.
From the graph 50 rand ≈ €5.70, therefore it can be deduced that 200 rand ≈ €22.80 (i.e. 4 × €5.70).

Exercise 17.1

1. Given that 80 km = 50 miles, draw a conversion graph up to 100 km. Using your graph estimate:
   a) how many miles is 50 km,
   b) how many kilometres is 80 miles,
   c) the speed in miles per hour (mph) equivalent to 100 km/h,
   d) the speed in km/h equivalent to 40 mph.

2. You can roughly convert temperature in degrees Celsius to degrees Fahrenheit by doubling the degrees Celsius and adding 30.
   Draw a conversion graph up to 50 °C. Use your graph to estimate the following:
   a) the temperature in °F equivalent to 25 °C,
   b) the temperature in °C equivalent to 100 °F,
   c) the temperature in °F equivalent to 0 °C,
   d) the temperature in °C equivalent to 200 °F.

3. Given that 0 °C = 32 °F and 50 °C = 122 °F, on the same graph as in question 2, draw a true conversion graph.
   i) Use the true graph to calculate the conversions in question 2.
   ii) Where would you say the rough conversion is most useful?

4. Long-distance calls from New York to Harare are priced at 85 cents/min off peak and $1.20/min at peak times.
   a) Draw, on the same axes, conversion graphs for the two different rates.
   b) From your graph estimate the cost of an 8 minute call made off peak.
   c) Estimate the cost of the same call made at peak rate.
   d) A call is to be made from a telephone box. If the caller has only $4 to spend, estimate how much more time he can talk for if he rings at off peak instead of at peak times.

5. A maths exam is marked out of 120. Draw a conversion graph to change the following marks to percentages.
   a) 80   b) 110   c) 54   d) 72

● **Speed, distance and time**

You may already be aware of the following formula:

$$\text{distance} = \text{speed} \times \text{time}$$

Rearranging the formula gives:

$$\text{speed} = \frac{\text{distance}}{\text{time}}$$

Where the speed is not constant:

$$\text{average speed} = \frac{\text{total distance}}{\text{total time}}$$

Algebra and graphs

Exercise 17.2

1. Find the average speed of an object moving:
   a) 30 m in 5 s
   b) 48 m in 12 s
   c) 78 km in 2 h
   d) 50 km in 2.5 h
   e) 400 km in 2 h 30 min
   f) 110 km in 2 h 12 min

2. How far will an object travel during:
   a) 10 s at 40 m/s
   b) 7 s at 26 m/s
   c) 3 hours at 70 km/h
   d) 4 h 15 min at 60 km/h
   e) 10 min at 60 km/h
   f) 1 h 6 min at 20 m/s?

3. How long will it take to travel:
   a) 50 m at 10 m/s
   b) 1 km at 20 m/s
   c) 2 km at 30 km/h
   d) 5 km at 70 m/s
   e) 200 cm at 0.4 m/s
   f) 1 km at 15 km/h?

Travel graphs

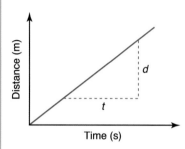

The graph of an object travelling at a constant speed is a straight line as shown (left).

$$\text{Gradient} = \frac{d}{t}$$

The units of the gradient are m/s, hence the gradient of a distance–time graph represents the speed at which the object is travelling.

Worked example

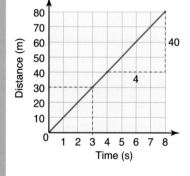

The graph (left) represents an object travelling at constant speed.

i) From the graph calculate how long it took to cover a distance of 30 m.
   The time taken to travel 30 m is 3 seconds.

ii) Calculate the gradient of the graph.
    Taking two points on the line, gradient $= \frac{40}{4} = 10$.

iii) Calculate the speed at which the object was travelling.
     Gradient of a distance–time graph = speed.
     Therefore the speed is 10 m/s.

Exercise 17.3

1. Draw a distance–time graph for the first 10 seconds of an object travelling at 6 m/s.

2. Draw a distance–time graph for the first 10 seconds of an object travelling at 5 m/s. Use your graph to estimate:
   a) the time taken to travel 25 m,
   b) how far the object travels in 3.5 seconds.

Graphs in practical situations

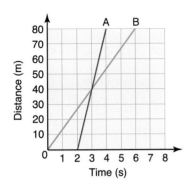

3. Two objects A and B set off from the same point and move in the same straight line. B sets off first, whilst A sets off 2 seconds later. Using the distance–time graph (left) estimate:
   a) the speed of each of the objects,
   b) how far apart the objects would be 20 seconds after the start.

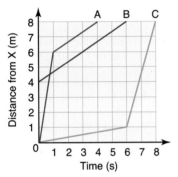

4. Three objects A, B and C move in the same straight line away from a point X. Both A and C change their speed during the journey, whilst B travels at the same constant speed throughout. From the distance–time graph (left) estimate:
   a) the speed of object B,
   b) the two speeds of object A,
   c) the average speed of object C,
   d) how far object C is from X 3 seconds from the start,
   e) how far apart objects A and C are 4 seconds from the start.

The graphs of two or more journeys can be shown on the same axes. The shape of the graph gives a clear picture of the movement of each of the objects.

Worked example The journeys of two cars, X and Y, travelling between A and B are represented on the distance–time graph (left). Car X and Car Y both reach point B 100 km from A at 11 00.

i) Calculate the speed of Car X between 07 00 and 08 00.

$$\text{speed} = \frac{\text{distance}}{\text{time}}$$

$$= \frac{60}{1} \text{ km/h} = 60 \text{ km/h}$$

ii) Calculate the speed of Car Y between 09 00 and 11 00.

$$\text{speed} = \frac{100}{2} \text{ km/h} = 50 \text{ km/h}$$

iii) Explain what is happening to Car X between 08 00 and 09 00.

No distance has been travelled, therefore Car X is stationary.

Algebra and graphs

Exercise 17.4

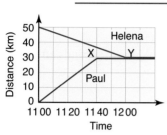

1. Two friends Paul and Helena arrange to meet for lunch at noon. They live 50 km apart and the restaurant is 30 km from Paul's home. The travel graph (left) illustrates their journeys.
   a) What is Paul's average speed between 11 00 and 11 40?
   b) What is Helena's average speed between 11 00 and 12 00?
   c) What does the line XY represent?

2. A car travels at a speed of 60 km/h for 1 hour. It then stops for 30 minutes and then continues at a constant speed of 80 km/h for a further 1.5 hours. Draw a distance–time graph for this journey.

3. A girl cycles for 1.5 hours at 10 km/h. She then stops for an hour and then travels for a further 15 km in 1 hour. Draw a distance–time graph of the girl's journey.

4. Two friends leave their houses at 16 00. The houses are 4 km apart and the friends travel towards each other on the same road. Fyodor walks at 7 km/h and Yin walks at 5 km/h.
   a) On the same axes, draw a distance–time graph of their journeys.
   b) From your graph estimate the time at which they meet.
   c) Estimate the distance from Fyodor's house to the point where they meet.

5. A train leaves a station P at 18 00 and travels to station Q 150 km away. It travels at a steady speed of 75 km/h. At 18 10 another train leaves Q for P at a steady speed of 100 km/h.
   a) On the same axes draw a distance–time graph to show both journeys.
   b) From the graph estimate the time at which both trains pass each other.
   c) At what distance from station Q do both trains pass each other?
   d) Which train arrives at its destination first?

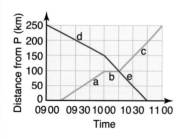

6. A train sets off from town P at 09 15 and heads towards town Q 250 km away. Its journey is split into the three stages a, b and c. At 09 00 a second train leaves town Q heading for town P. Its journey is split into the two stages d and e. Using the graph (left) calculate the following:
   a) the speed of the first train during stages a, b and c,
   b) the speed of the second train during stages d and e.

Speed–time graphs, acceleration and deceleration

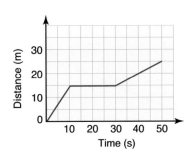

So far the graphs that have been dealt with have been similar to the one shown (left) i.e. distance–time graphs.

If the graph were of a girl walking it would indicate that initially she was walking at a constant speed of 1.5 m/s for 10 seconds, then she stopped for 20 seconds and finally she walked at a constant speed of 0.5 m/s for 20 seconds.

For a distance–time graph the following is true:
- a straight line represents constant speed,
- a horizontal line indicates no movement,
- the gradient of a line gives the speed.

This section also deals with the interpretation of travel graphs, but where the vertical axis represents the object's speed.

Worked example   The graph shows the speed of a car over a period of 16 seconds.

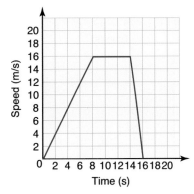

i) Explain the shape of the graph.

For the first 8 seconds the speed of the car is increasing uniformly with time. This means it is **accelerating** at a constant rate. Between 8 and 14 seconds the car is travelling at a constant speed of 16 m/s. Between 14 and 16 seconds the speed of the car decreases uniformly. This means that it is **decelerating** at a constant rate.

ii) Calculate the rate of acceleration during the first 8 seconds.

From a speed–time graph, the acceleration is found by calculating the gradient of the line. Therefore:

$$\text{acceleration} = \frac{16}{8} = 2 \text{ m/s}^2$$

iii) Calculate the rate of deceleration between 14 and 16 seconds:

$$\text{deceleration} = \frac{16}{2} = 8 \text{ m/s}^2$$

Algebra and graphs

Exercise 17.5 Using the graphs below, calculate the acceleration/deceleration in each case.

1.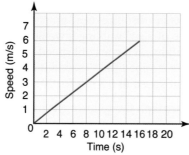

2.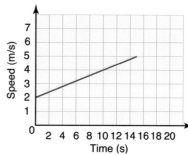

3.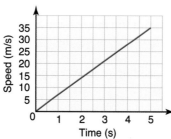

4.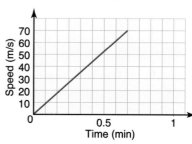

5.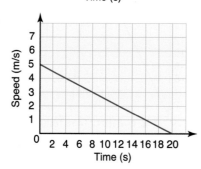

6.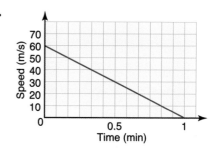

7. Sketch a graph to show an aeroplane accelerating from rest at a constant rate of 5 m/s² for 10 seconds.

8. A train travelling at 30 m/s starts to decelerate at a constant rate of 3 m/s². Sketch a speed–time graph showing the train's motion until it stops.

Exercise 17.6

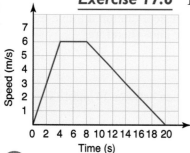

1. The graph (left) shows the speed–time graph of a boy running for 20 seconds. Calculate:
   a) the acceleration during the first four seconds,
   b) the acceleration during the second period of four seconds,
   c) the deceleration during the final twelve seconds.

Graphs in practical situations

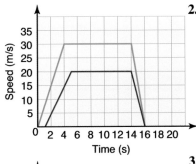

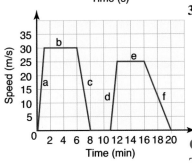

2. The speed–time graph (left) represents a cheetah chasing a gazelle.
   a) Does the top graph represent the cheetah or the gazelle?
   b) Calculate the cheetah's acceleration in the initial stages of the chase.
   c) Calculate the gazelle's acceleration in the initial stages of the chase.
   d) Calculate the cheetah's deceleration at the end.

3. The speed–time graph (left) represents a train travelling from one station to another.
   a) Calculate the acceleration during stage a.
   b) Calculate the deceleration during stage c.
   c) Calculate the deceleration during stage f.
   d) Describe the train's motion during stage b.
   e) Describe the train's motion 10 minutes from the start.

● **Area under a speed–time graph**

The area under a speed–time graph gives the distance travelled.

Worked example   The table below shows the speed of a train over a 30 second period.

Time (s)	0	5	10	15	20	25	30
Speed (m/s)	20	20	20	22.5	25	27.5	30

i) Plot a speed–time graph for the first 30 seconds.

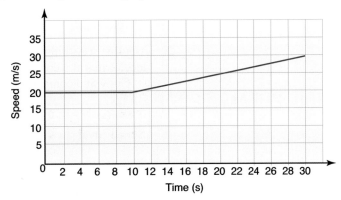

ii) Calculate the train's acceleration after the first 10 seconds.

$$\text{Acceleration} = \frac{10}{20} = \frac{1}{2} \text{ m/s}^2$$

iii) Calculate the distance travelled during the 30 seconds.

This is calculated by working out the area under the graph. The graph can be split into two regions as shown overleaf.

Algebra and graphs

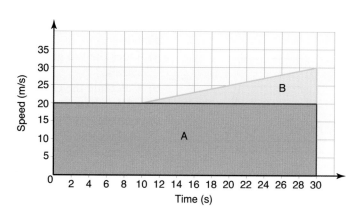

Distance represented by region A = (20 × 30) m
= 600 m
Distance represented by region B = ($\frac{1}{2}$ × 20 × 10) m
= 100 m
Total distance travelled = (600 + 100) m
= 700 m

Exercise 17.7

1. The table below gives the speed of a boat over a 10 second period.

Time (s)	0	2	4	6	8	10
Speed (m/s)	5	6	7	8	9	10

   a) Plot a speed–time graph for the 10 second period.
   b) Calculate the acceleration of the boat.
   c) Calculate the total distance travelled during the 10 seconds.

2. A cyclist travelling at 6 m/s applies the brakes and decelerates at a constant rate of 2 m/s².
   a) Copy and complete the table below.

Time (s)	0	0.5	1	1.5	2	2.5	3
Speed (m/s)	6						0

   b) Plot a speed–time graph for the 3 seconds shown in the table above.
   c) Calculate the distance travelled during the 3 seconds of deceleration.

3. A car accelerates as shown in the graph (left).
   a) Calculate the rate of acceleration in the first 40 seconds.
   b) Calculate the distance travelled over the 60 seconds shown.
   c) After what time had the motorist travelled half the distance?

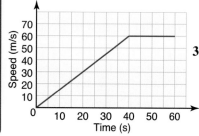

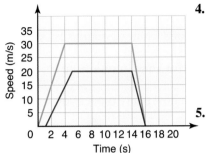

4. The graph (left) represents the cheetah and gazelle chase from question 2 in Exercise 17.6.
   a) Calculate the distance run by the cheetah during the chase.
   b) Calculate the distance run by the gazelle during the chase.

5. The graph (right) represents the train journey from question 3 in Exercise 17.6. Calculate, in km, the distance travelled during the 20 minutes shown.

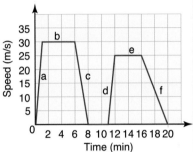

6. An aircraft accelerates uniformly from rest at a rate of 10 m/s^2 for 12 seconds before it takes off. Calculate the distance it travels along the runway.

7. The speed–time graph below depicts the motion of two motorbikes A and B over a 15 second period.

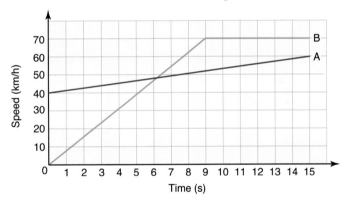

At the start of the graph motorbike A overtakes a stationary motorbike B. Assume they then travel in the same direction.
   a) Calculate motorbike A's acceleration over the 15 seconds in m/s^2.
   b) Calculate motorbike B's acceleration over the first 9 seconds in m/s^2.
   c) Calculate the distance travelled by A during the 15 seconds (give your answer to the nearest metre).
   d) Calculate the distance travelled by B during the 15 seconds (give your answer to the nearest metre).
   e) How far apart were the two motorbikes at the end of the 15 second period?

Algebra and graphs

Student assessment I

1. 1 euro had an exchange rate of 8 Chinese yuan and 200 Pakistani rupees.
   a) Draw a conversion graph for yuan to rupees up to 80 yuan.
   b) Estimate from your graph how many rupees you would get for 50 yuan.
   c) Estimate from your graph how many yuan you would get for 1600 rupees.

2. A South African taxi driver has a fixed charge of 20 rand and then charges 6 rand per km.
   a) Draw a conversion graph to enable you to estimate the cost of the following taxi rides:
      i) 5 km
      ii) 8.5 km
   b) If a trip cost 80 rand, estimate from your graph the distance travelled.

3. An electricity account can work in two ways:
   - account A which involves a fixed charge of $5 and then a rate of 7c per unit,
   - account B which involves no fixed charge but a rate of 9.5c per unit.

   a) On the same axes draw a graph up to 400 units for each type of account, converting units used to cost.
   b) Use your graph to advise a customer on which account to use.

4. A car travels at 60 km/h for 1 hour. The driver then takes a 30 minute break. After her break, she continues at 80 km/h for 90 minutes.
   a) Draw a distance–time graph for her journey.
   b) Calculate the total distance travelled.

5. Two trains depart at the same time from cities M and N, which are 200 km apart. One train travels from M to N, the other from N to M. The train departing from M travels a distance of 60 km in the first hour, 120 km in the next 1.5 hours and then the rest of the journey at 40 km/h. The train departing from N travels the whole distance at a speed of 100 km/h. Assuming all speeds are constant:
   a) draw a travel graph to show both journeys,
   b) estimate how far from city M the trains are when they pass each other,
   c) estimate how long after the start of the journey it is when the trains pass each other.

Graphs in practical situations

Student assessment 2

1. Absolute zero (0 K) is equivalent to −273 °C and 0 °C is equivalent to 273 K. Draw a conversion graph which will convert K into °C. Use your graph to estimate:
   a) the temperature in K equivalent to −40 °C,
   b) the temperature in °C equivalent to 100 K.

2. A Canadian plumber has a call-out charge of 70 Canadian dollars and then charges a rate of $50 per hour.
   a) Draw a conversion graph and estimate the cost of the following:
      i) a job lasting $4\frac{1}{2}$ hours,
      ii) a job lasting $6\frac{3}{4}$ hours.
   b) If a job cost $245, estimate from your graph how long it took to complete.

3. A boy lives 3.5 km from his school. He walks home at a constant speed of 9 km/h for the first 10 minutes. He then stops and talks to his friends for 5 minutes. He finally runs the rest of his journey home at a constant speed of 12 km/h.
   a) Illustrate this information on a distance–time graph.
   b) Use your graph to estimate the total time it took the boy to get home that day.

4. Below are four distance–time graphs A, B, C and D. Two of them are not possible.
   a) Which two graphs are impossible?
   b) Explain why the two you have chosen are not possible.

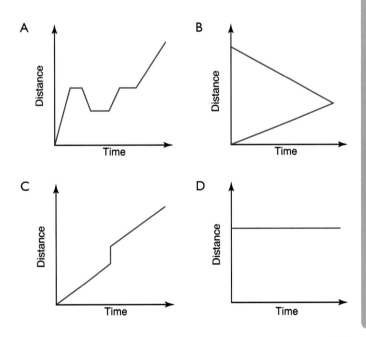

Algebra and graphs

Student assessment 3

1. The graph below is a speed–time graph for a car accelerating from rest.

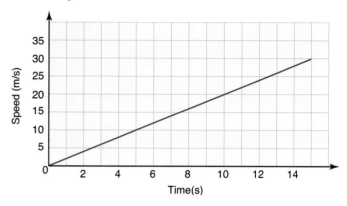

   a) Calculate the car's acceleration in m/s².
   b) Calculate, in metres, the distance the car travels in 15 seconds.
   c) How long did it take the car to travel half the distance?

2. The speed–time graph below represents a 100 m sprinter during a race.

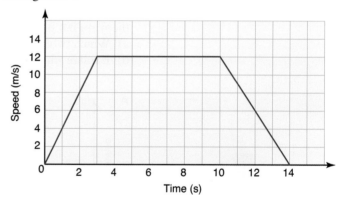

   a) Calculate the sprinter's acceleration during the first two seconds of the race.
   b) Calculate the sprinter's deceleration at the end of the race.
   c) Calculate the distance the sprinter ran in the first 10 seconds.
   d) Calculate the sprinter's time for the 100 m race. Give your answer to 3 s.f.

Graphs in practical situations

3. A motorcyclist accelerates uniformly from rest to 50 km/h in 8 seconds. He then accelerates to 110 km/h in a further 6 seconds.
   a) Draw a speed–time graph for the first 14 seconds.
   b) Use your graph to find the total distance the motorcyclist travels. Give your answer in metres.

4. The graph shows the speed of a car over a period of 50 seconds.

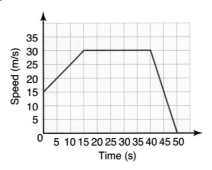

   a) Calculate the car's acceleration in the first 15 seconds.
   b) Calculate the distance travelled whilst the car moved at constant speed.
   c) Calculate the total distance travelled.

Student assessment 4

1. The graph below is a speed–time graph for a car decelerating to rest.

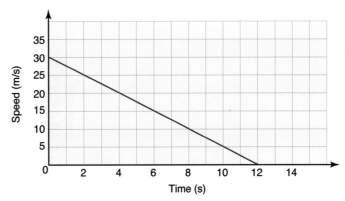

   a) Calculate the car's deceleration in m/s².
   b) Calculate, in metres, the distance the car travels in 12 seconds.
   c) How long did it take the car to travel half the distance?

Algebra and graphs

2. The graph below shows the speeds of two cars A and B over a 15 second period.

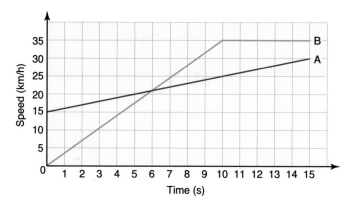

   a) Calculate the acceleration of car A in m/s².
   b) Calculate the distance travelled in metres during the 15 seconds by car A.
   c) Calculate the distance travelled in metres during the 15 seconds by car B.

3. A motor cycle accelerates uniformly from rest to 30 km/h in 3 seconds. It then accelerates to 150 km/h in a further 6 seconds.
   a) Draw a speed–time graph for the first 9 seconds.
   b) Use your graph to find the total distance the motor cycle travels. Give your answer in metres.

4. Two cars X and Y are travelling in the same direction. The speed–time graph (below) shows their speeds over 12 seconds.

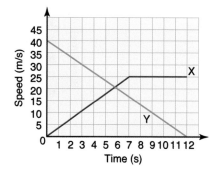

   a) Calculate the deceleration of Y during the 12 seconds.
   b) Calculate the distance travelled by Y in the 12 seconds.
   c) Calculate the total distance travelled by X in the 12 seconds.

18 Graphs of functions

You should be familiar with the work covered in Chapter 28 on straight-line graphs before embarking on this chapter.

● Quadratic functions

The general expression for a quadratic function takes the form $ax^2 + bx + c$ where a, b and c are constants. Some examples of quadratic functions are given below.

$$y = 2x^2 + 3x - 12 \qquad y = x^2 - 5x + 6 \qquad y = 3x^2 + 2x - 3$$

If a graph of a quadratic function is plotted, the smooth curve produced is called a **parabola**, e.g.

$$y = x^2$$

x	−4	−3	−2	−1	0	1	2	3	4
y	16	9	4	1	0	1	4	9	16

$$y = -x^2$$

x	−4	−3	−2	−1	0	1	2	3	4
y	−16	−9	−4	−1	0	−1	−4	−9	−16

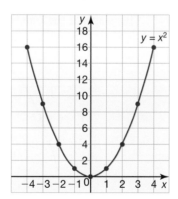
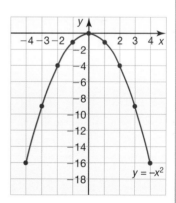

Algebra and graphs

Worked examples **a)** Plot a graph of the function $y = x^2 - 5x + 6$ for $0 \leq x \leq 5$.

A table of values for x and y is given below:

x	0	1	2	3	4	5
y	6	2	0	0	2	6

These can then be plotted to give the graph:

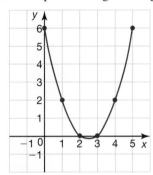

b) Plot a graph of the function $y = -x^2 + x + 2$ for $-3 \leq x \leq 4$.

Drawing up a table of values gives:

x	-3	-2	-1	0	1	2	3	4
y	-10	-4	0	2	2	0	-4	-10

The graph of the function is given below:

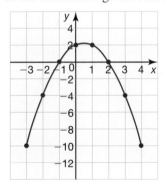

Exercise 18.1 For each of the following quadratic functions, construct a table of values and then draw the graph.

1. $y = x^2 + x - 2$, $-4 \leq x \leq 3$
2. $y = -x^2 + 2x + 3$, $-3 \leq x \leq 5$
3. $y = x^2 - 4x + 4$, $-1 \leq x \leq 5$
4. $y = -x^2 - 2x - 1$, $-4 \leq x \leq 2$
5. $y = x^2 - 2x - 15$, $-4 \leq x \leq 6$

6. $y = 2x^2 - 2x - 3$, $\quad -2 \leq x \leq 3$
7. $y = -2x^2 + x + 6$, $\quad -3 \leq x \leq 3$
8. $y = 3x^2 - 3x - 6$, $\quad -2 \leq x \leq 3$
9. $y = 4x^2 - 7x - 4$, $\quad -1 \leq x \leq 3$
10. $y = -4x^2 + 4x - 1$, $\quad -2 \leq x \leq 3$

● **Graphical solution of a quadratic equation**

Worked example i) Draw a graph of $y = x^2 - 4x + 3$ for $-2 \leq x \leq 5$.

x	-2	-1	0	1	2	3	4	5
y	15	8	3	0	-1	0	3	8

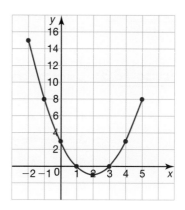

ii) Use the graph to solve the equation $x^2 - 4x + 3 = 0$.
To solve the equation it is necessary to find the values of x when $y = 0$, i.e. where the graph crosses the x-axis. These points occur when $x = 1$ and $x = 3$ and are therefore the solutions.

Exercise 18.2 Solve each of the quadratic functions below by plotting a graph for the ranges of x stated.

1. $x^2 - x - 6 = 0$, $\quad -4 \leq x \leq 4$
2. $-x^2 + 1 = 0$, $\quad -4 \leq x \leq 4$
3. $x^2 - 6x + 9 = 0$, $\quad 0 \leq x \leq 6$
4. $-x^2 - x + 12 = 0$, $\quad -5 \leq x \leq 4$
5. $x^2 - 4x + 4 = 0$, $\quad -2 \leq x \leq 6$
6. $2x^2 - 7x + 3 = 0$, $\quad -1 \leq x \leq 5$
7. $-2x^2 + 4x - 2 = 0$, $\quad -2 \leq x \leq 4$
8. $3x^2 - 5x - 2 = 0$, $\quad -1 \leq x \leq 3$

Algebra and graphs

In the previous worked example, as $y = x^2 - 4x + 3$, a solution could be found to the equation $x^2 - 4x + 3 = 0$ by reading off where the graph crossed the x-axis. The graph can, however, also be used to solve other quadratic equations.

Worked example Use the graph of $y = x^2 - 4x + 3$ to solve the equation $x^2 - 4x + 1 = 0$.

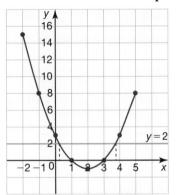

$x^2 - 4x + 1 = 0$ can be rearranged to give:

$$x^2 - 4x + 3 = 2$$

Using the graph of $y = x^2 - 4x + 3$ and plotting the line $y = 2$ on the same graph gives the graph shown on the left.

Where the curve and the line cross gives the solution to $x^2 - 4x + 3 = 2$ and hence also $x^2 - 4x + 1 = 0$.
Therefore the solutions to $x^2 - 4x + 1 = 0$ are $x \approx 0.3$ and $x \approx 3.7$.

Exercise 18.3 Using the graphs which you drew in Exercise 18.2, solve the following quadratic equations. Show your method clearly.

1. $x^2 - x - 4 = 0$
2. $-x^2 - 1 = 0$
3. $x^2 - 6x + 8 = 0$
4. $-x^2 - x + 9 = 0$
5. $x^2 - 4x + 1 = 0$
6. $2x^2 - 7x = 0$
7. $-2x^2 + 4x = -1$
8. $3x^2 = 2 + 5x$

● The reciprocal function

Worked example Draw the graph of $y = \dfrac{2}{x}$ for $-4 \leqslant x \leqslant 4$.

x	−4	−3	−2	−1	0	1	2	3	4
y	−0.5	−0.7	−1	−2	—	2	1	0.7	0.5

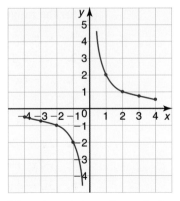

This is a reciprocal function giving a hyperbola.

Exercise 18.4

1. Plot the graph of the function $y = \dfrac{1}{x}$ for $-4 \leq x \leq 4$.

2. Plot the graph of the function $y = \dfrac{3}{x}$ for $-4 \leq x \leq 4$.

3. Plot the graph of the function $y = \dfrac{5}{2x}$ for $-4 \leq x \leq 4$.

● **Types of graph**

Graphs of functions of the form ax^n take different forms depending on the values of a and n. The different types of line produced also have different names, as described below.

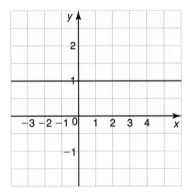
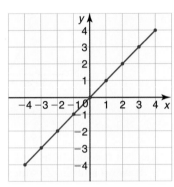
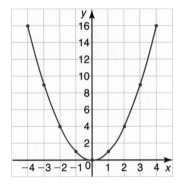

If $a = 1$ and $n = 0$, then $f(x) = x^0$. This is a **linear** function giving a **straight line**.

If $a = 1$ and $n = 1$, then $f(x) = x^1$. This is a **linear** function giving a **straight line**.

If $a = 1$ and $n = 2$, then $f(x) = x^2$. This is a **quadratic** function giving a **parabola**.

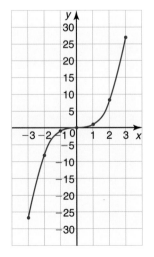
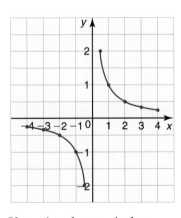
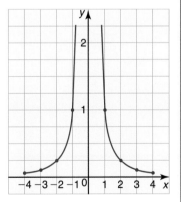

If $a = 1$ and $n = 3$, then $f(x) = x^3$. This is a **cubic** function giving a **cubic curve**.

If $a = 1$ and $n = -1$, then $f(x) = x^{-1}$ or $f(x) = \dfrac{1}{x}$. This is a **reciprocal** function giving a **hyperbola**.

If $a = 1$ and $n = -2$, then $f(x) = x^{-2}$ or $f(x) = \dfrac{1}{x^2}$. This is a **reciprocal** function, shown on the graph above.

Algebra and graphs

Worked example Draw a graph of the function $y = 2x^2$ for $-3 \leqslant x \leqslant 3$.

x	−3	−2	−1	0	1	2	3
y	18	8	2	0	2	8	18

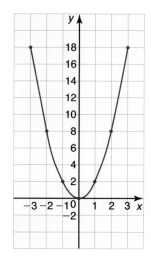

Exercise 18.5 For each of the functions given below:
i) draw up a table of values for x and $f(x)$,
ii) plot the graph of the function.

1. $f(x) = 3x + 2$, $\quad -3 \leqslant x \leqslant 3$
2. $f(x) = \frac{1}{2}x + 4$, $\quad -3 \leqslant x \leqslant 3$
3. $f(x) = -2x - 3$, $\quad -4 \leqslant x \leqslant 2$
4. $f(x) = 2x^2 - 1$, $\quad -3 \leqslant x \leqslant 3$
5. $f(x) = 0.5x^2 + x - 2$, $\quad -5 \leqslant x \leqslant 3$
6. $f(x) = 3x^2 - 2x + 1$, $\quad -2 \leqslant x \leqslant 2$
7. $f(x) = 2x^3$, $\quad -2 \leqslant x \leqslant 2$
8. $f(x) = \frac{1}{2}x^3 - 2x + 3$, $\quad -3 \leqslant x \leqslant 3$
9. $f(x) = 3x^{-1}$, $\quad -3 \leqslant x \leqslant 3$
10. $f(x) = 2x^{-2}$, $\quad -3 \leqslant x \leqslant 3$
11. $f(x) = \frac{1}{x^2} + 3x$, $\quad -3 \leqslant x \leqslant 3$

Exponential functions

Functions of the form $y = a^x$ are known as **exponential** functions. Plotting an exponential function is done in the same way as for other functions.

Worked example   Plot the graph of the function $y = 2^x$ for $-3 \leqslant x \leqslant 3$.

x	−3	−2	−1	0	1	2	3
y	0.125	0.25	0.5	1	2	4	8

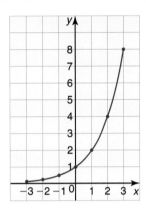

Exercise 18.6   For each of the functions below:
  i) draw up a table of values of x and $f(x)$,
  ii) plot a graph of the function.

1. $f(x) = 3^x$, $-3 \leqslant x \leqslant 3$
2. $f(x) = 1^x$, $-3 \leqslant x \leqslant 3$
3. $f(x) = 2^x + 3$, $-3 \leqslant x \leqslant 3$
4. $f(x) = 2^x + x$, $-3 \leqslant x \leqslant 3$
5. $f(x) = 2^x - x$, $-3 \leqslant x \leqslant 3$
6. $f(x) = 3^x - x^2$, $-3 \leqslant x \leqslant 3$

Gradients of curves

The gradient of a straight line is constant and is calculated by considering the coordinates of two of the points on the line and then carrying out the calculation $\dfrac{y_2 - y_1}{x_2 - x_1}$ as shown below:

$$\text{Gradient} = \frac{4-2}{4-0}$$

$$= \frac{1}{2}$$

Algebra and graphs

The gradient of a curve, however, is not constant: its slope changes. To calculate the gradient of a curve at a specific point, the following steps need to be taken:

- draw a tangent to the curve at that point,
- calculate the gradient of the tangent.

Worked example   For the function $y = 2x^2$, calculate the gradient of the curve at the point where $x = 1$.

On a graph of the function $y = 2x^2$, identify the point on the curve where $x = 1$ and then draw a tangent to that point. This gives:

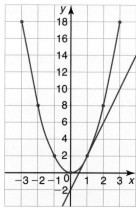

Two points on the tangent are identified in order to calculate its gradient.

$$\text{Gradient} = \frac{10-(-2)}{3-0}$$
$$= \frac{12}{3}$$
$$= 4$$

Therefore the gradient of the function $y = 2x^2$ when $x = 1$ is 4.

Exercise 18.7

For each of the functions below:
  i) plot a graph,
  ii) calculate the gradient of the function at the specified point.

1. $y = x^2$, $-4 \leq x \leq 4$,   gradient where $x = 1$
2. $y = \frac{1}{2}x^2$, $-4 \leq x \leq 4$,   gradient where $x = -2$
3. $y = x^3$, $-3 \leq x \leq 3$,   gradient where $x = 1$
4. $y = x^3 - 3x^2$, $-4 \leq x \leq 4$,   gradient where $x = -2$
5. $y = 4x^{-1}$, $-4 \leq x \leq 4$,   gradient where $x = -1$
6. $y = 2^x$, $-3 \leq x \leq 3$,   gradient where $x = 0$

● Solving equations by graphical methods

As shown earlier in this chapter, if a graph of a function is plotted, then it can be used to solve equations.

Worked examples

a) i) Plot a graph of $y = 3x^2 - x - 2$ for $-3 \leq x \leq 3$.

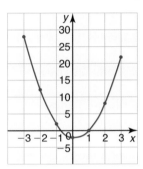

x	-3	-2	-1	0	1	2	3
y	28	12	2	-2	0	8	22

ii) Use the graph to solve the equation $3x^2 - x - 2 = 0$.

To solve the equation, $y = 0$. Therefore where the curve intersects the x-axis gives the solution to the equation.
 i.e. $3x^2 - x - 2 = 0$ when $x = -0.7$ and 1

iii) Use the graph to solve the equation $3x^2 - 7 = 0$

To be able to use the original graph, this equation needs to be manipulated in such a way that one side of the equation becomes:
 $3x^2 - x - 2$.
Manipulating $3x^2 - 7 = 0$ gives:
 $3x^2 - x - 2 = -x + 5$ (subtracting x from both sides, and adding 5 to both sides)
Hence finding where the curve $y = 3x^2 - x - 2$ intersects the line $y = -x + 5$ gives the solution to the equation $3x^2 - 7 = 0$.

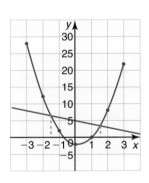

Therefore the solutions to $3x^2 - 7 = 0$ are $x \approx -1.5$ and 1.5.

Algebra and graphs

b) i) Plot a graph of $y = \frac{1}{x} + x$ for $-4 \leq x \leq 4$.

x	−4	−3	−2	−1	0	1	2	3	4
y	−4.25	−3.3	−2.5	−2	—	2	2.5	3.3	4.25

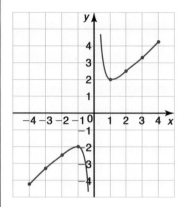

ii) Use the graph to explain why $\frac{1}{x} + x = 0$ has no solution.

For $\frac{1}{x} + x = 0$, the graph will need to intersect the x-axis. From the plot opposite, it can be seen that the graph does not intersect the x-axis and hence the equation

$\frac{1}{x} + x = 0$ has no solution.

iii) Use the graph to find the solution to $x^2 - x = 1$.

This equation needs to be manipulated in such a way that one side becomes $\frac{1}{x} + x$.

Manipulating $x^2 - x = 1$ gives:

$x - 1 = \frac{1}{x}$ (dividing both sides by x)

$2x - 1 = \frac{1}{x} + x$ (adding x to both sides)

Hence finding where the curve $y = \frac{1}{x} + x$ intersects the line $y = 2x - 1$ will give the solution to the equation $x^2 - x = 1$.

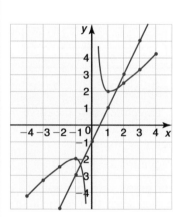

Therefore the solutions to the equation $x^2 - x = 1$ are $x \approx -0.6$ and 1.6.

Exercise 18.8

1. a) Plot the function $y = \frac{1}{2}x^2 + 1$ for $-4 \leq x \leq 4$.
    b) Showing your method clearly, use the graph to solve the equation $\frac{1}{2}x^2 = 4$.

2. a) Plot the function $y = x^3 + x - 2$ for $-3 \leq x \leq 3$.
    b) Showing your method clearly, use the graph to solve the equation $x^3 = 7 - x$.

3. a) Plot the function $y = 2x^3 - x^2 + 3$ for $-2 \leq x \leq 2$.
    b) Showing your method clearly, use the graph to solve the equation $2x^3 - 7 = 0$.

4. a) Plot the function $y = \dfrac{2}{x^2} - x$ for $-4 \leqslant x \leqslant 4$.
   b) Showing your method clearly, use the graph to solve the equation $4x^3 - 10x^2 + 2 = 0$.

5. a) Plot the function $y = 2^x - x$ for $-2 \leqslant x \leqslant 5$.
   b) Showing your method clearly, use the graph to solve the equation $2^x = 2x + 2$.

6. A tap is dripping at a constant rate into a container. The level (l cm) of the water in the container, is given by the equation $l = 2^t - 1$ where t hours is the time taken.
   a) Calculate the level of the water after 3 hours.
   b) Calculate the level of the water in the container at the start.
   c) Calculate the time taken for the level of the water to reach 31 cm.
   d) Plot a graph showing the level of the water over the first 6 hours.
   e) From your graph, estimate the time taken for the water to reach a level of 45 cm.

7. Draw a graph of $y = 4^x$ for values of x between -1 and 3. Use your graph to find approximate solutions to the following equations:
   a) $4^x = 30$
   b) $4^x = \dfrac{1}{2}$

8. Draw a graph of $y = 2^x$ for values of x between -2 and 5. Use your graph to find approximate solutions to the following equations:
   a) $2^x = 20$
   b) $2^{(x+2)} = 40$
   c) $2^{-x} = 0.2$

9. During an experiment it is found that harmful bacteria grow at an exponential rate with respect to time. The approximate population of the bacteria P is modelled by the equation $P = 4^t + 100$, where t is the time in hours.

   a) Calculate the approximate number of harmful bacteria at the start of the experiment.
   b) Calculate the number of harmful bacteria after 5 hours. Give your answer to 3 significant figures.
   c) Draw a graph of $P = 4^t + 100$, for values of t from 0 to 6.
   d) Estimate from your graph the time taken for the bacteria population to reach 600.

Algebra and graphs

10. The population of a type of insect is falling at an exponential rate. The population P is known to be modelled by the equation $P = 1000 \times (\frac{1}{2})^t$, where t is the time in weeks.

    a) Copy and complete the following table of results, giving each value of P to the nearest whole number.

t	0	1	2	3	4	5	6	7	8	9	10
P			250								

    b) Plot a graph for the table of results above.
    c) Estimate from your graph the population of insects after $3\frac{1}{2}$ weeks.

Student assessment 1

1. Sketch the graphs of the following functions:
   a) $y = x^2$
   b) $y = -x^2$

2. a) Copy and complete the table below for the function $y = x^2 + 8x + 15$

x	−7	−6	−5	−4	−3	−2	−1	0	1	2
y		3			3					

   b) Plot a graph of the function.

3. Plot a graph of each of the functions below between the given limits of x.
   a) $y = -x^2 - 2x - 1$, $-3 \leqslant x \leqslant 3$
   b) $y = x^2 + 2x - 7$, $-5 \leqslant x \leqslant 2$

4. a) Plot the graph of the quadratic function $y = x^2 + 9x + 20$ for $-7 \leqslant x \leqslant -2$.
   b) Showing your method clearly, use your graph to solve the equation $x^2 = -9x - 14$.

5. a) Plot the graph of $y = \frac{1}{x}$ for $-4 \leqslant x \leqslant 4$.
   b) Showing your method clearly, use your graph to solve the equation $1 = -x^2 + 3x$.

Student assessment 2

1. Sketch the graph of the function $y = \frac{1}{x}$.

2. a) Copy and complete the table below for the function $y = -x^2 - 7x - 12$.

x	−7	−6	−5	−4	−3	−2	−1	0	1	2
y		−6				−2				

   b) Plot a graph of the function.

3. Plot a graph of each of the functions below between the given limits of x.
   a) $y = x^2 - 3x - 10$, $-3 \leqslant x \leqslant 6$
   b) $y = -x^2 - 4x - 4$, $-5 \leqslant x \leqslant 1$

4. a) Plot the graph of the quadratic equation $y = -x^2 - x + 15$ for $-6 \leqslant x \leqslant 4$.
   b) Showing your method clearly, use your graph to solve the following equations:
      i) $10 = x^2 + x$ ii) $x^2 = x + 5$

5. a) Plot the graph of $y = \frac{2}{x}$ for $-4 \leqslant x \leqslant 4$.
   b) Showing your method clearly, use your graph to solve the equation $x^2 + x = 2$.

Student assessment 3

1. a) Name the types of graph shown below:

   i) 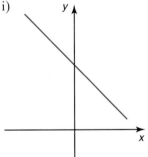   ii)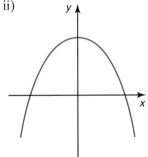

   b) Give a **possible** equation for each of the graphs drawn.

2. For each of the functions below:
   i) draw up a table of values,
   ii) plot a graph of the function.
      a) $f(x) = x^2 + 3x$, $-5 \leqslant x \leqslant 2$
      b) $f(x) = \frac{1}{x} + 3x$, $-3 \leqslant x \leqslant 3$

Algebra and graphs

3. a) Plot the function $y = \frac{1}{2}x^3 + 2x^2$ for $-5 \leqslant x \leqslant 2$.
   b) Calculate the gradient of the curve when:
      i) $x = 1$ ii) $x = -1$

4. a) Plot a graph of the function $y = 2x^2 - 5x - 5$ for $-2 \leqslant x \leqslant 5$.
   b) Use the graph to solve the equation $2x^2 - 5x - 5 = 0$.
   c) Showing your method clearly, use the graph to solve the equation $2x^2 - 3x = 10$.

Student assessment 4

1. a) Name the types of graph shown below:
      i)   ii)

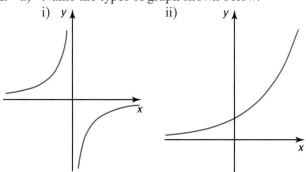

   b) Give a **possible** equation for each of the graphs drawn.

2. For each of the functions below:
   i) draw up a table of values,
   ii) plot a graph of the function.
      a) $f(x) = 2^x + x$, $-3 \leqslant x \leqslant 3$
      b) $f(x) = 3^x - x^2$, $-3 \leqslant x \leqslant 3$

3. a) Plot the function $y = -x^3 - 4x^2 + 5$ for $-5 \leqslant x \leqslant 2$.
   b) Calculate the gradient of the curve when:
      i) $x = 0$ ii) $x = -2$

4. a) Copy and complete the table below for the function $y = \dfrac{1}{x^2} - 5$.

x	-3	-2	-1	-0.5	-0.25	0	0.25	0.5	1	2	3
y				-1		—					

   b) Plot a graph of the function.
   c) Use the graph to solve the equation $\dfrac{1}{x^2} = 5$.
   d) Showing your method clearly, use your graph to solve the equation $\dfrac{1}{x^2} + x^2 = 7$.

19 Functions

An expression such as $4x - 9$, in which the variable is x, is called 'a function of x'. Its numerical value depends on the value of x. We sometimes write $f(x) = 4x - 9$, or $f: x \mapsto 4x - 9$.

Worked examples **a)** For the function $f(x) = 3x - 5$, evaluate:
  i) $f(2)$ ii) $f(0)$

$$f(2) = 3 \times 2 - 5 \qquad f(0) = 3 \times 0 - 5$$
$$= 6 - 5 \qquad\qquad\quad = 0 - 5$$
$$= 1 \qquad\qquad\qquad\;\; = -5$$

  iii) $f(-2)$

$$f(-2) = 3 \times (-2) - 5$$
$$= -6 - 5$$
$$= -11$$

b) For the function $f: x \mapsto \dfrac{2x+6}{3}$, evaluate:
  i) $f(3)$ ii) $f(1.5)$

$$f(3) = \frac{2 \times 3 + 6}{3} \qquad f(1.5) = \frac{2 \times 1.5 + 6}{3}$$

$$= \frac{6+6}{3} \qquad\qquad\qquad = \frac{3+6}{3}$$

$$= 4 \qquad\qquad\qquad\qquad\;\; = 3$$

  iii) $f(-1)$

$$f(-1) = \frac{2 \times (-1) + 6}{3}$$

$$= \frac{-2+6}{3}$$

$$= \frac{4}{3}$$

c) For the function $f(x) = x^2 + 4$, evaluate:
  i) $f(2)$ ii) $f(6)$

$$f(2) = 2^2 + 4 \qquad f(6) = 6^2 + 4$$
$$= 4 + 4 \qquad\qquad\;\; = 36 + 4$$
$$= 8 \qquad\qquad\qquad = 40$$

  iii) $f(-1)$

$$f(-1) = -1^2 + 4$$
$$= 1 + 4$$
$$= 5$$

Algebra and graphs

Exercise 19.1

1. If $f(x) = 2x + 2$, calculate:
   a) $f(2)$ b) $f(4)$ c) $f(0.5)$ d) $f(1.5)$
   e) $f(0)$ f) $f(-2)$ g) $f(-6)$ h) $f(-0.5)$

2. If $f(x) = 4x - 6$, calculate:
   a) $f(4)$ b) $f(7)$ c) $f(3.5)$ d) $f(0.5)$
   e) $f(0.25)$ f) $f(-3)$ g) $f(-4.25)$ h) $f(0)$

3. If $g(x) = -5x + 2$, calculate:
   a) $g(0)$ b) $g(6)$ c) $g(4.5)$ d) $g(3.2)$
   e) $g(0.1)$ f) $g(-2)$ g) $g(-6.5)$ h) $g(-2.3)$

4. If $h(x) = -3x - 7$, calculate:
   a) $h(4)$ b) $h(6.5)$ c) $h(0)$ d) $h(0.4)$
   e) $h(-9)$ f) $h(-5)$ g) $h(-2)$ h) $h(-3.5)$

Exercise 19.2

1. If $f(x) = \dfrac{3x+2}{4}$, calculate:
   a) $f(2)$ b) $f(8)$ c) $f(2.5)$ d) $f(0)$
   e) $f(-0.5)$ f) $f(-6)$ g) $f(-4)$ h) $f(-1.6)$

2. If $g(x) = \dfrac{5x-3}{3}$, calculate:
   a) $g(3)$ b) $g(6)$ c) $g(0)$ d) $g(-3)$
   e) $g(-1.5)$ f) $g(-9)$ g) $g(-0.2)$ h) $g(-0.1)$

3. If $h: x \mapsto \dfrac{-6x+8}{4}$, calculate:
   a) $h(1)$ b) $h(0)$ c) $h(4)$ d) $h(1.5)$
   e) $h(-2)$ f) $h(-0.5)$ g) $h(-22)$ h) $h(-1.5)$

4. If $f(x) = \dfrac{-5x-7}{-8}$, calculate:
   a) $f(5)$ b) $f(1)$ c) $f(3)$ d) $f(-1)$
   e) $f(-7)$ f) $f(-\tfrac{3}{5})$ g) $f(-0.8)$ h) $f(0)$

Exercise 19.3

1. If $f(x) = x^2 + 3$, calculate:
   a) $f(4)$ b) $f(7)$ c) $f(1)$ d) $f(0)$
   e) $f(-1)$ f) $f(0.5)$ g) $f(-3)$ h) $f(\sqrt{2})$

2. If $f(x) = 3x^2 - 5$, calculate:
   a) $f(5)$ b) $f(8)$ c) $f(1)$ d) $f(0)$
   e) $f(-2)$ f) $f(\sqrt{3})$ g) $f(-\tfrac{1}{2})$ h) $f(-\tfrac{1}{3})$

3. If $g(x) = -2x^2 + 4$, calculate:
   a) $g(3)$ b) $g(\tfrac{1}{2})$ c) $g(0)$ d) $g(1.5)$
   e) $g(-4)$ f) $g(-1)$ g) $g(\sqrt{5})$ h) $g(-6)$

Functions

4. If $h(x) = \dfrac{-5x^2 + 15}{-2}$, calculate:
   a) $h(1)$ b) $h(4)$ c) $h(\sqrt{3})$ d) $h(0.5)$
   e) $h(0)$ f) $h(-3)$ g) $h\left(\dfrac{1}{\sqrt{2}}\right)$ h) $h(-2.5)$

5. If $f(x) = -6x(x - 4)$, calculate:
   a) $f(0)$ b) $f(2)$ c) $f(4)$ d) $f(0.5)$
   e) $f(-\tfrac{1}{2})$ f) $f(-\tfrac{1}{6})$ g) $f(-2.5)$ h) $f(\sqrt{2})$

6. If $g: x \mapsto \dfrac{(x+2)(x-4)}{-x}$, calculate:
   a) $g(1)$ b) $g(4)$ c) $g(8)$ d) $g(0)$
   e) $g(-2)$ f) $g(-10)$ g) $g(-\tfrac{3}{2})$ h) $g(-8)$

Exercise 19.4

1. If $f(x) = 2x + 1$, write the following in their simplest form:
   a) $f(x + 1)$ b) $f(2x - 3)$ c) $f(x^2)$
   d) $f\left(\dfrac{x}{2}\right)$ e) $f\left(\dfrac{x}{4} + 1\right)$ f) $f(x) - x$

2. If $g(x) = 3x^2 - 4$, write the following in their simplest form:
   a) $g(2x)$ b) $g\left(\dfrac{x}{4}\right)$ c) $g(\sqrt{2x})$
   d) $g(3x) + 4$ e) $g(x - 1)$ f) $g(2x + 2)$

3. If $f(x) = 4x^2 + 3x - 2$, write the following in their simplest form:
   a) $f(x) + 4$ b) $f(2x) + 2$ c) $f(x + 2) - 20$
   d) $f(x - 1) + 1$ e) $f\left(\dfrac{x}{2}\right)$ f) $f(3x + 2)$

● Inverse functions

The **inverse** of a function is its reverse, i.e. it 'undoes' the function's effects. The inverse of the function $f(x)$ is written as $f^{-1}(x)$. To find the inverse of a function:
- rewrite the function replacing $f(x)$ with y
- interchange x and y
- rearrange the equation to make y the subject.

Worked examples a) Find the inverse of each of the following functions:

i) $f(x) = x + 2$
$y = x + 2$
$x = y + 2$
$y = x - 2$
So $f^{-1}(x) = x - 2$

ii) $g(x) = 2x - 3$
$y = 2x - 3$
$x = 2y - 3$
$y = \dfrac{x + 3}{2}$
So $g^{-1}(x) = \dfrac{x + 3}{2}$

Algebra and graphs

b) If $f(x) = \dfrac{x-3}{3}$ calculate:

i) $f^{-1}(2)$ ii) $f^{-1}(-3)$

First calculate the inverse function $f^{-1}(x)$:

$y = \dfrac{x-3}{3}$

$x = \dfrac{y-3}{3}$

$y = 3x + 3$

So $f^{-1}(x) = 3x + 3$

i) $f^{-1}(2) = 3(2) + 3 = 9$
ii) $f^{-1}(-3) = 3(-3) + 3 = -6$

Exercise 19.5 Find the inverse of each of the following functions:

1. a) $f(x) = x + 3$ b) $f(x) = x + 6$
   c) $f(x) = x - 5$ d) $g(x) = x$
   e) $h(x) = 2x$ f) $p(x) = \dfrac{x}{3}$

2. a) $f(x) = 4x$ b) $f(x) = 2x + 5$
   c) $f(x) = 3x - 6$ d) $f(x) = \dfrac{x+4}{2}$
   e) $g(x) = \dfrac{3x-2}{4}$ f) $g(x) = \dfrac{8x+7}{5}$

3. a) $f(x) = \tfrac{1}{2}x + 3$ b) $g(x) = \tfrac{1}{4}x - 2$
   c) $h(x) = 4(3x - 6)$ d) $p(x) = 6(x + 3)$
   e) $q(x) = -2(-3x + 2)$ f) $f(x) = \tfrac{2}{3}(4x - 5)$

Exercise 19.6
1. If $f(x) = x - 4$, evaluate:
   a) $f^{-1}(2)$ b) $f^{-1}(0)$ c) $f^{-1}(-5)$

2. If $f(x) = 2x + 1$, evaluate:
   a) $f^{-1}(5)$ b) $f^{-1}(0)$ c) $f^{-1}(-11)$

3. If $g(x) = 6(x - 1)$, evaluate:
   a) $g^{-1}(12)$ b) $g^{-1}(3)$ c) $g^{-1}(6)$

4. If $g(x) = \dfrac{2x+4}{3}$, evaluate:
   a) $g^{-1}(4)$ b) $g^{-1}(0)$ c) $g^{-1}(-6)$

5. If $h(x) = \tfrac{1}{3}x - 2$, evaluate:
   a) $h^{-1}(-\tfrac{1}{2})$ b) $h^{-1}(0)$ c) $h^{-1}(-2)$

6. If $f(x) = \dfrac{4x-2}{5}$, evaluate:
   a) $f^{-1}(6)$ b) $f^{-1}(-2)$ c) $f^{-1}(0)$

Composite functions

Worked examples

a) If $f(x) = x + 2$ and $g(x) = x + 3$, find $fg(x)$.

$$fg(x) = f(x + 3)$$
$$= (x + 3) + 2$$
$$= x + 5$$

b) If $f(x) = 2x - 1$ and $g(x) = x - 2$, find $fg(x)$.

$$fg(x) = f(x - 2)$$
$$= 2(x - 2) - 1$$
$$= 2x - 4 - 1$$
$$= 2x - 5$$

c) If $f(x) = 2x + 3$ and $g(x) = 2x$, evaluate $fg(3)$.

$$fg(x) = f(2x)$$
$$= 2(2x) + 3$$
$$= 4x + 3$$
$$fg(3) = 4 \times 3 + 3$$
$$= 15$$

Exercise 19.7

1. Write a formula for $fg(x)$ in each of the following:
  a) $f(x) = x - 3$ $g(x) = x + 5$
  b) $f(x) = x + 4$ $g(x) = x - 1$
  c) $f(x) = x$ $g(x) = 2x$
  d) $f(x) = \frac{x}{2}$ $g(x) = 2x$

2. Write a formula for $pq(x)$ in each of the following:
  a) $p(x) = 2x$ $q(x) = x + 4$
  b) $p(x) = 3x + 1$ $q(x) = 2x$
  c) $p(x) = 4x + 6$ $q(x) = 2x - 1$
  d) $p(x) = -x + 4$ $q(x) = x + 2$

3. Write a formula for $jk(x)$ in each of the following:
  a) $j(x) = \frac{x-2}{4}$ $k(x) = 4x$
  b) $j(x) = 3x + 2$ $k(x) = \frac{x-3}{2}$
  c) $j(x) = \frac{2x+5}{3}$ $k(x) = \frac{1}{2}x + 1$
  d) $j(x) = \frac{1}{4}(x - 3)$ $k(x) = \frac{8x+2}{5}$

4. Evaluate $fg(2)$ in each of the following:
  a) $f(x) = x - 4$ $g(x) = x + 3$
  b) $f(x) = 2x$ $g(x) = -x + 6$
  c) $f(x) = 3x$ $g(x) = 6x + 1$
  d) $f(x) = \frac{x}{2}$ $g(x) = -2x$

Algebra and graphs

5. Evaluate gh(−4) in each of the following:
   a) $g(x) = 3x + 2$ $h(x) = -4x$
   b) $g(x) = \frac{1}{2}(3x - 1)$ $h(x) = \frac{2x}{5}$
   c) $g(x) = 4(-x + 2)$ $h(x) = \frac{2x + 6}{4}$
   d) $g(x) = \frac{4x + 4}{5}$ $h(x) = -\frac{1}{3}(-x + 5)$

Student assessment 1

1. For the function $f(x) = 5x - 1$, evaluate:
   a) f(2)   b) f(0)   c) f(−3)

2. For the function $g: x \mapsto \frac{3x - 2}{2}$, evaluate:
   a) g(4)   b) g(0)   c) g(−3)

3. For the function $f(x) = \frac{(x + 3)(x - 4)}{2}$, evaluate:
   a) f(0)   b) f(−3)   c) f(−6)

4. Find the inverse of each of the following functions:
   a) $f(x) = -x + 4$ b) $g(x) = \frac{3(x - 6)}{2}$

5. If $h(x) = \frac{3}{2}(-x + 3)$, evaluate:
   a) $h^{-1}(-3)$ b) $h^{-1}(\frac{3}{2})$

6. If $f(x) = 4x + 2$ and $g(x) = -x + 3$, find fg(x).

Student assessment 2

1. For the function $f(x) = 3x + 1$, evaluate:
   a) f(4)   b) f(−1)   c) f(0)

2. For the function $g: x \mapsto \frac{-x - 2}{3}$, evaluate:
   a) g(4)   b) g(−5)   c) g(1)

3. For the function $f(x) = x^2 - 3x$, evaluate:
   a) f(1)   b) f(3)   c) f(−3)

4. Find the inverse of the following functions:
   a) $f(x) = -3x + 9$ b) $g(x) = \frac{(x - 2)}{4}$

5. If $h(x) = -5(-2x + 4)$, evaluate:
   a) $h^{-1}(-10)$ b) $h^{-1}(0)$

6. If $f(x) = 8x + 2$ and $g(x) = 4x - 1$, find fg(x).

Topic 2 Mathematical investigations and ICT

● House of cards

The drawing shows a house of cards 3 layers high. 15 cards are needed to construct it.

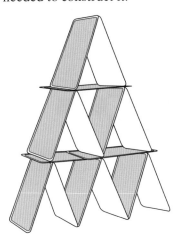

1. How many cards are needed to construct a house 10 layers high?
2. The world record is for a house 75 layers high. How many cards are needed to construct this house of cards?
3. Show that the general formula for a house n layers high requiring c cards is:
   $$c = \tfrac{1}{2}n(3n + 1)$$

● Chequered boards

A chessboard is an 8×8 square grid consisting of alternating black and white squares as shown:

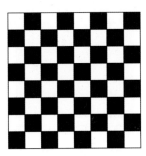

There are 64 unit squares of which 32 are black and 32 are white.

Algebra and graphs

Consider boards of different sizes. The examples below show rectangular boards, each consisting of alternating black and white unit squares.

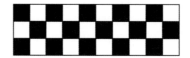

Total number of unit squares is 30
Number of black squares is 15
Number of white squares is 15

Total number of unit squares is 21
Number of black squares is 10
Number of white squares is 11

1. Investigate the number of black and white unit squares on different rectangular boards. Note: For consistency you may find it helpful to always keep the bottom right-hand square the same colour.
2. What are the numbers of black and white squares on a board $m \times n$ units?

● Modelling: Stretching a spring

A spring is attached to a clamp stand as shown below.

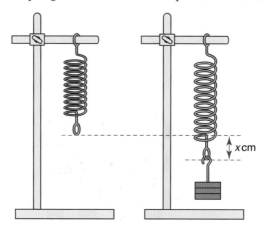

Different weights are attached to the end of the spring. The mass (m) in grams is noted as is the amount by which the spring stretches (x) in centimetres as shown on the right.

The data collected is shown in the table below:

Mass (g)	50	100	150	200	250	300	350	400	450	500
Extension (cm)	3.1	6.3	9.5	12.8	15.4	18.9	21.7	25.0	28.2	31.2

1. Plot a graph of mass against extension.
2. Describe the approximate relationship between the mass and the extension.
3. Draw a line of best fit through the data.
4. Calculate the equation of the line of best fit.
5. Use your equation to predict what the length of the spring would be for a mass of 275 g.
6. Explain why it is unlikely that the equation would be useful to find the extension if a mass of 5 kg was added to the spring.

● ICT activity 1

For each question, use a graphing package to plot the inequalities on the same pair of axes. Leave unshaded the region which satisfies all of them simultaneously.

1. $y \leqslant x$ $y > 0$ $x \leqslant 3$
2. $x + y > 3$ $y \leqslant 4$ $y - x > 2$
3. $2y + x \leqslant 5$ $y - 3x - 6 < 0$ $2y - x > 3$

● ICT activity 2

You have seen that it is possible to solve some exponential equations by applying the laws of indices.

Use a graphics calculator and appropriate graphs to solve the following exponential equations:

1. $4^x = 40$
2. $3^x = 17$
3. $5^{x-1} = 6$
4. $3^{-x} = 0.5$

Topic 3: Geometry

Syllabus

E3.1

Use and interpret the geometrical terms: point, line, parallel, bearing, right angle, acute, obtuse and reflex angles, perpendicular, similarity and congruence.
Use and interpret vocabulary of triangles, quadrilaterals, circles, polygons and simple solid figures including nets.

E3.2

Measure lines and angles.
Construct a triangle given the three sides using ruler and pair of compasses only.
Construct other simple geometrical figures from given data using ruler and protractor as necessary.
Construct angle bisectors and perpendicular bisectors using straight edge and pair of compasses only.

E3.3

Read and make scale drawings.

E3.4

Calculate lengths of similar figures.
Use the relationships between areas of similar triangles, with corresponding results for similar figures and extension to volumes and surface areas of similar solids.

E3.5

Recognise rotational and line symmetry (including order of rotational symmetry) in two dimensions.

Recognise symmetry properties of the prism (including cylinder) and the pyramid (including cone).
Use the following symmetry properties of circles:
- equal chords are equidistant from the centre
- the perpendicular bisector of a chord passes through the centre
- tangents from an external point are equal in length.

E3.6

Calculate unknown angles using the following geometrical properties:
- angles at a point
- angles at a point on a straight line and intersecting straight lines
- angles formed within parallel lines
- angle properties of triangles and quadrilaterals
- angle properties of regular polygons
- angle in a semi-circle
- angle between tangent and radius of a circle
- angle properties of irregular polygons
- angle at the centre of a circle is twice the angle at the circumference
- angles in the same segment are equal
- angles in opposite segments are supplementary; cyclic quadrilaterals.

E3.7

Use the following loci and the method of intersecting loci for sets of points in two dimensions which are:
- at a given distance from a given point
- at a given distance from a given straight line
- equidistant from two given points
- equidistant from two given intersecting straight lines.

Contents

Chapter 20　Geometrical vocabulary (E3.1)
Chapter 21　Geometrical constructions and scale drawings (E3.2 and E3.3)
Chapter 22　Similarity (E3.4)
Chapter 23　Symmetry (E3.5)
Chapter 24　Angle properties (E3.6)
Chapter 25　Loci (E3.7)

The Greeks

Many of the great Greek mathematicians came from the Greek Islands, from cities such as Ephesus or Miletus (which are in present day Turkey) or from Alexandria in Egypt. This section briefly mentions some of the Greek mathematicians of 'The Golden Age'. You may wish to find out more about them.

Thales of Alexandria invented the 365 day calendar and predicted the dates of eclipses of the Sun and the moon.

Pythagoras of Samos founded a school of mathematicians and worked with geometry. His successor as leader was Theano, the first woman to hold a major role in mathematics.

Eudoxus of Asia Minor (Turkey) worked with irrational numbers like pi and discovered the formula for the volume of a cone.

Euclid of Alexandria formed what would now be called a university department. His book became the set text in schools and universities for 2000 years.

Apollonius of Perga (Turkey) worked on, and gave names to, the parabola, the hyperbola and the ellipse.

Archimedes (287–212 BCE)

Archimedes is accepted today as the greatest mathematician of all time. However he was so far ahead of his time that his influence on his contemporaries was limited by their lack of understanding.

20 Geometrical vocabulary

● Angles and lines

Different types of angle have different names:

acute angles lie between 0° and 90°
right angles are exactly 90°
obtuse angles lie between 90° and 180°
reflex angles lie between 180° and 360°

To find the shortest distance between two points, you measure the length of the **straight line** which joins them.

Two lines which meet at right angles are **perpendicular** to each other.

So in the diagram below CD is **perpendicular** to AB, and AB is perpendicular to CD.

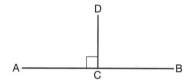

If the lines AD and BD are drawn to form a triangle, the line CD can be called the **height** or **altitude** of the triangle ABD.

Parallel lines are straight lines which can be continued to infinity in either direction without meeting.

Railway lines are an example of parallel lines. Parallel lines are marked with arrows as shown:

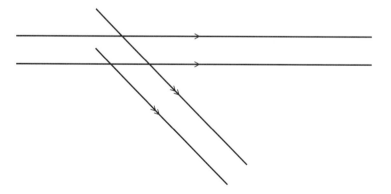

Triangles

Triangles can be described in terms of their sides or their angles, or both.

An **acute-angled** triangle has all its angles less than 90°.

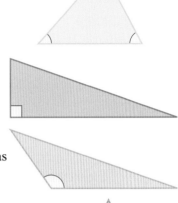

A **right-angled** triangle has an angle of 90°.

An **obtuse-angled** triangle has one angle greater than 90°.

An **isosceles** triangle has two sides of equal length, and the angles opposite the equal sides are equal.

An **equilateral** triangle has three sides of equal length and three equal angles.

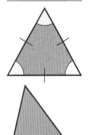

A **scalene** triangle has three sides of different lengths and all three angles are different.

Congruent triangles

Congruent triangles are **identical**. They have corresponding sides of the same length, and corresponding angles which are equal.

Triangles are congruent if any of the following can be proved:
- Three corresponding sides are equal (S S S);
- Two corresponding sides and the included angle are equal (S A S);
- Two angles and the corresponding side are equal (A S A);
- Each triangle has a right angle, and the hypotenuse and a corresponding side are equal in length.

Geometry

NB: All diagrams are not drawn to scale.

Similar triangles

If the angles of two triangles are the same, then their corresponding sides will also be in proportion to each other. When this is the case, the triangles are said to be **similar**.

In the diagram below, triangle ABC is similar to triangle XYZ. Similar shapes are covered in more detail in Chapter 22.

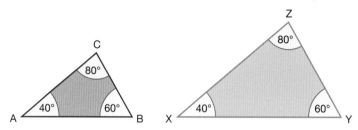

Exercise 20.1

1. In the diagrams below, identify pairs of congruent triangles. Give reasons for your answers.

a)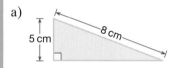

b)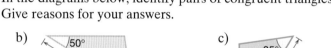

c)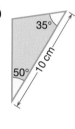

d)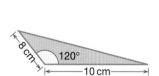

e)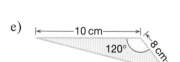

f)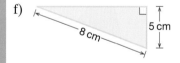

g)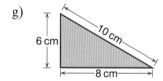

h)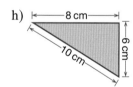

Circles

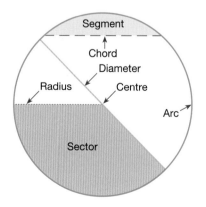

● Quadrilaterals

A **quadrilateral** is a plane shape consisting of four angles and four sides. There are several types of quadrilateral. The main ones, and their properties, are described below.

Two pairs of parallel sides.
All sides are equal.
All angles are equal.
Diagonals intersect at right angles.

Two pairs of parallel sides.
Opposite sides are equal.
All angles are equal.

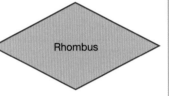

Two pairs of parallel sides.
All sides are equal.
Opposite angles are equal.
Diagonals intersect at right angles.

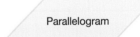

Two pairs of parallel sides.
Opposite sides are equal.
Opposite angles are equal.

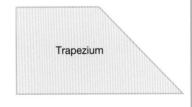

One pair of parallel sides.
An isosceles trapezium has one pair of parallel sides and the other pair of sides are equal in length.

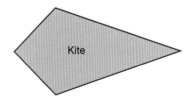

Two pairs of equal sides.
One pair of equal angles.
Diagonals intersect at right angles.

Geometry

Exercise 20.2 **1.** Copy and complete the following table. The first line has been started for you.

	Rectangle	Square	Parallelogram	Kite	Rhombus	Equilateral triangle
Opposite sides equal in length	Yes	Yes				
All sides equal in length						
All angles right angles						
Both pairs of opposite sides parallel						
Diagonals equal in length						
Diagonals intersect at right angles						
All angles equal						

● **Polygons**

Any closed figure made up of straight lines is called a **polygon**.

If the sides are the same length and the interior angles are equal, the figure is called a **regular polygon**.

The names of the common polygons are:

    3 sides   **tri**angle
    4 sides   **quad**rilateral
    5 sides   **penta**gon
    6 sides   **hexa**gon
    7 sides   **hepta**gon
    8 sides   **octa**gon
    10 sides   **deca**gon
    12 sides   **dodeca**gon

Two polygons are said to be **similar** if

a) their angles are the same
b) corresponding sides are in proportion.

● Nets

The diagram below is the **net** of a cube. It shows the faces of the cube opened out into a two-dimensional plan. The net of a three-dimensional shape can be folded up to make that shape.

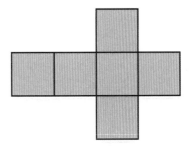

Exercise 20.3 Draw the following on squared paper:

1. Two other possible nets of a cube
2. The net of a cuboid (rectangular prism)
3. The net of a triangular prism
4. The net of a cylinder
5. The net of a square-based pyramid
6. The net of a tetrahedron

Student assessment 1

1. Are the angles below acute, obtuse, reflex or right angles?

   a)   b)   c)   d)

   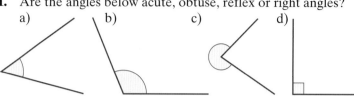

2. Draw and label two pairs of intersecting parallel lines.

3. Identify the types of triangles below in two ways (for example, obtuse-angled scalene triangle):

   a)   b)

4. Draw a circle of radius 3 cm. Mark on it:
   a) a diameter   b) a chord   c) a sector.

5. Draw a rhombus and write down three of its properties.

6. On squared paper, draw the net of a trianglular prism.

Geometrical constructions and scale drawings

Constructing triangles

Triangles can be drawn accurately by using a ruler and a pair of compasses. This is called **constructing** a triangle.

Worked example The sketch shows the triangle ABC.

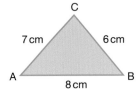

Construct the triangle ABC given that:

AB = 8 cm, BC = 6 cm and AC = 7 cm

- Draw the line AB using a ruler:

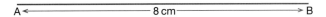

- Open up a pair of compasses to 6 cm. Place the compass point on B and draw an arc:

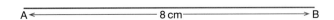

Note that every point on the arc is 6 cm away from B.

Geometry

- Open up the pair of compasses to 7 cm. Place the compass point on A and draw another arc, with centre A and radius 7 cm, ensuring that it intersects with the first arc. Every point on the second arc is 7 cm from A. Where the two arcs intersect is point C, as it is both 6 cm from B and 7 cm from A.
- Join C to A and C to B:

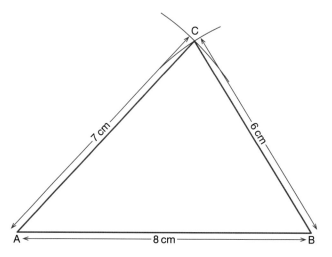

Exercise 21.1 Using only a ruler and a pair of compasses, construct the following triangles:

1. △ABC where AB = 10 cm, AC = 7 cm and BC = 9 cm
2. △LMN where LM = 4 cm, LN = 8 cm and MN = 5 cm
3. △PQR, an equilateral triangle of side length 7 cm
4. a) △ABC where AB = 8 cm, AC = 4 cm and BC = 3 cm
   b) Is this triangle possible? Explain your answer.

Constructing simple geometric figures

Worked example Using a ruler and a protractor only, construct the parallelogram ABCD in which AB = 6 cm, AD = 3 cm and angle DAB = 40°.

- Draw a line 6 cm long and label it AB.
- Place the protractor on A and mark an angle of 40°.

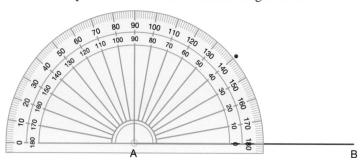

- Draw a line from A through the marked point.

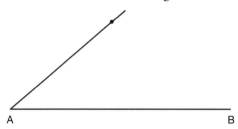

- Place the protractor on B and mark an angle of 40° reading from the inner scale. Draw a line from B through the marked point.

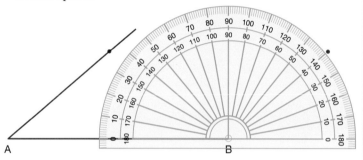

- Measure 3 cm from A and mark the point D.
- Measure 3 cm from B and mark the point C.
- Join D to C.

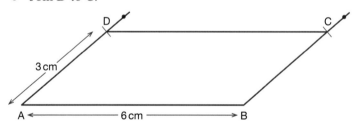

Exercise 21.2

1. Using only a ruler and a protractor, construct the triangle PQR in which PQ = 4 cm, angle RPQ = 115° and angle RQP = 30°.

2. Using only a ruler and a protractor, construct the trapezium ABCD in which AB = 8 cm, AD = 4 cm, angle DAB = 60° and angle ABC = 90°.

● Bisecting lines and angles

The word **bisect** means 'to divide in half'. Therefore, to bisect an angle means to divide an angle in half. Similarly, to bisect a line means to divide a line in half. A **perpendicular bisector** to a line is another line which divides it in half and meets the original line at right angles. To bisect either a line or an angle involves the use of a pair of compasses.

Geometry

Worked examples **a)** A line AB is drawn below. Construct the perpendicular bisector to AB.

- Open a pair of compasses to more than half the distance AB.
- Place the compass point on A and draw arcs above and below AB.
- With the same radius, place the compass point on B and draw arcs above and below AB. Note that the two pairs of arcs should intersect (see diagram below).
- Draw a line through the two points where the arcs intersect:

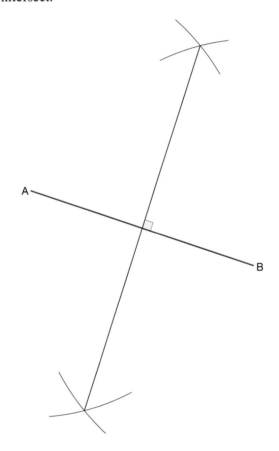

The line drawn is known as the perpendicular bisector of AB, as it divides AB in half and also meets it at right angles.

b) Using a pair of compasses, bisect the angle ABC below:

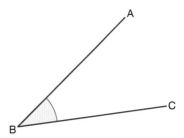

- Open a pair of compasses and place the point on B. Draw two arcs such that they intersect the arms of the angle:

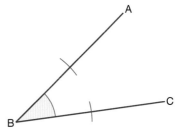

- Place the compasses in turn on the points of intersection, and draw another pair of arcs of the same radius. Ensure that they intersect.
- Draw a line through B and the point of intersection of the two arcs. This line bisects angle ABC.

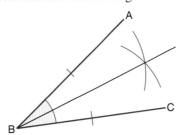

Geometry

Exercise 21.3

1. Draw a triangle similar to the one shown below:

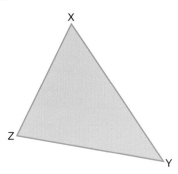

   Construct the perpendicular bisector of each of the sides of your triangle.
   Use a pair of compasses to draw a circle using the point where the three perpendicular bisectors cross as the centre and passing through the points X, Y and Z.
   This is called the **circumcircle** of the triangle.

2. Draw a triangle similar to the one shown below:

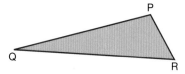

   By construction, draw a circle to pass through points P, Q and R.

Geometrical constructions and scale drawings

● **Scale drawings**

Scale drawings are used when an accurate diagram, drawn in proportion, is needed. Common uses of scale drawings include maps and plans. The use of scale drawings involves understanding how to scale measurements.

Worked examples

a) A map is drawn to a scale of 1 : 10 000. If two objects are 1 cm apart on the map, how far apart are they in real life? Give your answer in metres.

A scale of 1 : 10 000 means that 1 cm on the map represents 10 000 cm in real life.
Therefore the distance   = 10 000 cm
                          = 100 m

b) A model boat is built to a scale of 1 : 50. If the length of the real boat is 12 m, calculate the length of the model boat in cm.

A scale of 1 : 50 means that 50 cm on the real boat is 1 cm on the model boat.

12 m = 1200 cm

Therefore the length of the model boat   = 1200 ÷ 50 cm
                                         = 24 cm

c) i) Construct, to a scale of 1 : 1, a triangle ABC such that AB = 6 cm, AC = 5 cm and BC = 4 cm.

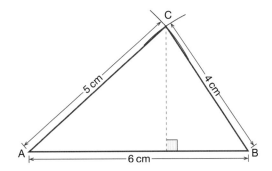

ii) Measure the perpendicular length of C from AB.
Perpendicular length is 3.3 cm.

iii) Calculate the area of the triangle.

$$\text{Area} = \frac{\text{base length} \times \text{perpendicular height}}{2}$$

$$\text{Area} = \frac{6 \times 3.3}{2} \text{ cm} = 9.9 \text{ cm}^2$$

Geometry

Exercise 21.4

1. In the following questions, both the scale to which a map is drawn and the distance between two objects on the map are given.
   Find the real distance between the two objects, giving your answer in metres.
   a) 1 : 10 000   3 cm
   b) 1 : 10 000   2.5 cm
   c) 1 : 20 000   1.5 cm
   d) 1 : 8000   5.2 cm

2. In the following questions, both the scale to which a map is drawn and the true distance between two objects are given.
   Find the distance between the two objects on the map, giving your answer in cm.
   a) 1 : 15 000   1.5 km
   b) 1 : 50 000   4 km
   c) 1 : 10 000   600 m
   d) 1 : 25 000   1.7 km

3. A rectangular pool measures 20 m by 36 m as shown below:

   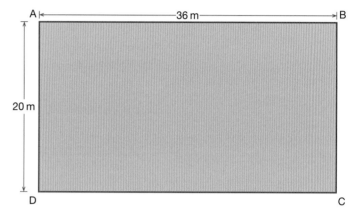

   a) Construct a scale drawing of the pool, using 1 cm for every 4 m.
   b) A boy swims across the pool in such a way that his path is the perpendicular bisector of BD. Show, by construction, the path that he takes.
   c) Work out the distance the boy swam.

4. A triangular enclosure is shown in the diagram below:

   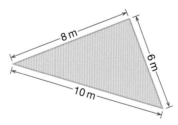

   a) Using a scale of 1 cm for each metre, construct a scale drawing of the enclosure.
   b) Calculate the true area of the enclosure.

5. Three radar stations A, B and C pick up a distress signal from a boat at sea.

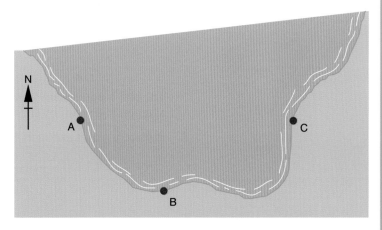

C is 24 km due East of A, AB = 12 km and BC = 18 km. The signal indicates that the boat is equidistant from all three radar stations.
a) By construction and using a scale of 1 cm for every 3 km, locate the position of the boat.
b) What is the boat's true distance from each radar station?

6. A plan view of a field is shown below:

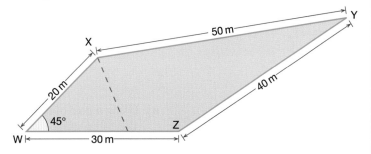

a) Using a scale of 1 cm for every 5 m, construct a scale drawing of the field.
b) A farmer divides the field by running a fence from X in such a way that it bisects angle WXY. By construction, show the position of the fence on your diagram.
c) Work out the length of fencing used.

Student assessment I

1. Construct △ABC such that AB = 8 cm, AC = 6 cm and BC = 12 cm.

Geometry

2. Three players, P, Q and R, are approaching a football. Their positions relative to each other are shown below:

   P●

   R
   ●

   Q●

   The ball is equidistant from all three players. Copy the diagram and show, by construction, the position of the ball.

3. A plan of a living room is shown below:

   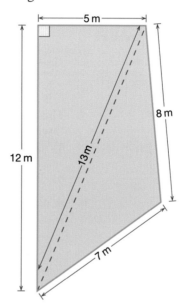

   a) Using a pair of compasses, construct a scale drawing of the room using 1 cm for every metre.
   b) Using a set square if necessary, calculate the total area of the actual living room.

Student assessment 2

1. a) Draw an angle of 320°.
   b) Using a pair of compasses, bisect the angle.

2. In the following questions, both the scale to which a map is drawn and the true distance between two objects are given. Find the distance between the two objects on the map, giving your answer in cm.
   a) 1 : 20 000   4.4 km   b) 1 : 50 000   12.2 km

3. a) Construct a regular hexagon with sides of length 3 cm.
   b) Calculate its area, showing your method clearly.

22 Similarity

NB: All diagrams are not drawn to scale.

Similar shapes

Two polygons are said to be **similar** if a) they are equi-angular and b) corresponding sides are in proportion.

For triangles, being equi-angular implies that corresponding sides are in proportion. The converse is also true.

In the diagrams (left) $\triangle ABC$ and $\triangle PQR$ are similar.

For similar figures the ratios of the lengths of the sides are the same and represent the **scale factor**, i.e.

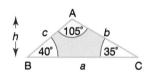

$$\frac{p}{a} = \frac{q}{b} = \frac{r}{c} = k \text{ (where } k \text{ is the scale factor of enlargement)}$$

The heights of similar triangles are proportional also:

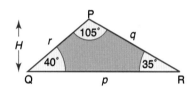

$$\frac{H}{h} = \frac{p}{a} = \frac{q}{b} = \frac{r}{c} = k$$

The ratio of the areas of similar triangles (the **area factor**) is equal to the square of the scale factor.

$$\frac{\text{Area of } \triangle PQR}{\text{Area of } \triangle ABC} = \frac{\frac{1}{2} H \times p}{\frac{1}{2} h \times a} = \frac{H}{h} \times \frac{p}{a} = k \times k = k^2$$

Exercise 22.1

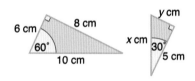

1. a) Explain why the two triangles (left) are similar.
   b) Calculate the scale factor which reduces the larger triangle to the smaller one.
   c) Calculate the value of x and the value of y.

2. Which of the triangles below are similar?

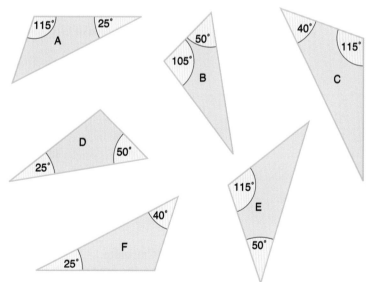

Geometry

3. The triangles below are similar.

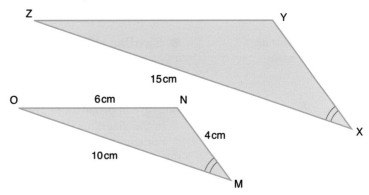

   a) Calculate the length XY.
   b) Calculate the length YZ.

4. In the triangle (right) calculate the lengths of sides *p*, *q* and *r*.

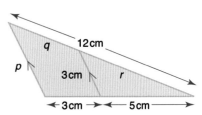

5. In the trapezium (right) calculate the lengths of the sides *e* and *f*.

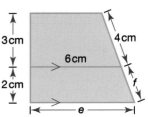

6. The triangles PQR and LMN are similar.

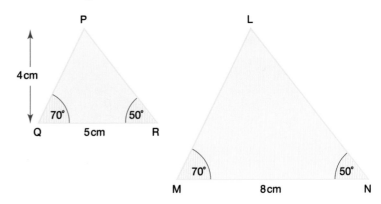

   Calculate:
   a) the area of △PQR
   b) the scale factor of enlargement
   c) the area of △LMN.

7. The triangles ABC and XYZ below are similar.

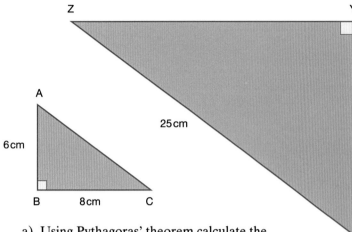

a) Using Pythagoras' theorem calculate the length of AC.
b) Calculate the scale factor of enlargement.
c) Calculate the area of △XYZ.

8. The triangle ADE shown (left) has an area of 12 cm².

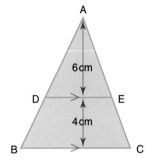

a) Calculate the area of △ABC.
b) Calculate the length BC.

9. The parallelograms below are similar.

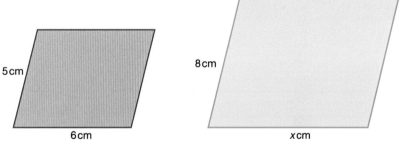

Calculate the length of the side marked x.

10. The diagram below shows two rhombuses.

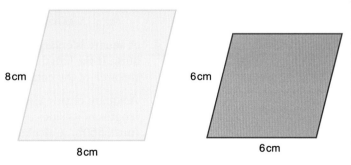

Explain, giving reasons, whether the two rhombuses are definitely similar.

Geometry

11. The diagram (right) shows a trapezium within a trapezium. Explain, giving reasons, whether the two trapezia are definitely similar.

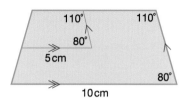

Exercise 22.2

1. In the hexagons below, hexagon B is an enlargement of hexagon A by a scale factor of 2.5.

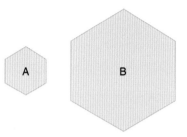

If the area of A is $8\,cm^2$, calculate the area of B.

2. P and Q are two regular pentagons. Q is an enlargement of P by a scale factor of 3. If the area of pentagon Q is $90\,cm^2$, calculate the area of P.

3. The diagram below shows four triangles A, B, C and D. Each is an enlargement of the previous one by a scale factor of 1.5

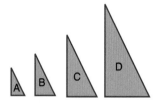

a) If the area of C is $202.5\,cm^2$, calculate the area of:
   i) triangle D   ii) triangle B   iii) triangle A.
b) If the triangles were to continue in this sequence, which letter triangle would be the first to have an area greater than $15\,000\,cm^2$?

4. A square is enlarged by increasing the length of its sides by 10%. If the length of its sides was originally 6 cm, calculate the area of the enlarged square.

5. A square of side length 4 cm is enlarged by increasing the lengths of its sides by 25% and then increasing them by a further 50%. Calculate the area of the final square.

6. An equilateral triangle has an area of $25\,cm^2$. If the lengths of its sides are reduced by 15%, calculate the area of the reduced triangle.

Similarity

Area and volume of similar shapes

Earlier in the chapter we found the following relationship between the scale factor and the area factor of enlargement:

Area factor = (scale factor)²

A similar relationship can be stated for volumes of similar shapes:

i.e. Volume factor = (scale factor)³

Exercise 22.3

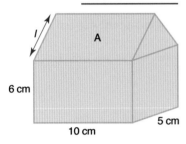

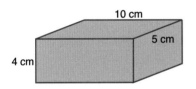

1. The diagram (left) shows a scale model of a garage. Its width is 5 cm, its length 10 cm and the height of its walls 6 cm.
   a) If the width of the real garage is 4 m, calculate:
      i) the length of the real garage
      ii) the real height of the garage wall.
   b) If the apex of the roof of the real garage is 2 m above the top of the walls, use Pythagoras' theorem to find the real slant length l.
   c) What is the area of the roof section A on the model?

2. A cuboid has dimensions as shown in the diagram (left):
   If the cuboid is enlarged by a scale factor of 2.5, calculate:
   a) the total surface area of the original cuboid
   b) the total surface area of the enlarged cuboid
   c) the volume of the original cuboid
   d) the volume of the enlarged cuboid.

3. A cube has side length 3 cm.
   a) Calculate its total surface area.
   b) If the cube is enlarged and has a total surface area of 486 cm², calculate the scale factor of enlargement.
   c) Calculate the volume of the enlarged cube.

4. Two cubes P and Q are of different sizes. If n is the ratio of their corresponding sides, express in terms of n:
   a) the ratio of their surface areas
   b) the ratio of their volumes.

5. The cuboids A and B shown below are similar.

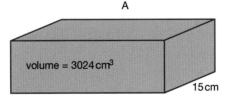

Calculate the volume of cuboid B.

Geometry

6. Two similar troughs X and Y are shown below.

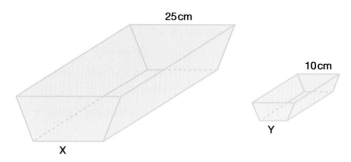

If the capacity of X is 10 litres, calculate the capacity of Y.

Exercise 22.4

1. The two cylinders L and M shown below are similar.

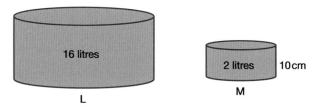

If the height of cylinder M is 10 cm, calculate the height of cylinder L.

2. A square-based pyramid (below) is cut into two shapes by a cut running parallel to the base and made half-way up.

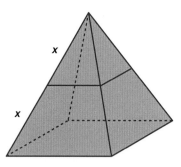

   a) Calculate the ratio of the volume of the smaller pyramid to that of the original one.
   b) Calculate the ratio of the volume of the small pyramid to that of the truncated base.

3. The two cones A and B (left) are similar. Cone B is an enlargement of A by a scale factor of 4.

   If the volume of cone B is 1024 cm³, calculate the volume of cone A.

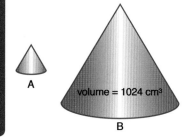

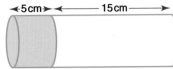

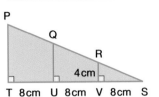

4. a) Stating your reasons clearly, decide whether the two cylinders shown (left) are similar or not.
   b) What is the ratio of the curved surface area of the shaded cylinder to that of the unshaded cylinder?

5. The diagram (left) shows a triangle.
   a) Calculate the area of $\triangle$RSV.
   b) Calculate the area of $\triangle$QSU.
   c) Calculate the area of $\triangle$PST.

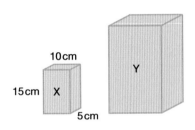

6. The area of an island on a map is $30\,cm^2$. The scale used on the map is 1:100 000.
   a) Calculate the area in square kilometres of the real island.
   b) An airport on the island is on a rectangular piece of land measuring 3 km by 2 km. Calculate the area of the airport on the map in cm^2.

7. The two packs of cheese X and Y (left) are similar.

   The total surface area of pack Y is four times that of pack X.

   Calculate:

   a) the dimensions of pack Y
   b) the mass of pack X if pack Y has a mass of 800 g.

Student assessment 1

1. Which of the triangles below are similar?

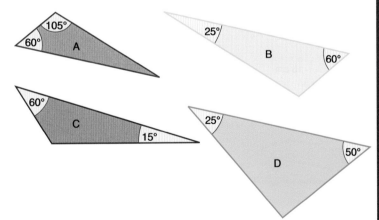

Geometry

2. Triangles P and Q (below) are similar. Express the ratio of their areas in the form, area of P : area of Q.

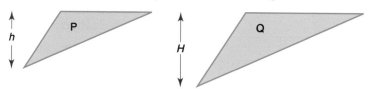

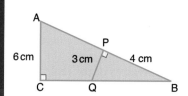

3. Using the triangle (left),
   a) explain whether △ABC and △PBQ are similar,
   b) calculate the length QB,
   c) calculate the length BC,
   d) calculate the length AP.

4. The vertical height of the large square-based solid pyramid (below) is 30 m. Its mass is 16 000 tonnes. If the mass of the smaller (shaded) pyramid is 2000 tonnes, calculate its vertical height.

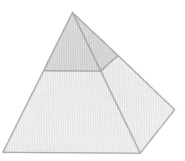

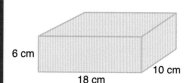

5. The cuboid (left) undergoes a reduction by a scale factor of 0.6.
   a) Draw a sketch of the reduced cuboid labelling its dimensions clearly.
   b) What is the volume of the new cuboid?
   c) What is the total surface area of the new cuboid?

6. Cuboids V and W (below) are similar.

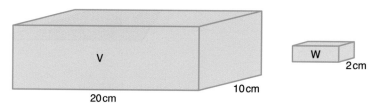

If the volume of cuboid V is 1600 cm³, calculate:
   a) the volume of cuboid W,
   b) the total surface area of cuboid V,
   c) the total surface area of cuboid W.

7. An island has an area of 50 km². What would be its area on a map of scale 1 : 20 000?

8. A box in the shape of a cube has a surface area of 2400 cm². What would be the volume of a similar box enlarged by a scale factor of 1.5?

Student assessment 2

1. The two triangles (below) are similar.

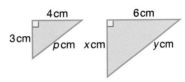

   a) Using Pythagoras' theorem, calculate the value of p.
   b) Calculate the values of x and y.

2. Cones M and N (left) are similar.

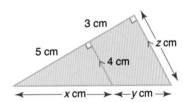

   a) Express the ratio of their surface areas in the form, area of M : area of N.
   b) Express the ratio of their volumes in the form, volume of M : volume of N.

3. Calculate the values of x, y and z in the triangle below.

4. The tins A and B (left) are similar. The capacity of tin B is three times that of tin A. If the label on tin A has an area of 75 cm², calculate the area of the label on tin B.

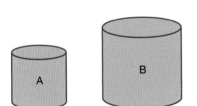

5. A cube of side 4 cm is enlarged by a scale factor of 2.5.
   a) Calculate the volume of the enlarged cube.
   b) Calculate the surface area of the enlarged cube.

6. The two troughs X and Y (left) are similar. The scale factor of enlargement from Y to X is 4. If the capacity of trough X is 1200 cm³, calculate the capacity of trough Y.

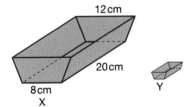

7. The rectangular floor plan of a house measures 8 cm by 6 cm. If the scale of the plan is 1 : 50, calculate:
   a) the dimensions of the actual floor,
   b) the area of the actual floor in m².

8. The volume of the cylinder (left) is 400 cm³. Calculate the volume of a similar cylinder formed by enlarging the one shown by a scale factor 2.

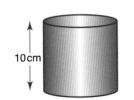

23 Symmetry

NB: All diagrams are not drawn to scale.

Symmetry and three-dimensional shapes

A **line of symmetry** divides a two-dimensional (flat) shape into two congruent (identical) shapes.
e.g.

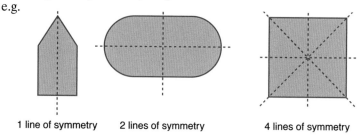

1 line of symmetry   2 lines of symmetry   4 lines of symmetry

A **plane of symmetry** divides a three-dimensional (solid) shape into two congruent solid shapes.
e.g.

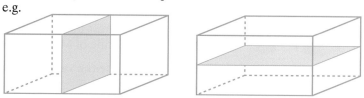

A cuboid has at least three planes of symmetry, two of which are shown above.

A shape has **reflective symmetry** if it has one or more lines or planes of symmetry.

A two-dimensional shape has **rotational symmetry** if, when rotated about a central point, it fits its outline. The number of times it fits its outline during a complete revolution is called the **order of rotational symmetry**.
e.g.

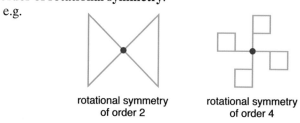

rotational symmetry of order 2   rotational symmetry of order 4

A three-dimensional shape has **rotational symmetry** if, when rotated about a central axis, it looks the same at certain intervals.
e.g.

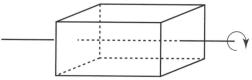

This cuboid has rotational symmetry of order 2 about the axis shown.

Symmetry

Exercise 23.1

1. Draw each of the solid shapes below twice, then:
   i) on each drawing of the shape, draw a different plane of symmetry,
   ii) state how many planes of symmetry the shape has in total.

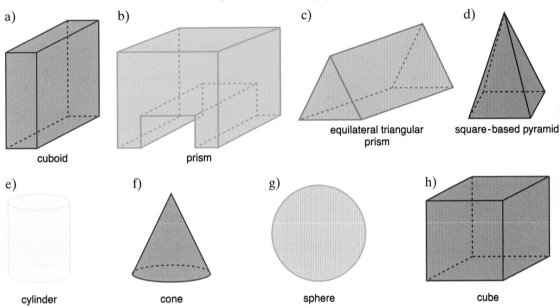

2. For each of the solid shapes shown below determine the order of rotational symmetry about the axis shown.

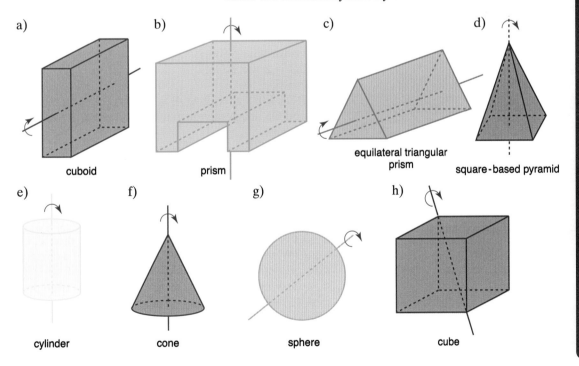

Geometry

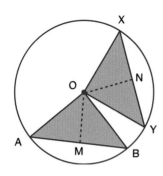

Circle properties

Equal chords and perpendicular bisectors

If chords AB and XY are of equal length, then, since OA, OB, OX and OY are radii, the triangles OAB and OXY are congruent isosceles triangles. It follows that:
- the section of a line of symmetry OM through △OAB is the same length as the section of a line of symmetry ON through △OXY,
- OM and ON are perpendicular bisectors of AB and XY respectively.

Exercise 23.2

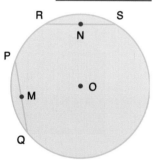

1. In the diagram (left) O is the centre of the circle, PQ and RS are chords of equal length and M and N are their respective midpoints.
   a) What kind of triangle is △POQ?
   b) Describe the line ON in relation to RS.
   c) If ∠POQ is 80°, calculate ∠OQP.
   d) Calculate ∠ORS.
   e) If PQ is 6 cm calculate the length OM.
   f) Calculate the diameter of the circle.

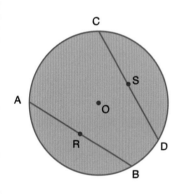

2. In the diagram (left) O is the centre of the circle. AB and CD are equal chords and the points R and S are their midpoints respectively.
   State whether the statements below are true or false, giving reasons for your answers.
   a) ∠COD = 2 × ∠AOR
   b) OR = OS
   c) If ∠ROB is 60° then △AOB is equilateral.
   d) OR and OS are perpendicular bisectors of AB and CD respectively.

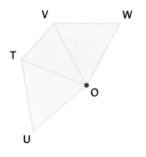

3. Using the diagram (left) state whether the following statements are true or false, giving reasons for your answer.
   a) If △VOW and △TOU are isosceles triangles, then T, U, V and W would all lie on the circumference of a circle with its centre at O.
   b) If △VOW and △TOU are congruent isosceles triangles, then T, U, V and W would all lie on the circumference of a circle with its centre at O.

23 Symmetry

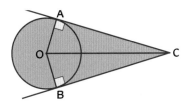

Tangents from an external point

Triangles OAC and OBC are congruent since ∠OAC and ∠OBC are right angles, OA = OB because they are both radii, and OC is common to both triangles. Hence AC = BC.

In general, therefore, tangents being drawn to the same circle from an external point are equal in length.

Exercise 23.3

1. Copy each of the diagrams below and calculate the size of the angle marked $x°$ in each case. Assume that the lines drawn from points on the circumference are tangents.

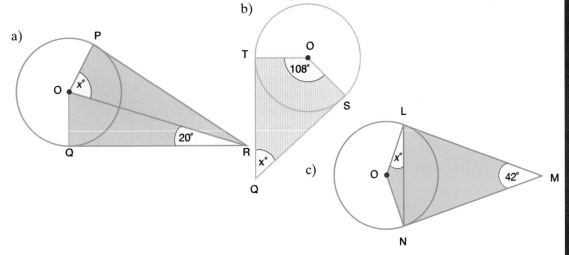

2. Copy each of the diagrams below and calculate the length of the side marked y cm in each case. Assume that the lines drawn from points on the circumference are tangents.

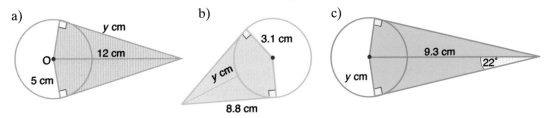

Student assessment I

1. Draw a shape with exactly:
   a) one line of symmetry,
   b) two lines of symmetry,
   c) three lines of symmetry.

2. Draw and name a shape with:
   a) two planes of symmetry,
   b) four planes of symmetry.

Geometry

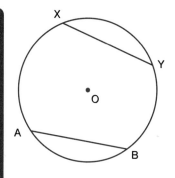

3. In the diagram (left) O is the centre of the circle and the lengths AB and XY are equal. Prove that △AOB and △XOY are congruent.

4. In the diagram (below) that PQ and QR are both tangents to the circle. Calculate the size of the angle marked $x°$.

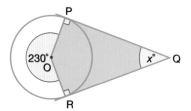

5. Calculate the diameter of the circle (right) given that LM and MN are both tangents to the circle, O is its centre and OM = 18 mm.

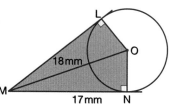

Student assessment 2

1. Draw a two-dimensional shape with exactly:
   a) rotational symmetry of order 2,
   b) rotational symmetry of order 4,
   c) rotational symmetry of order 6.

2. Draw and name a three-dimensional shape with the following orders of rotational symmetry. Mark the position of the axis of symmetry clearly.
   a) Order 2
   b) Order 3
   c) Order 8

3. In the diagram (left), OM and ON are perpendicular bisectors of AB and XY respectively. OM = ON. Prove that AB and XY are chords of equal length.

4. In the diagram (right), XY and YZ are both tangents to the circle with centre O.
   a) Calculate ∠OZX.
   b) Calculate the length XZ.

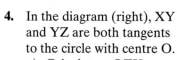

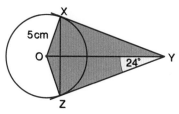

5. In the diagram (left), LN and MN are both tangents to the circle centre O. If ∠LNO is 35°, calculate the circumference of the circle.

24 Angle properties

NB: All diagrams are not drawn to scale.

Angles at a point and on a line

One complete revolution is equivalent to a rotation of 360° about a point. Similarly, half a complete revolution is equivalent to a rotation of 180° about a point. These facts can be seen clearly by looking at either a circular angle measurer or a semi-circular protractor.

Worked examples **a)** Calculate the size of the angle x in the diagram below:

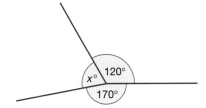

The sum of all the angles around a point is 360°. Therefore:

$$120 + 170 + x = 360$$
$$x = 360 - 120 - 170$$
$$x = 70$$

Therefore angle x is 70°.
Note that the size of the angle x is **calculated** and **not measured**.

b) Calculate the size of angle a in the diagram below:

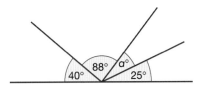

The sum of all the angles at a point on a straight line is 180°. Therefore:

$$40 + 88 + a + 25 = 180$$
$$a = 180 - 40 - 88 - 25$$
$$a = 27$$

Therefore angle a is 27°.

Angles formed within parallel lines

When two straight lines cross, it is found that the angles opposite each other are the same size. They are known as **vertically opposite angles**. By using the fact that angles at a point on a straight line add up to 180°, it can be shown why vertically opposite angles must always be equal in size.

$a + b = 180°$
$c + b = 180°$

Therefore, a is equal to c.

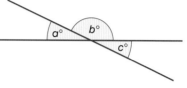

When a line intersects two parallel lines, as in the diagram below, it is found that certain angles are the same size.

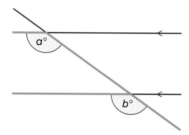

The angles a and b are equal and are known as **corresponding angles**. Corresponding angles can be found by looking for an 'F' formation in a diagram.

A line intersecting two parallel lines also produces another pair of equal angles, known as **alternate angles**. These can be shown to be equal by using the fact that both vertically opposite and corresponding angles are equal.

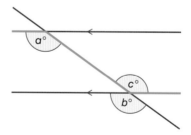

In the diagram above, $a = b$ (corresponding angles). But $b = c$ (vertically opposite). It can therefore be deduced that $a = c$.

Angles a and c are alternate angles. These can be found by looking for a 'Z' formation in a diagram.

Angle properties

Exercise 24.1 In each of the following questions, some of the angles are given. Deduce, giving your reasons, the size of the other labelled angles.

1.
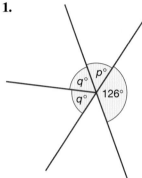

2.
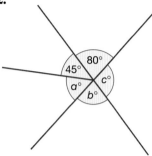

3.
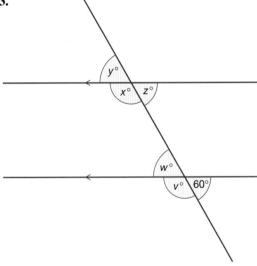

4.
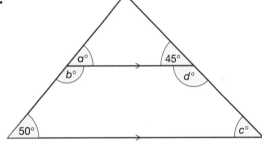

5.
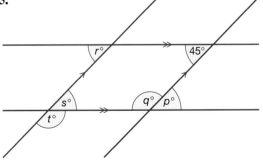

6.
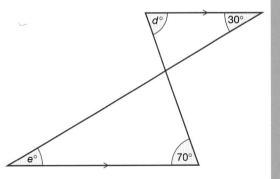

Geometry

7.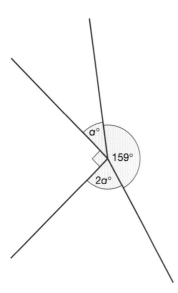

8.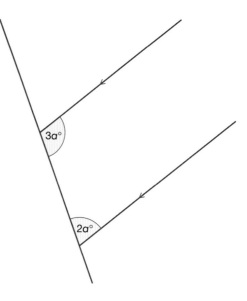

Angles in a triangle
The sum of the angles in a triangle is 180°.

Worked example   Calculate the size of the angle x in the triangle below:

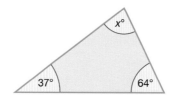

$37 + 64 + x = 180$
$x = 180 - 37 - 64$
Therefore angle x is 79°.

Exercise 24.2   **1.** For each of the triangles below, use the information given to calculate the size of angle x.

a) 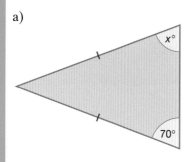   b)   c)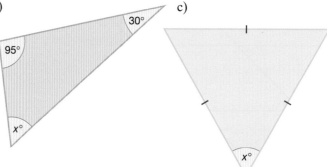

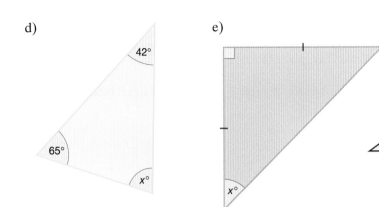

2. In each of the diagrams below, calculate the size of the labelled angles.

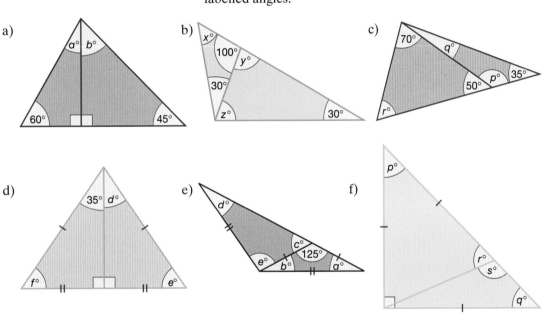

Angles in a quadrilateral

In the quadrilaterals below, a straight line is drawn from one of the corners (vertices) to the opposite corner. The result is to split the quadrilaterals into two triangles.

   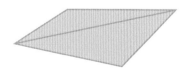

Geometry

The sum of the angles of a triangle is 180°. Therefore, as a quadrilateral can be drawn as two triangles, the sum of the four angles of any quadrilateral must be 360°.

Worked example Calculate the size of angle p in the quadrilateral below:

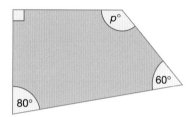

$90 + 80 + 60 + p = 360$
$p = 360 - 90 - 80 - 60$

Therefore angle p is 130°.

Exercise 24.3 In each of the diagrams below, calculate the size of the labelled angles.

1.

2.

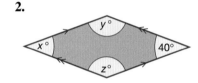

3.

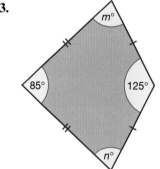

4.

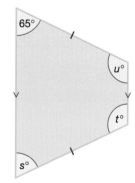

5.

6.

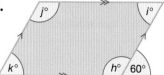

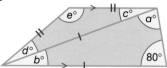

7. 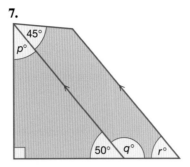 **8.**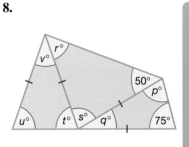

● Polygons

A **regular polygon** is distinctive in that all its sides are of equal length and all its angles are of equal size. Below are some examples of regular polygons.

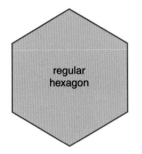  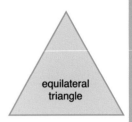

● The sum of the interior angles of a polygon

In the polygons below a straight line is drawn from each vertex to vertex A.

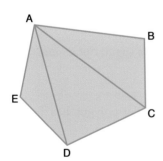 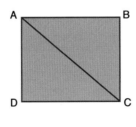 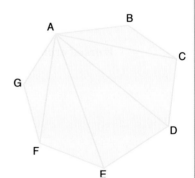

As can be seen, the number of triangles is always two less than the number of sides the polygon has, i.e. if there are n sides, there will be $(n-2)$ triangles.

Since the angles of a triangle add up to $180°$, the sum of the interior angles of a polygon is therefore $180(n-2)$ degrees.

Geometry

Worked example Find the sum of the interior angles of a regular pentagon and hence the size of each interior angle.

For a pentagon, $n = 5$.

Therefore the sum of the interior angles $= 180(5 - 2)°$
$= 180 \times 3°$
$= 540°$

For a regular pentagon the interior angles are of equal size.

Therefore each angle $= \dfrac{540°}{5} = 108°$.

The sum of the exterior angles of a polygon

The angles marked a, b, c, d, e and f in the diagram below represent the exterior angles of a regular hexagon.

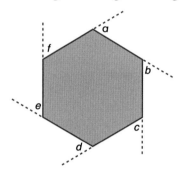

For any convex polygon the sum of the exterior angles is $360°$. If the polygon is regular and has n sides, then each exterior angle $= \dfrac{360°}{n}$.

Worked examples **a)** Find the size of an exterior angle of a regular nonagon.

$\dfrac{360°}{9} = 40°$

b) Calculate the number of sides a regular polygon has if each exterior angle is $15°$.

$n = \dfrac{360}{15}$
$= 24$

The polygon has 24 sides.

Exercise 24.4

1. Find the sum of the interior angles of the following polygons:
   a) a hexagon   b) a nonagon   c) a heptagon

2. Find the value of each interior angle of the following regular polygons:
   a) an octagon   b) a square
   c) a decagon   d) a dodecagon

3. Find the size of each exterior angle of the following regular polygons:
   a) a pentagon   b) a dodecagon   c) a heptagon

4. The exterior angles of regular polygons are given below. In each case calculate the number of sides the polygon has.
   a) 20°   b) 36°   c) 10°
   d) 45°   e) 18°   f) 3°

5. The interior angles of regular polygons are given below. In each case calculate the number of sides the polygon has.
   a) 108°   b) 150°   c) 162°
   d) 156°   e) 171°   f) 179°

6. Calculate the number of sides a regular polygon has if an interior angle is five times the size of an exterior angle.

The angle in a semi-circle

In the diagram below, if AB represents the diameter of the circle, then the angle at C is 90°.

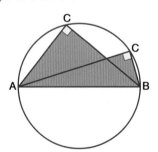

Geometry

Exercise 24.5 In each of the following diagrams, O marks the centre of the circle. Calculate the value of x in each case.

1.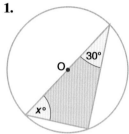

2.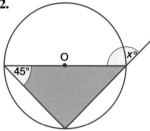

3.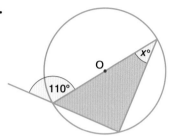

4.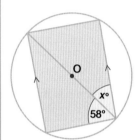

5.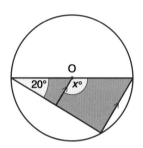

6.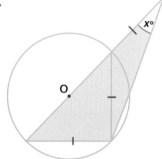

● **The angle between a tangent and a radius of a circle**

The angle between a tangent at a point and the radius to the same point on the circle is a right angle.

In the diagram below, triangles OAC and OBC are congruent as ∠OAC and ∠OBC are right angles, OA = OB because they are both radii and OC is common to both triangles.

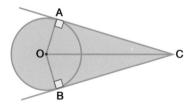

Exercise 24.6

In each of the following diagrams, O marks the centre of the circle. Calculate the value of x in each case.

1.

2.

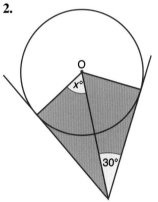

3.

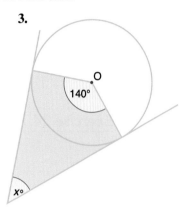

4.

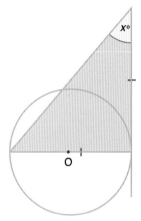

5.

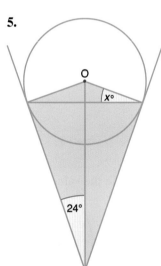

6.

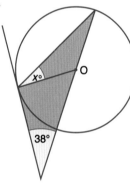

7.

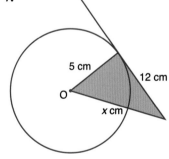

8.

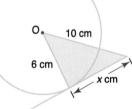

9.

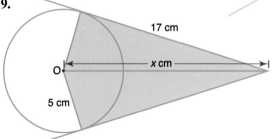

Geometry

Angle properties of irregular polygons

As explained earlier in this chapter, the sum of the interior angles of a polygon is given by $180(n - 2)°$, where n represents the number of sides of the polygon. The sum of the exterior angles of any polygon is 360°.

Both of these rules also apply to irregular polygons, i.e. those where the lengths of the sides and the sizes of the interior angles are not all equal.

Exercise 24.7

1. For the pentagon (right):
   a) calculate the value of x,
   b) calculate the size of each of the angles.

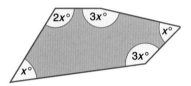

2. Find the size of each angle in the octagon (below).

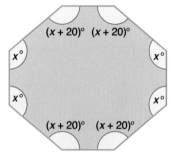

3. Calculate the value of x for the pentagon shown.

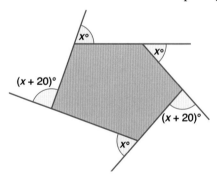

4. Calculate the size of each of the angles a, b, c, d and e in the hexagon (right).

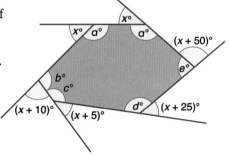

Angle properties

Angle at the centre of a circle

The angle subtended at the centre of a circle by an arc is twice the size of the angle on the circumference subtended by the same arc. Both diagrams below illustrate this theorem.

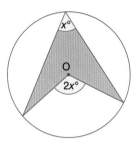
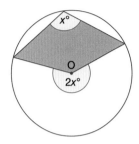

Exercise 24.8 In each of the following diagrams, O marks the centre of the circle. Calculate the size of the marked angles:

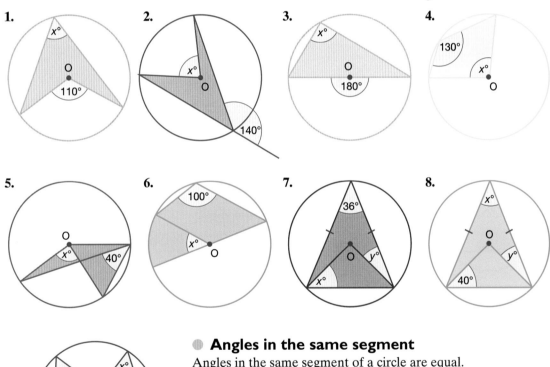

Angles in the same segment

Angles in the same segment of a circle are equal.

This can be explained simply by using the theorem that the angle subtended at the centre is twice the angle on the circumference. Looking at the diagram (left), if the angle at the centre is $2x°$, then each of the angles at the circumference must be equal to $x°$.

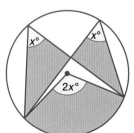

Geometry

Exercise 24.9 Calculate the marked angles in the following diagrams:

1.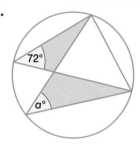
2.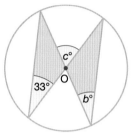
3.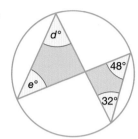
4.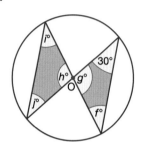
5.
6.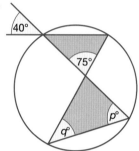

Angles in opposite segments

Points P, Q, R and S all lie on the circumference of the circle (below). They are called concyclic points. Joining the points P, Q, R and S produces a **cyclic quadrilateral**.

The opposite angles are **supplementary**, i.e. they add up to 180°. Since $p° + r° = 180°$ (supplementary angles) and $r° + t° = 180°$ (angles on a straight line) it follows that $p° = t°$.

Therefore the exterior angle of a cyclic quadrilateral is equal to the interior opposite angle.

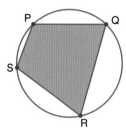
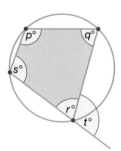

Exercise 24.10 Calculate the size of the marked angles in each of the following:

1.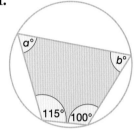

2.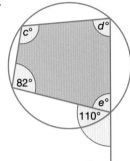

3.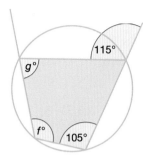

4.

5.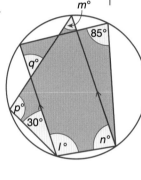

6.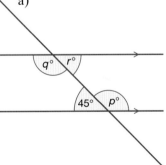

Student assessment 1

1. For the diagrams below, calculate the size of the labelled angles.

   a)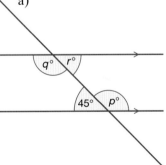

   b)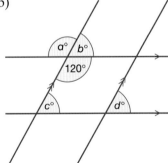

Geometry

2. For the diagrams below, calculate the size of the labelled angles.

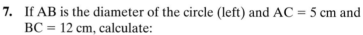

3. Find the size of each interior angle of a twenty-sided regular polygon.

4. What is the sum of the interior angles of a nonagon?

5. What is the sum of the exterior angles of a polygon?

6. What is the size of each exterior angle of a regular pentagon?

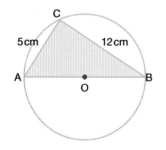

7. If AB is the diameter of the circle (left) and AC = 5 cm and BC = 12 cm, calculate:
  a) the size of angle ACB,
  b) the length of the radius of the circle.

In questions 8–11, O marks the centre of the circle. Calculate the size of the angle marked x in each case.

8. **9.** **10.** **11.**

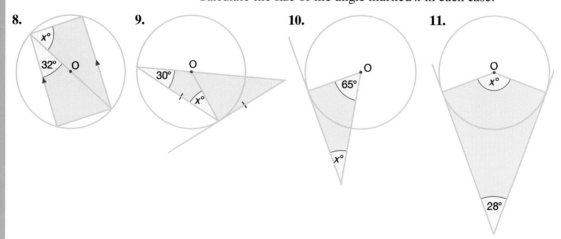

Student assessment 2

1. For the diagrams below, calculate the size of the labelled angles.

   a) 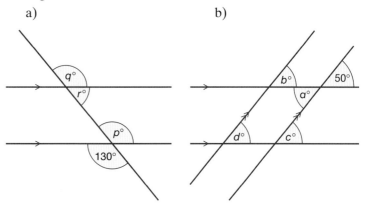 b)

2. For the diagrams below, calculate the size of the labelled angles.

   a) b)

   c) 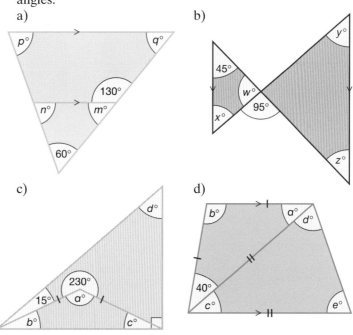 d)

3. Find the value of each interior angle of a regular polygon with 24 sides.

4. What is the sum of the interior angles of a regular dodecagon?

5. What is the size of an exterior angle of a regular dodecagon?

Geometry

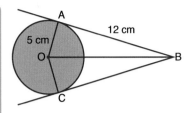

6. AB and BC are both tangents to the circle centre O (left). If OA = 5 cm and AB = 12 cm calculate:
   a) the size of angle OAB,
   b) the length OB.

7. If OA is a radius of the circle (below) and PB the tangent to the circle at A, calculate angle ABO.

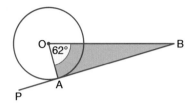

In question 8–11, O marks the centre of the circle. Calculate the size of the angle marked x in each case.

8. 9. 10. 11.

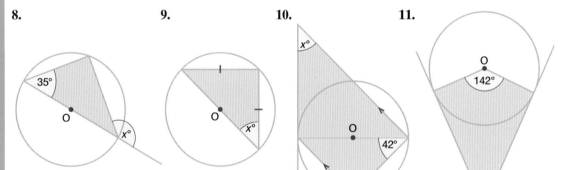

Student assessment 3

1. In the following diagrams, O marks the centre of the circle. Identify which angles are:
   i) supplementary angles,
   ii) right angles,
   iii) equal.

a) b) c) d)

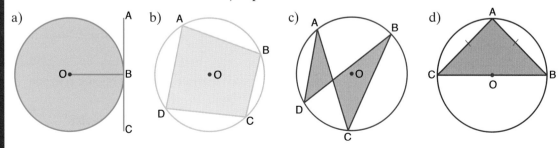

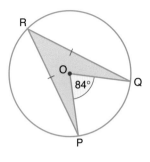

2. If ∠POQ = 84° and O marks the centre of the circle in the diagram (left), calculate the following:
   a) ∠PRQ
   b) ∠OQR

3. Calculate ∠DAB and ∠ABC in the diagram below.

4. If DC is a diameter, and O marks the centre of the circle, calculate angles BDC and DAB.

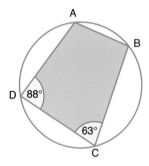
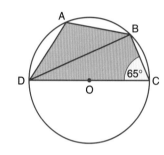

5. Calculate as many angles as possible in the diagram below. O marks the centre of the circle.

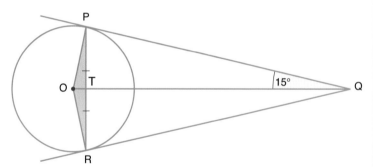

6. Calculate the values of c, d and e.

7. Calculate the values of f, g, h and i.

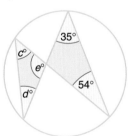
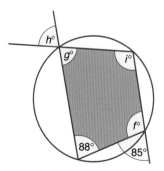

Geometry

Student assessment 4

1. In the following diagrams, O is the centre of the circle. Identify which angles are:
   i) supplementary angles,
   ii) right angles,
   iii) equal.

a)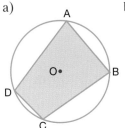
b)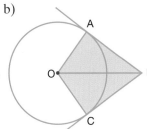
c)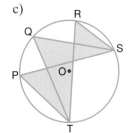
d)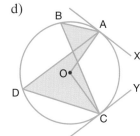

2. If ∠AOC is 72°, calculate ∠ABC.

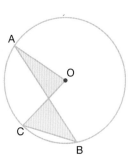

3. If ∠AOB = 130°, calculate ∠ABC, ∠OAB and ∠CAO.

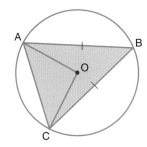

4. Show that ABCD is a cyclic quadrilateral.

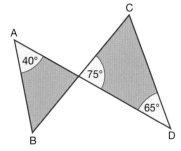

5. Calculate f and g.

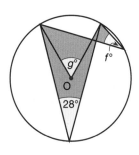

6. If $y = 22.5°$ calculate the value of x.

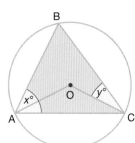

Loci

NB: All diagrams are not drawn to scale.

A **locus** (plural **loci**) refers to all the points which fit a particular description. These points can belong to either a region or a line, or both. The principal types of loci are explained below.

● The locus of the points which are at a given distance from a given point

In the diagram (left) it can be seen that the locus of all the points equidistant from a point A is the circumference of a circle centre A. This is due to the fact that all points on the circumference of a circle are equidistant from the centre.

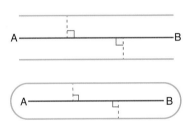

● The locus of the points which are at a given distance from a given straight line

In the diagram (left) it can be seen that the locus of the points equidistant from a straight line AB runs parallel to that straight line. It is important to note that the distance of the locus from the straight line is measured at right angles to the line. This diagram, however, excludes the ends of the line. If these two points are taken into consideration then the locus takes the form shown in the next diagram (left).

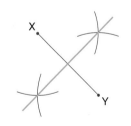

● The locus of the points which are equidistant from two given points

The locus of the points equidistant from points X and Y lies on the perpendicular bisector of the line XY.

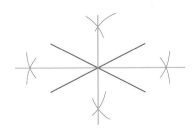

● The locus of the points which are equidistant from two given intersecting straight lines

The locus in this case lies on the bisectors of both pairs of opposite angles as shown left.

The application of the above cases will enable you to tackle problems involving loci at this level.

Geometry

Worked example The diagram shows a trapezoidal garden. Three of its sides are enclosed by a fence, and the fourth is adjacent to a house.

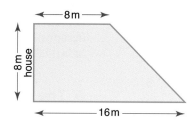

i) Grass is to be planted in the garden. However, it must be at least 2 m away from the house and at least 1 m away from the fence. Shade the region in which the grass can be planted.

The shaded region is therefore the locus of all the points which are both at least 2 m away from the house and at least 1 m away from the surrounding fence. Note that the boundary of the region also forms part of the locus of the points.

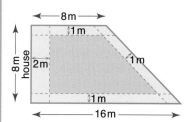

ii) Using the same garden as before, grass must now be planted according to the following conditions: it must be **more than** 2 m away from the house and **more than** 1 m away from the fence. Shade the region in which the grass can be planted.

The shape of the region is the same as in the first case; however, in this instance the boundary is not included in the locus of the points as the grass cannot be exactly 2 m away from the house or exactly 1 m away from the fence.

Note: If the boundary is included in the locus points, it is represented by a **solid** line. If it is not included then it is represented by a **dashed** (broken) line.

Exercise 25.1 Questions 1–4 are about a rectangular garden measuring 8 m by 6 m. For each question draw a scale diagram of the garden and identify the locus of the points which fit the criteria.

1. Draw the locus of all the points at least 1 m from the edge of the garden.

2. Draw the locus of all the points at least 2 m from each corner of the garden.

3. Draw the locus of all the points more than 3 m from the centre of the garden.

4. Draw the locus of all the points equidistant from the longer sides of the garden.

5. A port has two radar stations at P and Q which are 20 km apart. The radar at P is set to a range of 20 km, whilst the radar at Q is set to a range of 15 km.
    a) Draw a scale diagram to show the above information.
    b) Shade the region in which a ship must be sailing if it is only picked up by radar P. Label this region 'a'.
    c) Shade the region in which a ship must be sailing if it is only picked up by radar Q. Label this region 'b'.
    d) Identify the region in which a ship must be sailing if it is picked up by both radars. Label this region 'c'.

6. X and Y are two ship-to-shore radio receivers (left). They are 25 km apart.
    A ship sends out a distress signal. The signal is picked up by both X and Y. The radio receiver at X indicates that the ship is within a 30 km radius of X, whilst the radio receiver at Y indicates that the ship is within 20 km of Y. Draw a scale diagram and identify the region in which the ship must lie.

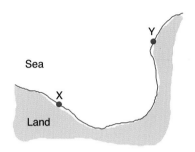

7. a) Mark three points L, M and N not in a straight line. By construction find the point which is equidistant from L, M and N.
    b) What would happen if L, M and N were on the same straight line?

8. Draw a line AB 8 cm long. What is the locus of a point C such that the angle ACB is always a right angle?

9. Draw a circle by drawing round a circular object (do not use a pair of compasses). By construction determine the position of the centre of the circle.

10. Three lionesses L_1, L_2 and L_3 have surrounded a gazelle. The three lionesses are equidistant from the gazelle. Draw a diagram with the lionesses in similar positions to those shown (left) and by construction determine the position (g) of the gazelle.

Exercise 25.2

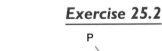

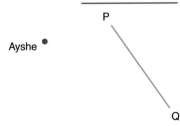

1. Three girls are playing hide and seek. Ayshe and Belinda are at the positions shown (left) and are trying to find Cristina. Cristina is on the opposite side of a wall PQ to her two friends.
    Assuming Ayshe and Belinda cannot see over the wall identify, by copying the diagram, the locus of points where Cristina could be if:
    a) Cristina can only be seen by Ayshe,
    b) Cristina can only be seen by Belinda,
    c) Cristina can not be seen by either of her two friends,
    d) Cristina can be seen by both of her friends.

Geometry

2. A security guard S is inside a building in the position shown. The building is inside a rectangular compound. If the building has three windows as shown, identify the locus of points in the compound which can be seen by the security guard.

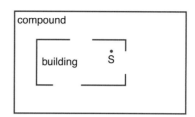

3. The circular cage shown (left) houses a snake. Inside the cage are three obstacles.
A rodent is placed inside the cage at R. From where it is lying, the snake can see the rodent.
Trace the diagram and identify the regions in which the snake could be lying.

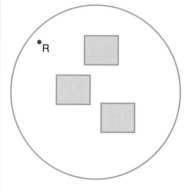

Exercise 25.3

1. A coin is rolled in a straight line on a flat surface as shown below.

Draw the locus of the centre of the coin O as the coin rolls along the surface.

2. The diameter of the disc (left) is the same as the width and height of each of the steps shown.
Copy the diagram and draw the locus of the centre of the disc as it rolls down the steps.

3. A stone is thrown vertically upwards. Draw the locus of its trajectory from the moment it leaves the person's hand to the moment it is caught again.

4. A stone is thrown at an angle of elevation of 45°. Sketch the locus of its trajectory.

5. X and Y are two fixed posts in the ground. The ends of a rope are tied to X and Y. A goat is attached to the rope by a ring on its collar which enables it to move freely along the rope's length.
Copy the diagram (left) and sketch the locus of points in which the goat is able to graze.

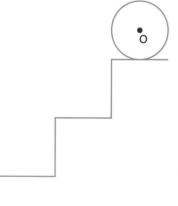

Loci

Student assessment 1

1. Pedro and Sara are on opposite sides of a building as shown (left). Their friend Raul is standing in a place such that he cannot be seen by either Pedro or Sara. Copy the above diagram and identify the locus of points at which Raul could be standing.

2. A rectangular garden measures 10 m by 6 m. A tree stands in the centre of the garden. Grass is to be planted according to the following conditions:
   - it must be at least 1 m from the edge of the garden,
   - it must be more than 2 m away from the centre of the tree.

   a) Make a scale drawing of the garden.
   b) Draw the locus of points in which the grass can be planted.

3. A rectangular rose bed in a park measures 8 m by 5 m as shown (left).
   The park keeper puts a low fence around the rose bed. The fence is at a constant distance of 2 m from the rose bed.
   a) Make a scale drawing of the rose bed.
   b) Draw the position of the fence.

4. A and B (left) are two radio beacons 80 km apart, either side of a shipping channel. A ship sails in such a way that it is always equidistant from A and B.
   Showing your method of construction clearly, draw the path of the ship.

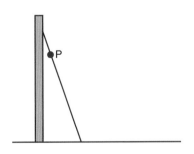

5. A ladder 10 m long is propped up against a wall as shown (left). A point P on the ladder is 2 m from the top.
   Make a scale drawing to show the locus of point P if the ladder were to slide down the wall. Note: several positions of the ladder will need to be shown.

6. The equilateral triangle PQR is rolled along the line shown. At first, corner Q acts as the pivot point until P reaches the line, then P acts as the pivot point until R reaches the line, and so on.

Showing your method clearly, draw the locus of point P as the triangle makes one full rotation, assuming there is no slipping.

Geometry

Student assessment 2

1. Jose, Katrina and Luis are standing at different points around a building as shown (left).
   Trace the diagram and show whether any of the three friends can see each other or not.

2. A rectangular courtyard measures 20 m by 12 m. A horse is tethered in the centre with a rope 7 m long. Another horse is tethered, by a rope 5 m long, to a rail which runs along the whole of the left-hand (shorter) side of the courtyard. This rope is able to run freely along the length of the rail. Draw a scale diagram of the courtyard and draw the locus of points which can be reached by both horses.

3. The view in the diagram (left) is of two walls which form part of an obstacle course. A girl decides to ride her bicycle in between the two walls in such a way that she is always equidistant from them.
   Copy the diagram and, showing your construction clearly, draw the locus of her path.

4. A ball is rolling along the line shown in the diagram (below). Copy the diagram and draw the locus of the centre, O, of the ball as it rolls.

5. A square ABCD is 'rolled' along the flat surface shown below. Initially corner C acts as a pivot point until B touches the surface, then B acts as a pivot point until A touches the surface, and so on.

Assuming there is no slipping, draw the locus of point A as the square makes one complete rotation. Show your method clearly.

Topic 3
Mathematical investigations and ICT

● Fountain borders

The Alhambra Palace in Granada, Spain has many fountains which pour water into pools. Many of the pools are surrounded by beautiful ceramic tiles. This investigation looks at the number of square tiles needed to surround a particular shape of pool.

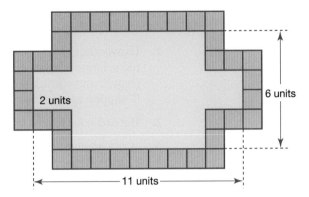

The diagram above shows a rectangular pool 11×6 units, in which a square of dimension 2×2 units is taken from each corner.

The total number of unit square tiles needed to surround the pool is 38.

The shape of the pools can be generalised as shown below:

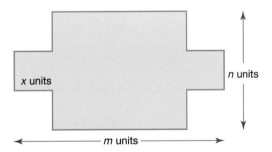

1. Investigate the number of unit square tiles needed for different sized pools. Record your results in an ordered table.

2. From your results write an algebraic rule in terms of m, n and x (if necessary) for the number of tiles T needed to surround a pool.

3. Justify, in words and using diagrams, why your rule works.

Tiled walls

Many cultures have used tiles to decorate buildings. Putting tiles on a wall takes skill. These days, to make sure that each tile is in the correct position, 'spacers' are used between the tiles.

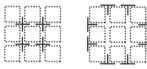

You can see from the diagrams that there are + shaped spacers (used where four tiles meet) and T shaped spacers (used at the edges of a pattern).

1. Draw other sized squares and rectangles, and investigate the relationship between the dimensions of the shape (length and width) and the number of + shaped and T shaped spacers.

2. Record your results in an ordered table.

3. Write an algebraic rule for the number of + shaped spacers c in a rectangle l tiles long by w tiles wide.

4. Write an algebraic rule for the number of T shaped spacers t in a rectangle l tiles long by w tiles wide.

ICT activity 1

In this activity you will be using a dynamic geometry package such as Cabri or Geogebra to demonstrate that for the triangle (below):

$$\frac{AB}{ED} = \frac{AC}{EC} = \frac{BC}{DC}$$

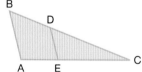

1. a) Using the geometry package construct the triangle ABC.
   b) Construct the line segment ED such that it is parallel to AB. (You will need to construct a line parallel to AB first and then attach the line segment ED to it.)
   c) Using a 'measurement' tool, measure each of the lengths AB, AC, BC, ED, EC and DC.
   d) Using a 'calculator' tool, calculate the ratios $\frac{AB}{ED}, \frac{AC}{EC}, \frac{BC}{DC}$.

2. Comment on your answers to question 1(d).

3. a) Grab vertex B and move it to a new position. What happens to the ratios you calculated in question 1(d)?
   b) Grab the vertices A and C in turn and move them to new positions. What happens to the ratios? Explain why this happens.

4. Grab point D and move it to a new position along the side BC. Explain, giving reasons, what happens to the ratios.

ICT activity 2

Using a geometry package, such as Cabri or Geogebra, demonstrate the following angle properties of a circle:

1. The angle subtended at the centre of a circle by an arc is twice the size of the angle on the circumference subtended by the same arc.

    The diagram below demonstrates the construction that needs to be formed:

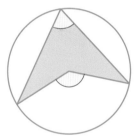

2. The angles in the same segment of a circle are equal.

3. The exterior angle of a cyclic quadrilateral is equal to the interior opposite angle.

Topic 4: Mensuration

Syllabus

E4.1
Use current units of mass, length, area, volume and capacity in practical situations and express quantities in terms of larger or smaller units.

E4.2
Carry out calculations involving the perimeter and area of a rectangle, triangle, parallelogram and trapezium and compound shapes derived from these.

E4.3
Carry out calculations involving the circumference and area of a circle.
Solve problems involving the arc length and sector area as fractions of the circumference and area of a circle.

E4.4
Carry out calculations involving the volume of a cuboid, prism and cylinder and the surface area of a cuboid and a cylinder.
Carry out calculations involving the surface area and volume of a sphere, pyramid and cone.

E4.5
Carry out calculations involving the areas and volumes of compound shapes.

Contents

Chapter 26 Measures (E4.1)
Chapter 27 Perimeter, area and volume (E4.2, E4.3, E4.4, E4.5)

The Egyptians

The Egyptians must have been very talented architects and surveyors to have planned the many large buildings and monuments built in that country thousands of years ago.

Evidence of the use of mathematics in Egypt in the Old Kingdom (about 2500BCE) is found on a wall near Meidum (south of Cairo) which gives guidelines for the slope of the stepped pyramid built there. The lines in the diagram are spaced at a distance of one cubit. A cubit is the distance from the tip of the finger to the elbow (about 50 cm). They show the use of that unit of measurement.

The earliest true mathematical documents date from about 1800BCE. The Moscow Mathematical Papyrus, the Egyptian Mathematical Leather Roll, the Kahun Papyri and the Berlin Papyrus all date to this period.

The Rhind Mathematical Papyrus, which was written in about 1650BCE, is said to be based on an older mathematical text. The Moscow Mathematical Papyrus and Rhind Mathematical Papyrus are so-called mathematical problem texts. They consist of a collection of mainly mensuration problems with solutions. These could have been written by a teacher for students to solve similar problems to the ones you will work on in this topic.

During the New Kingdom (about 1500−100BCE) papyri record land measurements. In the worker's village of Deir el-Medina several records have been found that record volumes of dirt removed while digging the underground tombs.

The Rhind Mathematical Papyrus

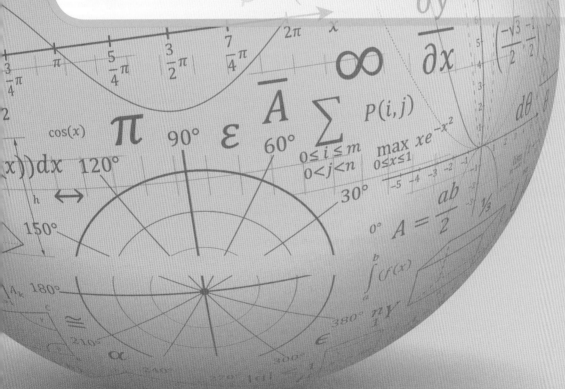

26 Measures

● **Metric units**

The metric system uses a variety of units for length, mass and capacity.

- The common units of length are: kilometre (km), metre (m), centimetre (cm) and millimetre (mm).
- The common units of mass are: tonne (t), kilogram (kg), gram (g) and milligram (mg).
- The common units of capacity are: litre (L or l) and millilitre (ml).

Note: 'centi' comes from the Latin *centum* meaning hundred (a centimetre is one hundredth of a metre); 'milli' comes from the Latin *mille* meaning thousand (a millimetre is one thousandth of a metre); 'kilo' comes from the Greek *khilloi* meaning thousand (a kilometre is one thousand metres).

It may be useful to have some practical experience of estimating lengths, volumes and capacities before starting the following exercises.

Exercise 26.1 Copy and complete the sentences below:

1. a) There are ... centimetres in one metre.
    b) There are ... millimetres in one metre.
    c) One metre is one ... of a kilometre.
    d) One milligram is one ... of a gram.
    e) One thousandth of a litre is one ....

2. Which of the units below would be used to measure the following?

    mm, cm, m, km, mg, g, kg, t, ml, litres

    a) your height
    b) the length of your finger
    c) the mass of a shoe
    d) the amount of liquid in a cup
    e) the height of a van
    f) the mass of a ship
    g) the capacity of a swimming pool
    h) the length of a highway
    i) the mass of an elephant
    j) the capacity of the petrol tank of a car

Converting from one unit to another

Length

1 km = 1000 m

Therefore 1 m = $\frac{1}{1000}$ km

1 m = 1000 mm

Therefore 1 mm = $\frac{1}{1000}$ m

1 m = 100 cm

Therefore 1 cm = $\frac{1}{100}$ m

1 cm = 10 mm

Therefore 1 mm = $\frac{1}{10}$ cm

Worked examples

a) Change 5.8 km into m.
Since 1 km = 1000 m,
5.8 km is 5.8 × 1000 m

5.8 km = 5800 m

b) Change 4700 mm to m.
Since 1 m is 1000 mm,
4700 mm is 4700 ÷ 1000 m

4700 mm = 4.7 m

c) Convert 2.3 km into cm.
2.3 km is 2.3 × 1000 m = 2300 m
2300 m is 2300 × 100 cm

2.3 km = 230 000 cm

Exercise 26.2

1. Put in the missing unit to make the following statements correct:
   a) 300 ... = 30 cm
   b) 6000 mm = 6 ...
   c) 3.2 m = 3200 ...
   d) 4.2 ... = 4200 mm
   e) 2.5 km = 2500 ...

2. Convert the following to millimetres:
   a) 8.5 cm
   b) 23 cm
   c) 0.83 m
   d) 0.05 m
   e) 0.004 m

3. Convert the following to metres:
   a) 560 cm
   b) 6.4 km
   c) 96 cm
   d) 0.004 km
   e) 12 mm

4. Convert the following to kilometres:
   a) 1150 m
   b) 250 000 m
   c) 500 m
   d) 70 m
   e) 8 m

Mensuration

Mass

1 tonne is 1000 kg

Therefore 1 kg = $\frac{1}{1000}$ tonne

1 kilogram is 1000 g

Therefore 1 g = $\frac{1}{1000}$ kg

1 g is 1000 mg

Therefore 1 mg = $\frac{1}{1000}$ g

Worked examples **a)** Convert 8300 kg to tonnes.
Since 1000 kg = 1 tonne, 8300 kg is 8300 ÷ 1000 tonnes

8300 kg = 8.3 tonnes

b) Convert 2.5 g to mg.

Since 1 g is 1000 mg, 2.5 g is 2.5 × 1000 mg

2.5 g = 2500 mg

Exercise 26.3 **1.** Convert the following:
  a) 3.8 g to mg   b) 28 500 kg to tonnes
  c) 4.28 tonnes to kg   d) 320 mg to g
  e) 0.5 tonnes to kg

Capacity

1 litre is 1000 millilitres

Therefore 1 ml = $\frac{1}{1000}$ litre

Exercise 26.4 **1.** Calculate the following and give the totals in ml:
  a) 3 litres + 1500 ml   b) 0.88 litre + 650 ml
  c) 0.75 litre + 6300 ml   d) 450 ml + 0.55 litre

2. Calculate the following and give the total in litres:
  a) 0.75 litre + 450 ml   b) 850 ml + 490 ml
  c) 0.6 litre + 0.8 litre   d) 80 ml + 620 ml + 0.7 litre

- ## Area and volume conversions

Converting between units for area and volume is not as straightforward as converting between units for length.

The diagram below shows a square of side length 1 m.

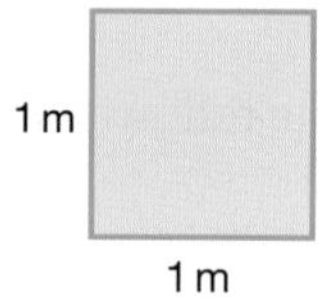

1 m

1 m

Area of the square = 1 m²
However, if the lengths of the sides are written in cm, each of the sides are 100 cm.

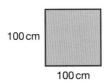

Area of the square = 100 × 100 = 10 000 cm²
Therefore an area of 1 m² = 10 000 cm².

Similarly a square of side length 1 cm is the same as a square of side length 10 mm. Therefore an area of 1 cm² is equivalent to an area of 100 mm².

The diagram below shows a cube of side length 1 m.

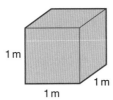

Volume of the cube = 1 m³
Once again, if the lengths of the sides are written in cm, each of the sides are 100 cm.

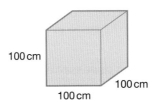

Volume of the cube = 100 × 100 × 100 = 1 000 000 cm³
Therefore a volume of 1 m³ = 1 000 000 cm³.

Similarly a cube of side length 1 cm is the same as a cube of side length 10 mm.

Therefore a volume of 1 cm³ is equivalent to a volume of 1000 mm³.

Exercise 26.5

1. Convert the following areas:
   a) 10 m² to cm²
   b) 2 m² to mm²
   c) 5 km² to m²
   d) 3.2 km² to m²
   e) 8.3 cm² to mm²

2. Convert the following areas:
   a) 500 cm² to m²
   b) 15 000 mm² to cm²
   c) 1000 m² to km²
   d) 40 000 mm² to m²
   e) 2 500 000 cm² to km²

Mensuration

3. Convert the following volumes:
   a) 2.5 m³ to cm³
   b) 3.4 cm³ to mm³
   c) 2 km³ to m³
   d) 0.2 m³ to cm³
   e) 0.03 m³ to mm³

4. Convert the following volumes:
   a) 150 000 cm³ to m³
   b) 24 000 mm³ to cm³
   c) 850 000 m³ to km³
   d) 300 mm³ to cm³
   e) 15 cm³ to m³

Student assessment 1

1. Convert the following lengths into the units indicated:
   a) 2.6 cm to mm
   b) 0.88 m to cm
   c) 6800 m to km
   d) 0.875 km to m

2. Convert the following masses into the units indicated:
   a) 4.2 g to mg
   b) 3940 g to kg
   c) 4.1 kg to g
   d) 0.72 tonnes to kg

3. Convert the following liquid measures into the units indicated:
   a) 1800 ml to litres
   b) 3.2 litres to ml
   c) 0.083 litre to ml
   d) 250 000 ml to litres

4. Convert the following areas:
   a) 56 cm² to mm²
   b) 2.05 m³ to cm³

5. Convert the following volumes:
   a) 8670 cm³ to m³
   b) 444 000 cm³ to m³

Student assessment 2

1. Convert the following lengths into the units indicated:
   a) 3100 mm to cm
   b) 6.4 km to m
   c) 0.4 cm to mm
   d) 460 mm to cm

2. Convert the following masses into the units indicated:
   a) 3.6 mg to g
   b) 550 mg to g
   c) 6500 g to kg
   d) 1510 kg to tonnes

3. Convert the following measures of capacity to the units indicated:
   a) 3400 ml to litres
   b) 6.7 litres to ml
   c) 0.73 litre to ml
   d) 300 000 ml to litres.

4. Convert the following areas:
   a) 0.03 m² to mm²
   b) 0.005 km² to m²

5. Convert the following volumes:
   a) 100 400 cm³ to m³
   b) 5005 m³ to km³

Perimeter, area and volume

NB: All diagrams are not drawn to scale.

● The perimeter and area of a rectangle

The **perimeter** of a shape is the distance around the outside of the shape. Perimeter can be measured in mm, cm, m, km, etc.

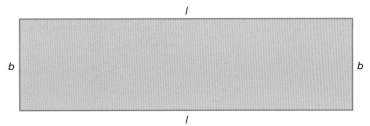

The perimeter of the rectangle above of length l and breadth b is therefore:

Perimeter = $l + b + l + b$

This can be rearranged to give:

Perimeter = $2l + 2b$

This in turn can be factorised to give:

Perimeter = $2(l + b)$

The **area** of a shape is the amount of surface that it covers. Area is measured in mm², cm², m², km², etc.

The area A of the rectangle above is given by the formula:

$A = lb$

Worked example Calculate the breadth of a rectangle of area 200 cm² and length 25 cm.

$A = lb$
$200 = 25b$
$b = 8$

So the breadth is 8 cm.

● The area of a triangle

Rectangle ABCD has a triangle CDE drawn inside it.

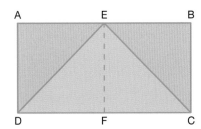

Mensuration

Point E is said to be a **vertex** of the triangle.
EF is the **height** or **altitude** of the triangle.
CD is the **length** of the rectangle, but is called the **base** of the triangle.

It can be seen from the diagram that triangle DEF is half the area of the rectangle AEFD.
Also triangle CFE is half the area of rectangle EBCF.
It follows that **triangle CDE is half the area of rectangle ABCD**.

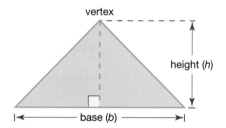

Area of a triangle $A = \frac{1}{2}bh$, where b is the base and h is the height.

Note: it does not matter which side is called the base, but the height must be measured at right angles from the base to the opposite vertex.

Exercise 27.1

1. Calculate the areas of the triangles below:

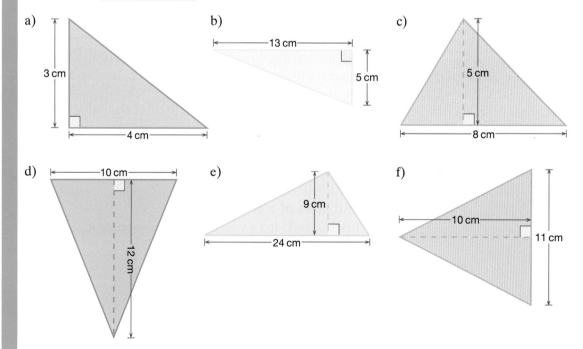

2. Calculate the areas of the shapes below:

a)

b)

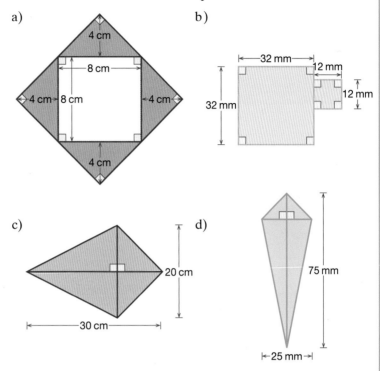

c)

d)

● The area of a parallelogram and a trapezium

A **parallelogram** can be rearranged to form a rectangle as shown below:

Therefore: area of parallelogram
= base length × perpendicular height.

Mensuration

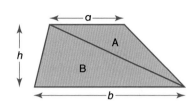

A **trapezium** can be visualised as being split into two triangles as shown on the left:

Area of triangle $A = \frac{1}{2} \times a \times h$

Area of triangle $B = \frac{1}{2} \times b \times h$

Area of the trapezium
$$= \text{area of triangle } A + \text{area of triangle } B$$
$$= \tfrac{1}{2}ah + \tfrac{1}{2}bh$$
$$= \tfrac{1}{2}h(a+b)$$

Worked examples

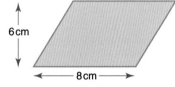

a) Calculate the area of the parallelogram (left):

Area = base length × perpendicular height
$$= 8 \times 6$$
$$= 48 \text{ cm}^2$$

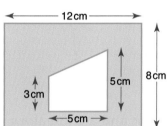

b) Calculate the shaded area in the shape (left):

Area of rectangle $= 12 \times 8$
$$= 96 \text{ cm}^2$$

Area of trapezium $= \tfrac{1}{2} \times 5(3+5)$
$$= 2.5 \times 8$$
$$= 20 \text{ cm}^2$$

Shaded area $= 96 - 20$
$$= 76 \text{ cm}^2$$

Exercise 27.2 Find the area of each of the following shapes:

1.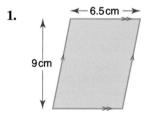

2.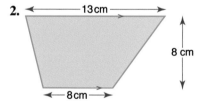

3. 4.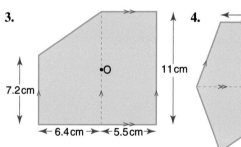

Exercise 27.3

1. Calculate *a*.

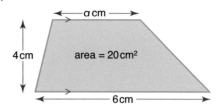

2. If the areas of this trapezium and parallelogram are equal, calculate *x*.

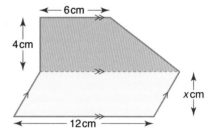

3. The end view of a house is as shown in the diagram (below). If the door has a width and height of 0.75 m and 2 m respectively. Calculate the area of brickwork.

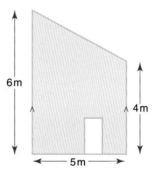

4. A garden in the shape of a trapezium is split into three parts: flower beds in the shape of a triangle and a parallelogram; and a section of grass in the shape of a trapezium, as shown below. The area of the grass is two and a half times the total area of flower beds. Calculate:
 a) the area of each flower bed,
 b) the area of grass,
 c) the value of *x*.

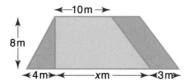

Mensuration

The circumference and area of a circle

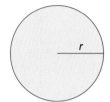 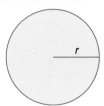

The circumference is $2\pi r$.   The area is πr^2.

$C = 2\pi r$ $A = \pi r^2$

Worked examples

a) Calculate the circumference of this circle, giving your answer to 3 s.f.

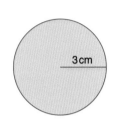

$C = 2\pi r$
$= 2\pi \times 3 = 18.8$

The circumference is 18.8 cm.

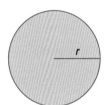

b) If the circumference of this circle is 12 cm, calculate the radius, giving your answer to 3 s.f.

$C = 2\pi r$

$r = \dfrac{C}{2\pi}$

$r = \dfrac{12}{2\pi} = 1.91$

The radius is 1.91 cm.

c) Calculate the area of this circle, giving your answer to 3 s.f.

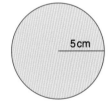

$A = \pi r^2$
$= \pi \times 5^2 = 78.5$

The area is 78.5 cm².

d) The area of a circle is 34 cm², calculate the radius, giving your answer to 3 s.f.

$A = \pi r^2$

$r = \sqrt{\dfrac{A}{\pi}}$

$r = \sqrt{\dfrac{34}{\pi}} = 3.29$

The radius is 3.29 cm.

Perimeter, area and volume

Exercise 27.4

1. Calculate the circumference of each circle, giving your answer to 3 s.f.

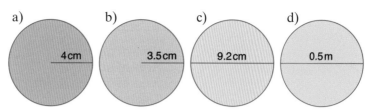

a) 4 cm   b) 3.5 cm   c) 9.2 cm   d) 0.5 m

2. Calculate the area of each of the circles in question 1. Give your answers to 3 s.f.

3. Calculate the radius of a circle when the circumference is:
   a) 15 cm
   b) π cm
   c) 4 m
   d) 8 mm

4. Calculate the diameter of a circle when the area is:
   a) 16 cm²
   b) 9π cm²
   c) 8.2 m²
   d) 14.6 mm²

Exercise 27.5

1. The wheel of a car has an outer radius of 25 cm. Calculate:
   a) how far the car has travelled after one complete turn of the wheel,
   b) how many times the wheel turns for a journey of 1 km.

2. If the wheel of a bicycle has a diameter of 60 cm, calculate how far a cyclist will have travelled after the wheel has rotated 100 times.

3. A circular ring has a cross-section as shown (left). If the outer radius is 22 mm and the inner radius 20 mm, calculate the cross-sectional area of the ring.

4.

   3 cm

   Four circles are drawn in a line and enclosed by a rectangle as shown. If the radius of each circle is 3 cm, calculate:
   a) the area of the rectangle,
   b) the area of each circle,
   c) the unshaded area within the rectangle.

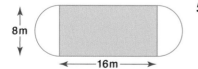

8 m, 16 m

5. A garden is made up of a rectangular patch of grass and two semi-circular vegetable patches. If the dimensions of the rectangular patch are 16 m (length) and 8 m (width) respectively, calculate:
   a) the perimeter of the garden,
   b) the total area of the garden.

Mensuration

The surface area of a cuboid and a cylinder

To calculate the surface area of a **cuboid** start by looking at its individual faces. These are either squares or rectangles. The surface area of a cuboid is the sum of the areas of its faces.

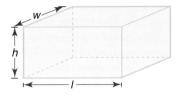

Area of top = wl
Area of front = lh
Area of one side = wh
Total surface area
$= 2wl + 2lh + 2wh$
$= 2(wl + lh + wh)$

Area of bottom = wl
Area of back = lh
Area of other side = wh

For the surface area of a **cylinder** it is best to visualise the net of the solid: it is made up of one rectangular piece and two circular pieces.

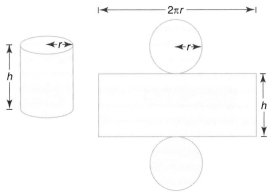

Area of circular pieces = $2 \times \pi r^2$
Area of rectangular piece = $2\pi r \times h$
Total surface area = $2\pi r^2 + 2\pi rh$
$= 2\pi r(r + h)$

Worked examples

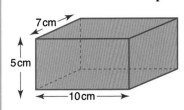

a) Calculate the surface area of the cuboid shown (left).

Total area of top and bottom = $2 \times 7 \times 10 = 140$ cm²
Total area of front and back = $2 \times 5 \times 10 = 100$ cm²
Total area of both sides = $2 \times 5 \times 7 = 70$ cm²

Total surface area = 310 cm²

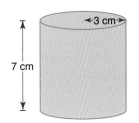

b) If the height of a cylinder is 7 cm and the radius of its circular top is 3 cm, calculate its surface area.

Total surface area = $2\pi r(r + h)$
$= 2\pi \times 3 \times (3 + 7)$
$= 6\pi \times 10$
$= 60\pi$
$= 188$ cm² (3 s.f.)

The total surface area is 188 cm².

Perimeter, area and volume

Exercise 27.6

1. Calculate the surface area of each of the following cuboids:
   a) $l = 12$ cm, $w = 10$ cm, $h = 5$ cm
   b) $l = 4$ cm, $w = 6$ cm, $h = 8$ cm
   c) $l = 4.2$ cm, $w = 7.1$ cm, $h = 3.9$ cm
   d) $l = 5.2$ cm, $w = 2.1$ cm, $h = 0.8$ cm

2. Calculate the height of each of the following cuboids:
   a) $l = 5$ cm, $w = 6$ cm, surface area = 104 cm²
   b) $l = 2$ cm, $w = 8$ cm, surface area = 112 cm²
   c) $l = 3.5$ cm, $w = 4$ cm, surface area = 118 cm²
   d) $l = 4.2$ cm, $w = 10$ cm, surface area = 226 cm²

3. Calculate the surface area of each of the following cylinders:
   a) $r = 2$ cm, $h = 6$ cm
   b) $r = 4$ cm, $h = 7$ cm
   c) $r = 3.5$ cm, $h = 9.2$ cm
   d) $r = 0.8$ cm, $h = 4.3$ cm

4. Calculate the height of each of the following cylinders. Give your answers to 1 d.p.
   a) $r = 2.0$ cm, surface area = 40 cm²
   b) $r = 3.5$ cm, surface area = 88 cm²
   c) $r = 5.5$ cm, surface area = 250 cm²
   d) $r = 3.0$ cm, surface area = 189 cm²

Exercise 27.7

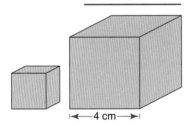

1. Two cubes (left) are placed next to each other. The length of each of the edges of the larger cube is 4 cm. If the ratio of their surface areas is 1 : 4, calculate:
   a) the surface area of the small cube,
   b) the length of an edge of the small cube.

2. A cube and a cylinder have the same surface area. If the cube has an edge length of 6 cm and the cylinder a radius of 2 cm calculate:
   a) the surface area of the cube,
   b) the height of the cylinder.

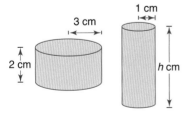

3. Two cylinders (left) have the same surface area. The shorter of the two has a radius of 3 cm and a height of 2 cm, and the taller cylinder has a radius of 1 cm. Calculate:
   a) the surface area of one of the cylinders,
   b) the height of the taller cylinder.

4. Two cuboids have the same surface area. The dimensions of one of them are: length = 3 cm, width = 4 cm and height = 2 cm.
   Calculate the height of the other cuboid if its length is 1 cm and its width is 4 cm.

Mensuration

The volume of a prism

A prism is any three-dimensional object which has a constant cross-sectional area.

Below are a few examples of some of the more common types of prism.

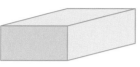
Rectangular prism (cuboid)

Circular prism (cylinder)

Triangular prism

When each of the shapes is cut parallel to the shaded face, the cross-section is constant and the shape is therefore classified as a prism.

Volume of a prism = area of cross-section × length

Worked examples

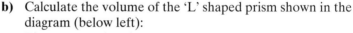

a) Calculate the volume of the cylinder in the diagram (left):

Volume = cross-sectional area × length
$= \pi \times 4^2 \times 10$
Volume = 503 cm³ (3 s.f.)

b) Calculate the volume of the 'L' shaped prism shown in the diagram (below left):

The cross-sectional area can be split into two rectangles:

Area of rectangle A = 5 × 2
= 10 cm²
Area of rectangle B = 5 × 1
= 5 cm²
Total cross-sectional area = (10 cm² + 5 cm²) = 15 cm²
Volume of prism = 15 × 5
= 75 cm³

Exercise 27.8

1. Calculate the volume of each of the following cuboids, where w, l and h represent the width, length and height respectively.
   a) $w = 2$ cm, $l = 3$ cm, $h = 4$ cm
   b) $w = 6$ cm, $l = 1$ cm, $h = 3$ cm
   c) $w = 6$ cm, $l = 23$ mm, $h = 2$ cm
   d) $w = 42$ mm, $l = 3$ cm, $h = 0.007$ m

2. Calculate the volume of each of the following cylinders, where r represents the radius of the circular face and h the height of the cylinder.
   a) $r = 4$ cm, $h = 9$ cm
   b) $r = 3.5$ cm, $h = 7.2$ cm
   c) $r = 25$ mm, $h = 10$ cm
   d) $r = 0.3$ cm, $h = 17$ mm

Perimeter, area and volume

3. Calculate the volume of each of the following triangular prisms, where b represents the base length of the triangular face, h its perpendicular height and l the length of the prism.
   a) $b = 6$ cm, $h = 3$ cm, $l = 12$ cm
   b) $b = 4$ cm, $h = 7$ cm, $l = 10$ cm
   c) $b = 5$ cm, $h = 24$ mm, $l = 7$ cm
   d) $b = 62$ mm, $h = 2$ cm, $l = 0.01$ m

4. Calculate the volume of each of the following prisms. All dimensions are given in centimetres.

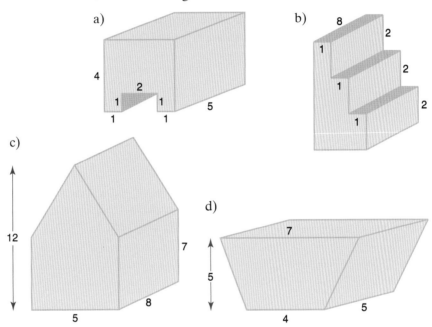

Exercise 27.9

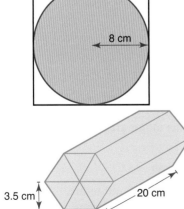

1. The diagram shows a plan view of a cylinder inside a box the shape of a cube. If the radius of the cylinder is 8 cm, calculate:
   a) the height of the cube,
   b) the volume of the cube,
   c) the volume of the cylinder,
   d) the percentage volume of the cube not occupied by the cylinder.

2. A chocolate bar is made in the shape of a triangular prism. The triangular face of the prism is equilateral and has an edge length of 4 cm and a perpendicular height of 3.5 cm. The manufacturer also sells these in special packs of six bars arranged as a hexagonal prism.
   If the prisms are 20 cm long, calculate:
   a) the cross-sectional area of the pack,
   b) the volume of the pack.

Mensuration

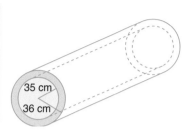

3. A cuboid and a cylinder have the same volume. The radius and height of the cylinder are 2.5 cm and 8 cm respectively. If the length and width of the cuboid are each 5 cm, calculate its height to 1 d.p.

4. A section of steel pipe is shown in the diagram. The inner radius is 35 cm and the outer radius 36 cm. Calculate the volume of steel used in making the pipe if it has a length of 130 m.

Arc length

An **arc** is part of the circumference of a circle between two radii.

Its length is proportional to the size of the angle ϕ between the two radii. The length of the arc as a fraction of the circumference of the whole circle is therefore equal to the fraction that ϕ is of 360°.

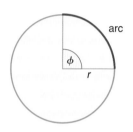

$$\text{Arc length} = \frac{\phi}{360} \times 2\pi r$$

Worked examples

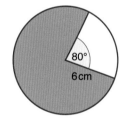

a) Find the length of the minor arc in the circle (right). Give your answer to 3 s.f.

$$\text{Arc length} = \frac{80}{360} \times 2 \times \pi \times 6$$

$$= 8.38 \text{ cm}$$

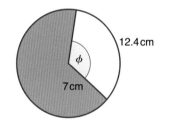

b) In the circle (left), the length of the minor arc is 12.4 cm and the radius is 7 cm.

   i) Calculate the angle ϕ.

$$\text{Arc length} = \frac{\phi}{360} \times 2\pi r$$

$$12.4 = \frac{\phi}{360} \times 2 \times \pi \times 7$$

$$\frac{12.4 \times 360}{2 \times \pi \times 7} = \phi$$

$$\phi = 101.5° \text{ (1 d.p.)}$$

   ii) Calculate the length of the major arc.

$$C = 2\pi r$$
$$= 2 \times \pi \times 7 = 44.0 \text{ cm (3 s.f.)}$$

Major arc = circumference − minor arc
$$= (44.0 - 12.4) = 31.6 \text{ cm}$$

Perimeter, area and volume

Exercise 27.10

1. For each of the following, give the length of the arc to 3 s.f. O is the centre of the circle.

a)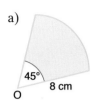
b)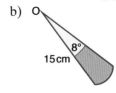
c)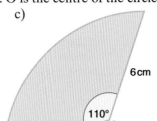
d)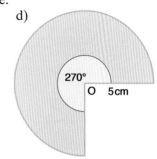

2. A sector is the region of a circle enclosed by two radii and an arc. Calculate the angle ϕ for each of the following sectors. The radius r and arc length a are given in each case.
   a) $r = 14$ cm, $\quad a = 8$ cm
   b) $r = 4$ cm, $\quad a = 16$ cm
   c) $r = 7.5$ cm, $\quad a = 7.5$ cm
   d) $r = 6.8$ cm, $\quad a = 13.6$ cm

3. Calculate the radius r for each of the following sectors. The angle ϕ and arc length a are given in each case.
   a) $\phi = 75°$, $\quad a = 16$ cm
   b) $\phi = 300°$, $\quad a = 24$ cm
   c) $\phi = 20°$, $\quad a = 6.5$ cm
   d) $\phi = 243°$, $\quad a = 17$ cm

Exercise 27.11

1. Calculate the perimeter of each of these shapes.

a)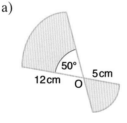
b)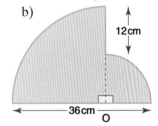

2. A shape (left) is made from two sectors arranged in such a way that they share the same centre. The radius of the smaller sector is 7 cm and the radius of the larger sector is 10 cm. If the angle at the centre of the smaller sector is 30° and the arc length of the larger sector is 12 cm, calculate:
   a) the arc length of the smaller sector,
   b) the total perimeter of the two sectors,
   c) the angle at the centre of the larger sector.

Mensuration

3. For the diagram (right), calculate:
   a) the radius of the smaller sector,
   b) the perimeter of the shape,
   c) the angle ϕ.

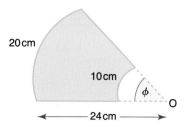

The area of a sector

A **sector** is the region of a circle enclosed by two radii and an arc. Its area is proportional to the size of the angle ϕ between the two radii. The area of the sector as a fraction of the area of the whole circle is therefore equal to the fraction that ϕ is of 360°.

$$\text{Area of sector} = \frac{\phi}{360} \times \pi r^2$$

Worked examples

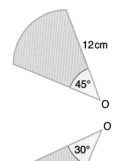

a) Calculate the area of the sector (left), giving your answer to 3 s.f.

$$\begin{aligned}\text{Area} &= \frac{\phi}{360} \times \pi r^2 \\ &= \frac{45}{360} \times \pi \times 12^2 \\ &= 56.5 \text{ cm}^2\end{aligned}$$

b) Calculate the radius of the sector (left), giving your answer to 3 s.f.

$$\begin{aligned}\text{Area} &= \frac{\phi}{360} \times \pi r^2 \\ 50 &= \frac{30}{360} \times \pi \times r^2 \\ \frac{50 \times 360}{30\pi} &= r^2 \\ r &= 13.8\end{aligned}$$

The radius is 13.8 cm.

Exercise 27.12

1. Calculate the area of each of the following sectors, using the values of the angles ϕ and radius r in each case.
   a) $\phi = 60°$, $r = 8$ cm
   b) $\phi = 120°$, $r = 14$ cm
   c) $\phi = 2°$, $r = 18$ cm
   d) $\phi = 320°$, $r = 4$ cm

Perimeter, area and volume

2. Calculate the radius for each of the following sectors, using the values of the angle ϕ and the area A in each case.
   a) $\phi = 40°$, $A = 120$ cm^2
   b) $\phi = 12°$, $A = 42$ cm^2
   c) $\phi = 150°$, $A = 4$ cm^2
   d) $\phi = 300°$, $A = 400$ cm^2

3. Calculate the value of the angle ϕ, to the nearest degree, for each of the following sectors, using the values of A and r in each case.
   a) $r = 12$ cm, $A = 60$ cm^2
   b) $r = 26$ cm, $A = 0.02$ m^2
   c) $r = 0.32$ m, $A = 180$ cm^2
   d) $r = 38$ mm, $A = 16$ cm^2

Exercise 27.13

1. A rotating sprinkler is placed in one corner of a garden (below). If it has a reach of 8 m and rotates through an angle of 30°, calculate the area of garden not being watered.

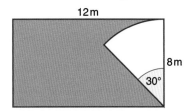

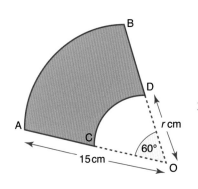

2. Two sectors AOB and COD share the same centre O. The area of AOB is three times the area of COD. Calculate:
   a) the area of sector AOB,
   b) the area of sector COD,
   c) the radius r cm of sector COD.

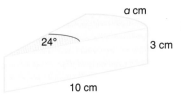

3. A circular cake is cut. One of the slices is shown. Calculate:
   a) the length a cm of the arc,
   b) the total surface area of all the sides of the slice,
   c) the volume of the slice.

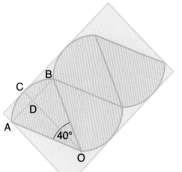

4. The diagram shows a plan view of four tiles in the shape of sectors placed in the bottom of a box. C is the midpoint of the arc AB and intersects the chord AB at point D. If the length OB is 10 cm, calculate:
   a) the length OD,
   b) the length CD,
   c) the area of the sector AOB,
   d) the length and width of the box,
   e) the area of the base of the box not covered by the tiles.

Mensuration

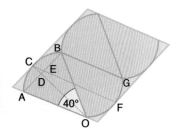

5. The tiles in question 4 are repackaged and are now placed in a box, the base of which is a parallelogram. Given that C and F are the midpoints of arcs AB and OG respectively, calculate:
   a) the angle OCF,
   b) the length CE,
   c) the length of the sides of the box,
   d) the area of the base of the box not covered by the tiles.

● **The volume of a sphere**

Volume of sphere $= \frac{4}{3}\pi r^3$

Worked examples a) Calculate the volume of the sphere (left), giving your answer to 3 s.f.

Volume of sphere $= \frac{4}{3}\pi r^3$
$= \frac{4}{3} \times \pi \times 3^3$
$= 113.1$

The volume is 113 cm³.

b) Given that the volume of a sphere is 150 cm³, calculate its radius to 3 s.f.

$V = \frac{4}{3}\pi r^3$

$r^3 = \frac{3V}{4\pi}$

$r^3 = \frac{3 \times 150}{4 \times \pi}$

$r = \sqrt[3]{35.8} = 3.30$

The radius is 3.30 cm.

Exercise 27.14 1. Calculate the volume of each of the following spheres. The radius r is given in each case.
   a) $r = 6$ cm   b) $r = 9.5$ cm
   c) $r = 8.2$ cm   d) $r = 0.7$ cm

2. Calculate the radius of each of the following spheres. Give your answers in centimetres and to 1 d.p. The volume V is given in each case.
   a) $V = 130$ cm³   b) $V = 720$ cm³
   c) $V = 0.2$ m³   d) $V = 1000$ mm³

Perimeter, area and volume

Exercise 27.15

1. Given that sphere B has twice the volume of sphere A, calculate the radius of sphere B. Give your answer to 1 d.p.

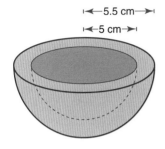

2. Calculate the volume of material used to make the hemispherical bowl on the left, if the inner radius of the bowl is 5 cm and its outer radius 5.5 cm.

3. The volume of the material used to make the sphere and hemispherical bowl (right) are the same. Given that the radius of the sphere is 7 cm and the inner radius of the bowl is 10 cm, calculate, to 1 d.p., the outer radius r cm of the bowl.

4. A ball is placed inside a box into which it will fit tightly. If the radius of the ball is 10 cm, calculate:
   a) the volume of the ball,
   b) the volume of the box,
   c) the percentage volume of the box not occupied by the ball.

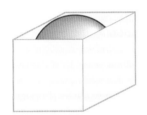

5. A steel ball is melted down to make eight smaller identical balls. If the radius of the original steel ball was 20 cm, calculate to the nearest millimetre the radius of each of the smaller balls.

6. A steel ball of volume 600 cm³ is melted down and made into three smaller balls A, B and C. If the volumes of A, B and C are in the ratio 7 : 5 : 3, calculate to 1 d.p. the radius of each of A, B and C.

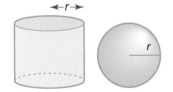

7. The cylinder and sphere shown (left) have the same radius and the same height. Calculate the ratio of their volumes, giving your answer in the form,
volume of cylinder : volume of sphere.

● **The surface area of a sphere**

Surface area of sphere = $4\pi r^2$

Exercise 27.16

1. Calculate the surface area of each of the following spheres when:
   a) $r = 6$ cm
   b) $r = 4.5$ cm
   c) $r = 12.25$ cm
   d) $r = \dfrac{1}{\sqrt{\pi}}$ cm

Mensuration

2. Calculate the radius of each of the following spheres, given the surface area in each case.
   a) $A = 50 \text{ cm}^2$
   b) $A = 16.5 \text{ cm}^2$
   c) $A = 120 \text{ mm}^2$
   d) $A = \pi \text{ cm}^2$

3. Sphere A has a radius of 8 cm and sphere B has a radius of 16 cm. Calculate the ratio of their surface areas in the form $1 : n$.

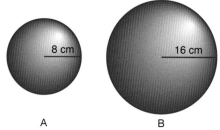

4. A hemisphere of diameter 10 cm is attached to a cylinder of equal diameter as shown.
   If the total length of the shape is 20 cm, calculate:
   a) the surface area of the hemisphere,
   b) the length of the cylinder,
   c) the surface area of the whole shape.

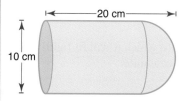

5. A sphere and a cylinder both have the same surface area and the same height of 16 cm.
   Calculate:
   a) the surface area of the sphere,
   b) the radius of the cylinder.

● The volume of a pyramid

A pyramid is a three-dimensional shape in which each of its faces must be plane. A pyramid has a polygon for its base and the other faces are triangles with a common vertex, known as the **apex**. Its individual name is taken from the shape of the base.

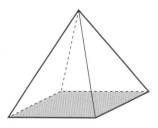
Square-based pyramid

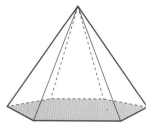
Hexagonal-based pyramid

Volume of any pyramid
$= \frac{1}{3} \times \text{area of base} \times \text{perpendicular height}$

27 Perimeter, area and volume

Worked examples

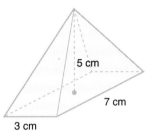

a) A rectangular-based pyramid has a perpendicular height of 5 cm and base dimensions as shown. Calculate the volume of the pyramid.

$$\text{Volume} = \tfrac{1}{3} \times \text{base area} \times \text{height}$$
$$= \tfrac{1}{3} \times 3 \times 7 \times 5$$
$$= 35$$

The volume is 35 cm³.

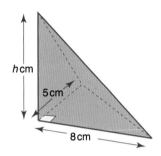

b) The pyramid shown has a volume of 60 cm³. Calculate its perpendicular height h cm.

$$\text{Volume} = \tfrac{1}{3} \times \text{base area} \times \text{height}$$
$$\text{Height} = \frac{3 \times \text{volume}}{\text{base area}}$$
$$h = \frac{3 \times 60}{\tfrac{1}{2} \times 8 \times 5}$$
$$h = 9$$

The height is 9 cm.

Exercise 27.17 Find the volume of each of the following pyramids:

1.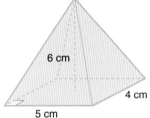

2. Base area = 50 cm²

3.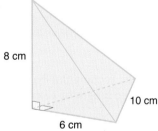

4.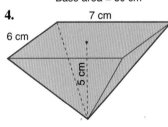

Exercise 27.18

1. Calculate the perpendicular height h cm for the pyramid (right), given that it has a volume of 168 cm³.

2. Calculate the length of the edge marked x cm, given that the volume of the pyramid (left) is 14 cm³.

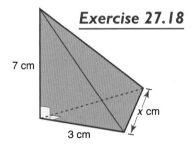

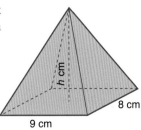

Mensuration

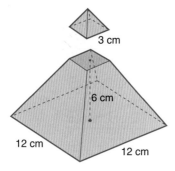

3. The top of a square-based pyramid (left) is cut off. The cut is made parallel to the base. If the base of the smaller pyramid has a side length of 3 cm and the vertical height of the truncated pyramid is 6 cm, calculate:
   a) the height of the original pyramid,
   b) the volume of the original pyramid,
   c) the volume of the truncated pyramid.

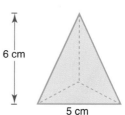

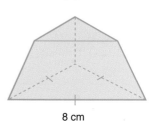

4. The top of a triangular-based pyramid (tetrahedron) is cut off. The cut is made parallel to the base. If the vertical height of the top is 6 cm, calculate:
   a) the height of the truncated piece,
   b) the volume of the small pyramid,
   c) the volume of the original pyramid.

● The surface area of a pyramid

The surface area of a pyramid is found simply by adding together the areas of all of its faces.

Exercise 27.19

1. Calculate the surface area of a regular tetrahedron with edge length 2 cm.

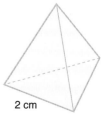

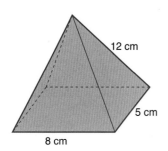

2. The rectangular-based pyramid shown (left) has a sloping edge length of 12 cm. Calculate its surface area.

3. Two square-based pyramids are glued together as shown (right). Given that all the triangular faces are identical, calculate the surface area of the whole shape.

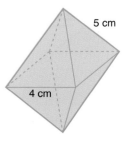

300

Perimeter, area and volume

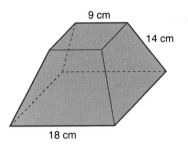

4. Calculate the surface area of the truncated square-based pyramid shown (left). Assume that all the sloping faces are identical.

5. The two pyramids shown below have the same surface area.

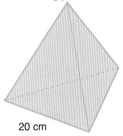
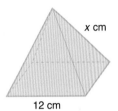

Calculate:
a) the surface area of the regular tetrahedron,
b) the area of one of the triangular faces on the square-based pyramid,
c) the value of x.

● The volume of a cone

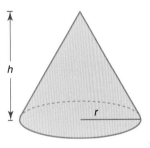

A cone is a pyramid with a circular base. The formula for its volume is therefore the same as for any other pyramid.

Volume $= \frac{1}{3} \times$ base area $\times$ height
$= \frac{1}{3}\pi r^2 h$

Worked examples a) Calculate the volume of the cone (left).

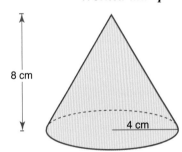

Volume $= \frac{1}{3}\pi r^2 h$
$= \frac{1}{3} \times \pi \times 4^2 \times 8$
$= 134.0$ (1 d.p.)

The volume is 134 cm³ (3 s.f.).

b) The sector below is assembled to form a cone as shown.

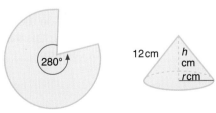

i) Calculate the base circumference of the cone.
   The base circumference of the cone is equal to the arc length of the sector.

   $$\text{Sector arc length} = \frac{\phi}{360} \times 2\pi r$$

   $$= \frac{280}{360} \times 2\pi \times 12 = 58.6 \text{ (3 s.f.)}$$

   So the base circumference is 58.6 cm.

ii) Calculate the base radius of the cone.
   The base of a cone is circular, therefore:

   $$C = 2\pi r$$

   $$r = \frac{C}{2\pi} = \frac{58.6}{2\pi}$$

   $$= 9.33 \text{ (3 s.f.)}$$

   So the radius is 9.33 cm.

iii) Calculate the vertical height of the cone.
   The vertical height of the cone can be calculated using Pythagoras' theorem on the right-angled triangle enclosed by the base radius, vertical height and the sloping face, as shown below.
   Note that the length of the sloping face is equal to the radius of the sector.

   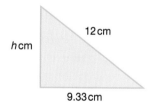

   $$12^2 = h^2 + 9.33^2$$
   $$h^2 = 12^2 - 9.33^2$$
   $$h^2 = 56.9$$
   $$h = 7.54 \text{ (3 s.f.)}$$

   So the height is 7.54 cm.

iv) Calculate the volume of the cone.

   $$\text{Volume} = \frac{1}{3} \times \pi r^2 h$$
   $$= \frac{1}{3} \times \pi \times 9.33^2 \times 7.54$$
   $$= 688 \text{ (3 s.f.)}$$

   So the volume is 688 cm³.

It is important to note that, although answers were given to 3 s.f. in each case, where the answer was needed in a subsequent calculation the exact value was used and not the rounded one. By doing this we avoid introducing rounding errors into the calculations.

Exercise 27.20

1. Calculate the volume of each of the following cones. Use the values for the base radius r and the vertical height h given in each case.
   a) $r = 3$ cm, $h = 6$ cm
   b) $r = 6$ cm, $h = 7$ cm
   c) $r = 8$ mm, $h = 2$ cm
   d) $r = 6$ cm, $h = 44$ mm

2. Calculate the base radius of each of the following cones. Use the values for the volume V and the vertical height h given in each case.
   a) $V = 600$ cm³, $h = 12$ cm
   b) $V = 225$ cm³, $h = 18$ mm
   c) $V = 1400$ mm³, $h = 2$ cm
   d) $V = 0.04$ m³, $h = 145$ mm

3. The base circumference C and the length of the sloping face l is given for each of the following cones. Calculate
   i) the base radius,
   ii) the vertical height,
   iii) the volume in each case.
   Give all answers to 3 s.f.
   a) $C = 50$ cm, $l = 15$ cm
   b) $C = 100$ cm, $l = 18$ cm
   c) $C = 0.4$ m, $l = 75$ mm
   d) $C = 240$ mm, $l = 6$ cm

Exercise 27.21

1. The two cones A and B shown below have the same volume. Using the dimensions shown and given that the base circumference of cone B is 60 cm, calculate the height h cm.

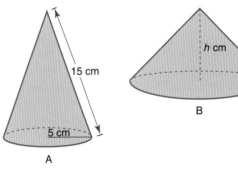

2. The sector shown is assembled to form a cone. Calculate:
   a) the base circumference of the cone,
   b) the base radius of the cone,
   c) the vertical height of the cone,
   d) the volume of the cone.

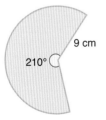

Mensuration

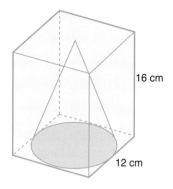

3. A cone is placed inside a cuboid as shown (left). If the base diameter of the cone is 12 cm and the height of the cuboid is 16 cm, calculate:
   a) the volume of the cuboid,
   b) the volume of the cone,
   c) the volume of the cuboid not occupied by the cone.

4. Two similar sectors are assembled into cones (below). Calculate:
   a) the volume of the smaller cone,
   b) the volume of the larger cone,
   c) the ratio of their volumes.

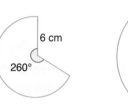 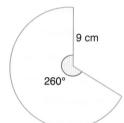

Exercise 27.22

1. An ice cream consists of a hemisphere and a cone (right). Calculate its total volume.

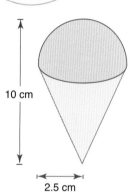

2. A cone is placed on top of a cylinder. Using the dimensions given (right), calculate the total volume of the shape.

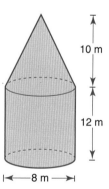

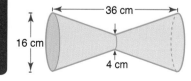

3. Two identical truncated cones are placed end to end as shown.
   Calculate the total volume of the shape.

Perimeter, area and volume

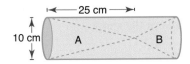

4. Two cones A and B are placed either end of a cylindrical tube as shown.
   Given that the volumes of A and B are in the ratio 2 : 1, calculate:
   a) the volume of cone A,
   b) the height of cone B,
   c) the volume of the cylinder.

● **The surface area of a cone**

The surface area of a cone comprises the area of the circular base and the area of the curved face. The area of the curved face is equal to the area of the sector from which it is formed.

Worked example   Calculate the total surface area of the cone shown (left).

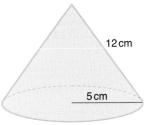

$$\text{Surface area of base} = \pi r^2 = 25\pi \text{ cm}^2$$

The curved surface area can best be visualised if drawn as a sector as shown in the diagram below left:

The radius of the sector is equivalent to the slant height of the cone. The curved perimeter of the sector is equivalent to the base circumference of the cone.

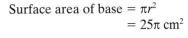

$$\frac{\phi}{360} = \frac{10\pi}{24\pi}$$

Therefore $\phi = 150°$

$$\text{Area of sector} = \frac{150}{360} \times \pi \times 12^2 = 60\pi \text{ cm}^2$$

$$\text{Total surface area} = 60\pi + 25\pi$$
$$= 85\pi$$
$$= 267 \text{ (3 s.f.)}$$

The total surface area is 267 cm².

Exercise 27.23   1. Calculate the surface area of each of the following cones:

a)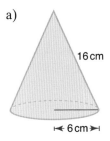

b)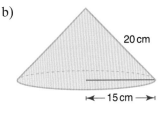

Mensuration

2. Two cones with the same base radius are stuck together as shown. Calculate the surface area of the shape.

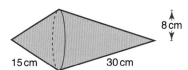

Student assessment I

1. Calculate the area of the shape below.

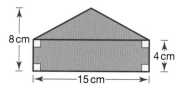

2. Calculate the circumference and area of each of the following circles. Give your answers to 3 s.f.
   a)   b)

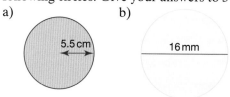

3. A semi-circular shape is cut out of the side of a rectangle as shown. Calculate the shaded area to 3 s.f.

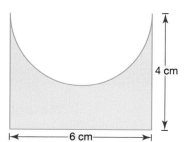

4. For the diagram (right), calculate the area of:
   a) the semi-circle,
   b) the trapezium,
   c) the whole shape.

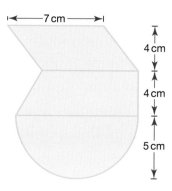

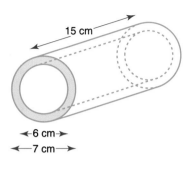

5. A cylindrical tube has an inner diameter of 6 cm, an outer diameter of 7 cm and a length of 15 cm. Calculate the following to 3 s.f.:
   a) the surface area of the shaded end,
   b) the inside surface area of the tube,
   c) the total surface area of the tube.

6. Calculate the volume of each of the following cylinders:
   a)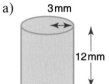
   b)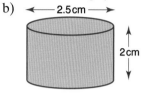

Student assessment 2

1. Calculate the area of this shape:

   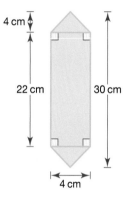

2. Calculate the circumference and area of each of the following circles. Give your answers to 3 s.f.

   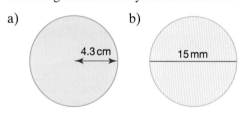

3. A rectangle of length 32 cm and width 20 cm has a semicircle cut out of two of its sides as shown (below). Calculate the shaded area to 3 s.f.

   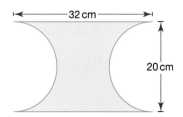

Mensuration

4. Calculate the area of:
   a) the semi-circle,
   b) the parallelogram,
   c) the whole shape.

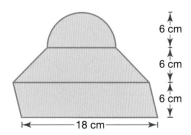

5. A prism in the shape of a hollowed-out cuboid has dimensions as shown. If the end is square, calculate the volume of the prism.

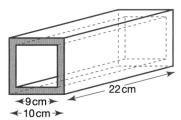

6. Calculate the surface area of each of the following cylinders:

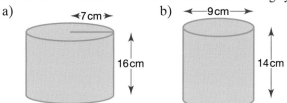

Student assessment 3

1. Calculate the arc length of each of the following sectors. The angle ϕ and radius r are given in each case.
   a) $\phi = 45°$
      $r = 15$ cm
   b) $\phi = 150°$
      $r = 13.5$ cm

2. Calculate the angle ϕ in each of the following sectors. The radius r and arc length a are given in each case.
   a) $r = 20$ mm
      $a = 95$ mm
   b) $r = 9$ cm
      $a = 9$ mm

3. Calculate the area of the sector shown below:

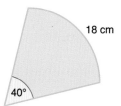

Perimeter, area and volume

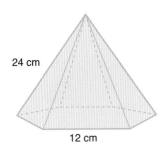

4. A sphere has a radius of 6.5 cm. Calculate to 3 s.f:
   a) its total surface area,
   b) its volume.

5. A pyramid with a base the shape of a regular hexagon is shown (left). If the length of each of its sloping edges is 24 cm, calculate:
   a) its total surface area,
   b) its volume.

Student assessment 4

1. Calculate the arc length of the following sectors. The angle ϕ and radius r are given in each case.
   a) $\phi = 255°$
      $r = 40$ cm
   b) $\phi = 240°$
      $r = 16.3$ mm

2. Calculate the angle ϕ in each of the following sectors. The radius r and arc length a are given in each case.
   a) $r = 40$ cm
      $a = 100$ cm
   b) $r = 20$ cm
      $a = 10$ mm

3. Calculate the area of the sector shown below:

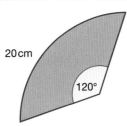

4. A hemisphere has a radius of 8 cm. Calculate to 1 d.p.:
   a) its total surface area,
   b) its volume.

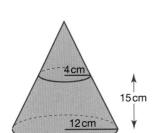

5. A cone has its top cut as shown (left). Calculate:
   a) the height of the large cone,
   b) the volume of the small cone,
   c) the volume of the truncated cone.

Student assessment 5

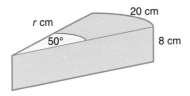

1. The prism (left) has a cross-sectional area in the shape of a sector.
   Calculate:
   a) the radius r cm,
   b) the cross-sectional area of the prism,
   c) the total surface area of the prism,
   d) the volume of the prism.

Mensuration

2. The cone and sphere shown (below) have the same volume.

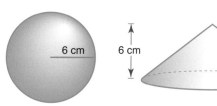

If the radius of the sphere and the height of the cone are both 6 cm, calculate:
a) the volume of the sphere,
b) the base radius of the cone,
c) the slant height x cm,
d) the surface area of the cone.

3. The top of a cone is cut off and a cylindrical hole is drilled out of the remaining truncated cone as shown (left). Calculate:
a) the height of the original cone,
b) the volume of the original cone,
c) the volume of the solid truncated cone,
d) the volume of the cylindrical hole,
e) the volume of the remaining truncated cone.

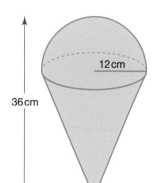

Student assessment 6

1. A metal object (left) is made from a hemisphere and a cone, both of base radius 12 cm. The height of the object when upright is 36 cm.
Calculate:
a) the volume of the hemisphere,
b) the volume of the cone,
c) the curved surface area of the hemisphere,
d) the total surface area of the object.

2. A regular tetrahedron (right) has edges of length 5 cm. Calculate:
a) the surface area of the tetrahedron,
b) the surface area of a tetrahedron with edge lengths of 10 cm.

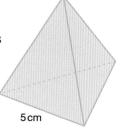

3. A regular tetrahedron and a sphere have the same surface area. If the radius of the sphere is 10 cm, calculate:
a) the area of one face of the tetrahedron,
b) the length of each edge of the tetrahedron.

(Hint: Use the trigonometric formula for the area of a triangle.)

Mathematical investigations and ICT

● **Metal trays**

A rectangular sheet of metal measures 30 cm by 40 cm.

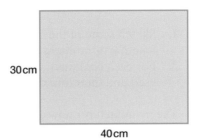

The sheet has squares of equal size cut from each corner. It is then folded to form a metal tray as shown.

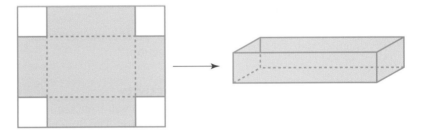

1. a) Calculate the length, width and height of the tray if a square of side length 1 cm is cut from each corner of the sheet of metal.
   b) Calculate the volume of this tray.

2. a) Calculate the length, width and height of the tray if a square of side length 2 cm is cut from each corner of the sheet of metal.
   b) Calculate the volume of this tray.

3. Using a spreadsheet if necessary, investigate the relationship between the volume of the tray and the size of the square cut from each corner. Enter your results in an ordered table.

4. Calculate, to 1 d.p., the side length of the square that produces the tray with the greatest volume.

5. State the greatest volume to the nearest whole number.

Mensuration

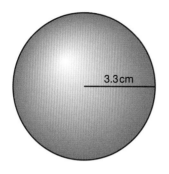

● Tennis balls

Tennis balls are spherical and have a radius of 3.3 cm.

A manufacturer wishes to make a cuboidal container with a lid that holds 12 tennis balls. The container is to be made of cardboard. The manufacturer wishes to use as little cardboard as possible.

1. Sketch some of the different containers that the manufacturer might consider.
2. For each container, calculate the total area of cardboard used and therefore decide on the most economical design.

The manufacturer now considers the possibility of using other flat-faced containers.

3. Sketch some of the different containers that the manufacturer might consider.
4. Investigate the different amounts of cardboard used for each design.
5. Which type of container would you recommend to the manufacturer?

● ICT activity

In this topic you will have seen that it is possible to construct a cone from a sector. The dimensions of the cone are dependent on the dimensions of the sector. In this activity you will be using a spreadsheet to investigate the maximum possible volume of a cone constructed from a sector of fixed radius.

Circles of radius 10 cm are cut from paper and used to construct cones. Different sized sectors are cut from the circles and then arranged to form a cone, e.g.

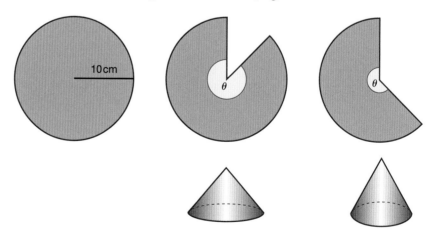

1. Using a spreadsheet similar to the one below, calculate the maximum possible volume, for a cone constructed from one of these circles:

	A	B	C	D	E	F
1	Angle of sector (θ)	Sector arc length (cm)	Base circumference of cone (cm)	Base radius of cone (cm)	Vertical height of cone (cm)	Volume of cone (cm³)
2	5	0.873	0.873	0.139	9.999	0.202
3	10	1.745	1.745	0.278	9.996	0.808
4	15	2.618	2.618	0.417	9.991	1.816
5	20					
6	25					
7	30					
8	Continue to 355°	Enter formulae here to calculate the results for each column				

2. Plot a graph to show how the volume changes as θ increases. Comment on your graph.

Topic 5: Coordinate geometry

Syllabus

E5.1
Demonstrate familiarity with Cartesian co-ordinates in two dimensions.

E5.2
Find the gradient of a straight line.
Calculate the gradient of a straight line from the co-ordinates of two points on it.

E5.3
Calculate the length and the co-ordinates of the midpoint of a straight line from the co-ordinates of its end points.

E5.4
Interpret and obtain the equation of a straight-line graph in the form $y = mx + c$.

E5.5
Determine the equation of a straight line parallel to a given line.

E5.6
Find the gradient of parallel and perpendicular lines.

Contents

Chapter 28   Straight-line graphs (E5.1, E5.2, E5.3, E5.4, E5.5, E5.6)

The French

In the middle of the seventeenth century there were three great French mathematicians, René Descartes, Blaise Pascal and Pierre de Fermat.

René Descartes was a philosopher and a mathematician. His book *The Meditations* asks 'How and what do I know?' His work in mathematics made a link between algebra and geometry. He thought that all nature could be explained in terms of mathematics. Although he was not considered as talented a mathematician as Pascal and Fermat, he has had greater influence on modern thought. The (x, y) coordinates we use are called Cartesian coordinates after Descartes.

Blaise Pascal (1623–1662) was a genius who studied geometry as a child. When he was 16 he stated and proved Pascal's Theorem, which relates any six points on any conic section. The Theorem is sometimes called the 'Cat's Cradle'. He founded probability theory and made contributions to the invention of calculus. He is best known for Pascal's Triangle.

René Descartes (1596–1650)

Pierre de Fermat (1601–1665) was a brilliant mathematician and, along with Descartes, one of the most influential. Fermat invented number theory and worked on calculus. He discovered probability theory with his friend Pascal. It can be argued that Fermat was at least Newton's equal as a mathematician.

Fermat's most famous discovery in number theory includes 'Fermat's Last Theorem'. This theorem is derived from Pythagoras' theorem which states that for a right-angled triangle, $x^2 = y^2 + z^2$ where x is the length of the hypotenuse. Fermat said that if the index (power) was greater than two and x, y, z are all whole numbers, then the equation was never true. (This theorem was only proved in 1995 by the English mathematician Andrew Wiles.)

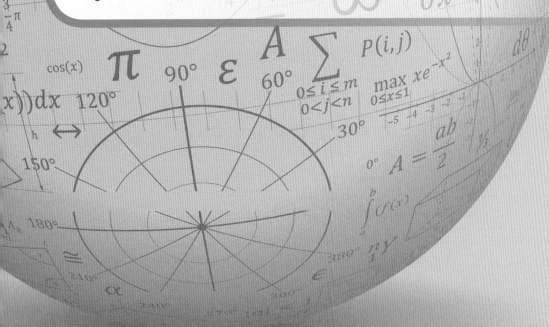

Straight-line graphs

● The gradient of a straight line

Lines are made of an infinite number of points. This chapter looks at those whose points form a straight line.

The graph below shows three straight lines.

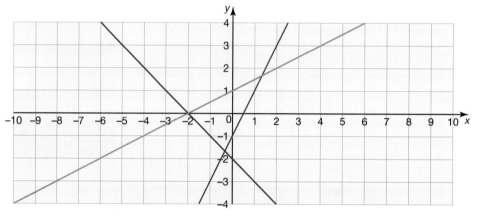

The lines have some properties in common (i.e. they are straight), but also have differences. One of their differences is that they have different slopes. The slope of a line is called its **gradient**.

The gradient of a straight line is constant, i.e. it does not change. The gradient of a straight line can be calculated by considering the coordinates of any two points on the line.

On the line below two points A and B have been chosen.

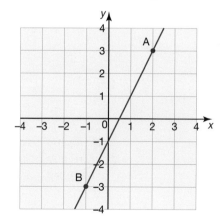

The coordinates of the points are A(2, 3) and B(−1, −3). The gradient is calculated using the following formula:

$$\text{Gradient} = \frac{\text{vertical distance between two points}}{\text{horizontal distance between two points}}$$

Graphically this can be represented as follows:

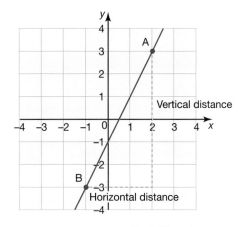

Therefore gradient $= \dfrac{3-(-3)}{2-(-1)} = \dfrac{6}{3} = 2$

In general therefore, if the two points chosen have coordinates (x_1, y_1) and (x_2, y_2) the gradient is calculated as:

$$\text{Gradient} = \frac{y_2 - y_1}{x_2 - x_1}$$

Worked example   Calculate the gradient of the line shown below.

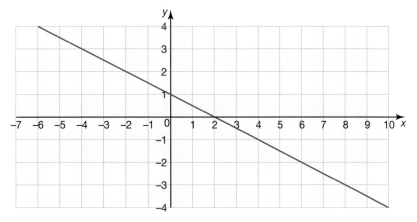

Choose two points on the line, e.g. $(-4, 3)$ and $(8, -3)$

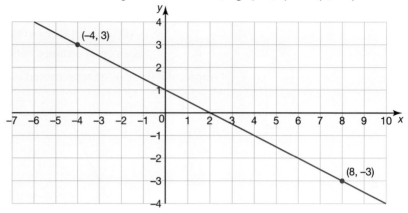

Let point 1 be $(-4, 3)$ and point 2 be $(8, -3)$.

$$\text{Gradient} = \frac{y_2 - y_1}{x_2 - x_1} = \frac{-3 - 3}{8 - (-4)}$$

$$= \frac{-6}{12} = -\frac{1}{2}$$

Note, the gradient is not affected by which point is chosen as point 1 and which is chosen as point 2. In the example above if point 1 was $(8, -3)$ and point 2 $(-4, 3)$, the gradient would be calculated as:

$$\text{Gradient} = \frac{y_2 - y_1}{x_2 - x_1} = \frac{3 - (-3)}{-4 - 8}$$

$$= \frac{6}{-12} = -\frac{1}{2}$$

To check whether or not the sign of the gradient is correct, the following guideline is useful.

 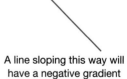

A line sloping this way will have a positive gradient   A line sloping this way will have a negative gradient

A large value for the gradient implies that the line steep. The line on the right below will have a greater value for the gradient than the line on the left as it is steeper.

Straight-line graphs

Exercise 28.1 **1.** For each of the following lines, select two points on the line and then calculate its gradient.

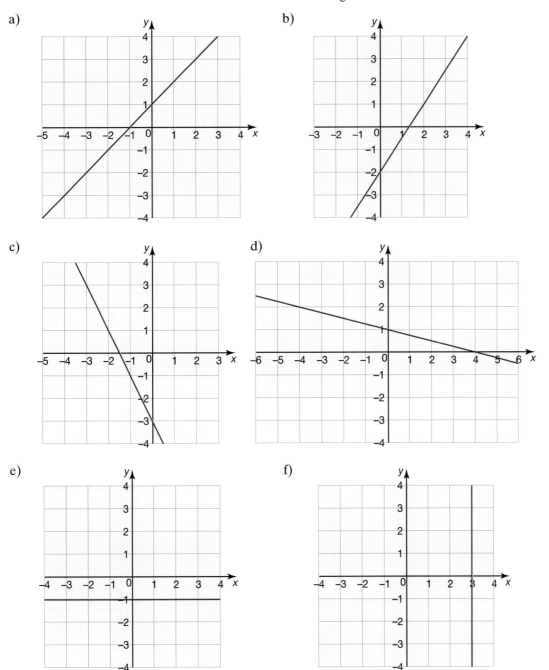

2. From your answers to question 1, what conclusion can you make about the gradient of any horizontal line?

3. From your answers to question 1, what conclusion can you make about the gradient of any vertical line?

4. The graph below shows six straight lines labelled A–F.

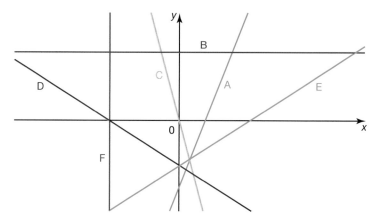

Six gradients are given below. Deduce which line has which gradient.

Gradient = $\frac{1}{2}$ Gradient is infinite   Gradient = 2

Gradient = −3   Gradient = 0   Gradient = $-\frac{1}{2}$

● The equation of a straight line

The coordinates of every point on a straight line all have a common relationship. This relationship when expressed algebraically as an equation in terms of x and/or y is known as the equation of the straight line.

Worked examples **a)** By looking at the coordinates of some of the points on the line below, establish the equation of the straight line.

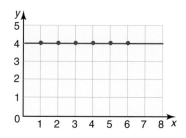

x	y
1	4
2	4
3	4
4	4
5	4
6	4

Some of the points on the line have been identified and their coordinates entered in a table above. By looking at the table it can be seen that the only rule all the points have in common is that $y = 4$.

Hence the equation of the straight line is $y = 4$.

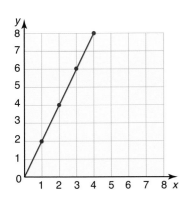

b) By looking at the coordinates of some of the points on the line (left), establish the equation of the straight line.

x	y
1	2
2	4
3	6
4	8

Once again, by looking at the table it can be seen that the relationship between the x- and y-coordinates is that each y coordinate is twice the corresponding x-coordinate. Hence the equation of the straight line is $y = 2x$.

Exercise 28.2 **1.** In each of the following identify the coordinates of some of the points on the line and use these to find the equation of the straight line.

a)

b)

c)

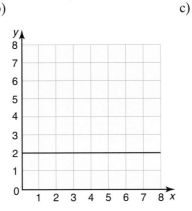

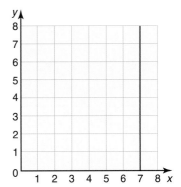

d)

e)

f)

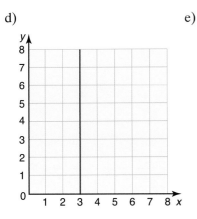

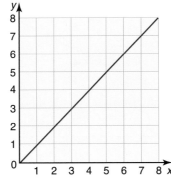

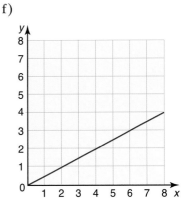

Coordinate geometry

g)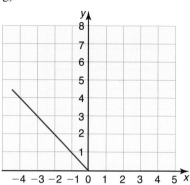

h)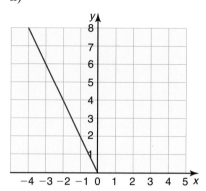

Exercise 28.3 1. In each of the following identify the coordinates of some of the points on the line and use these to find the equation of the straight line.

a)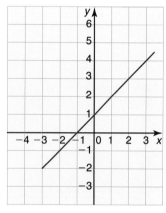

b)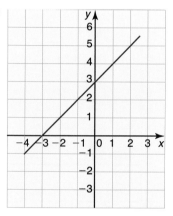

c)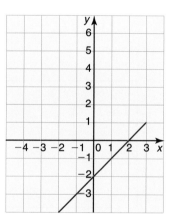

d)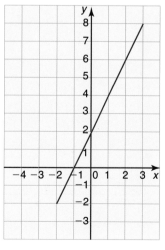

e)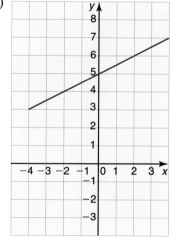

f)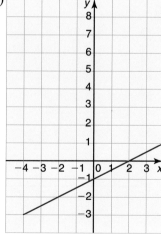

2. In each of the following identify the coordinates of some of the points on the line and use these to find the equation of the straight line.

a)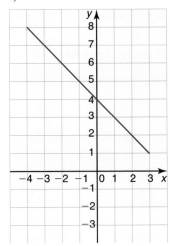

b)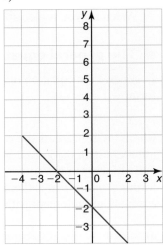

c)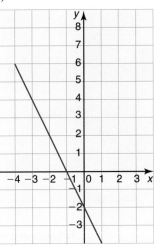

d)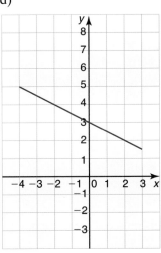

e)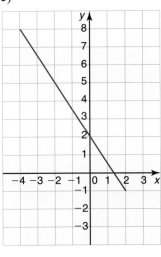

f)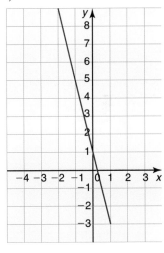

3. a) For each of the graphs in questions 1 and 2 calculate the gradient of the straight line.
   b) What do you notice about the gradient of each line and its equation?
   c) What do you notice about the equation of the straight line and where the line intersects the y-axis?

4. Copy the diagrams in question 1. Draw two lines on each diagram parallel to the given line.
   a) Write the equation of these new lines in the form $y = mx + c$.
   b) What do you notice about the equations of these new parallel lines?

5. In question 2 you have an equation for these lines in the form $y = mx + c$. Change the value of the intercept c and then draw the new line.
   What do you notice about this new line and the first line?

The general equation of a straight line

In general the equation of any straight line can be written in the form:

$$y = mx + c$$

where 'm' represents the gradient of the straight line and 'c' the intercept with the y-axis. This is shown in the diagram (left).

By looking at the equation of a straight line written in the form $y = mx + c$, it is therefore possible to deduce the line's gradient and intercept with the y-axis without having to draw it.

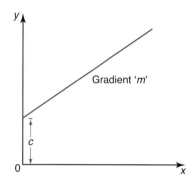

Gradient 'm'

Worked examples a) Calculate the gradient and y-intercept of the following straight lines:

i) $y = 3x - 2$ gradient $= 3$
                     y-intercept $= -2$
ii) $y = -2x + 6$ gradient $= -2$
                     y-intercept $= 6$

b) Calculate the gradient and y-intercept of the following straight lines:

i) $2y = 4x + 2$

This needs to be rearranged into **gradient-intercept** form (i.e. $y = mx + c$).
$y = 2x + 1$ gradient $= 2$
                    y-intercept $= 1$

ii) $y - 2x = -4$

Rearranging into gradient-intercept form, we have:
$y = 2x - 4$ gradient $= 2$
                    y-intercept $= -4$

iii) $-4y + 2x = 4$

Rearranging into gradient-intercept form, we have:
$y = \frac{1}{2}x - 1$ gradient $= \frac{1}{2}$
                           y-intercept $= -1$

iv) $\frac{y+3}{4} = -x + 2$

Rearranging into gradient-intercept form, we have:
$y + 3 = -4x + 8$
$y = -4x + 5$ gradient $= -4$
                    y-intercept $= 5$

Exercise 28.4 For the following linear equations, calculate both the gradient and y-intercept in each case.

1. a) $y = 2x + 1$ b) $y = 3x + 5$ c) $y = x - 2$
   d) $y = \frac{1}{2}x + 4$ e) $y = -3x + 6$ f) $y = -\frac{2}{3}x + 1$
   g) $y = -x$ h) $y = -x - 2$ i) $y = -(2x - 2)$

2. a) $y - 3x = 1$ b) $y + \frac{1}{2}x - 2 = 0$
   c) $y + 3 = -2x$ d) $y + 2x + 4 = 0$
   e) $y - \frac{1}{4}x - 6 = 0$ f) $-3x + y = 2$
   g) $2 + y = x$ h) $8x - 6 + y = 0$
   i) $-(3x + 1) + y = 0$

3. a) $2y = 4x - 6$ b) $2y = x + 8$
   c) $\frac{1}{2}y = x - 2$ d) $\frac{1}{4}y = -2x + 3$
   e) $3y - 6x = 0$ f) $\frac{1}{3}y + x = 1$
   g) $6y - 6 = 12x$ h) $4y - 8 + 2x = 0$
   i) $2y - (4x - 1) = 0$

4. a) $2x - y = 4$ b) $x - y + 6 = 0$
   c) $-2y = 6x + 2$ d) $12 - 3y = 3x$
   e) $5x - \frac{1}{2}y = 1$ f) $-\frac{2}{3}y + 1 = 2x$
   g) $9x - 2 = -y$ h) $-3x + 7 = -\frac{1}{2}y$
   i) $-(4x - 3) = -2y$

5. a) $\frac{y+2}{4} = \frac{1}{2}x$ b) $\frac{y-3}{x} = 2$ c) $\frac{y-x}{8} = 0$
   d) $\frac{2y-3x}{2} = 6$ e) $\frac{3y-2}{x} = -3$ f) $\frac{\frac{1}{2}y-1}{x} = -2$
   g) $\frac{3x-y}{2} = 6$ h) $\frac{6-2y}{3} = 2$ i) $\frac{-(x+2y)}{5x} = 1$

6. a) $\frac{3x-y}{y} = 2$ b) $\frac{-x+2y}{4} = y+1$
   c) $\frac{y-x}{x+y} = 2$ d) $\frac{1}{y} = \frac{1}{x}$
   e) $\frac{-(6x+y)}{2} = y+1$ f) $\frac{2x-3y+4}{4} = 4$

7. a) $\frac{y+1}{x} + \frac{3y-2}{2x} = -1$ b) $\frac{x}{y+1} + \frac{1}{2y+2} = 3$
   c) $\frac{-(-y+3x)}{-(6x-2y)} = 1$ d) $\frac{-(x-2y)-(-x-2y)}{4+x-y} = -2$

Coordinate geometry

Parallel lines and their equations

Lines that are parallel, by their very definition must have the same gradient. Similarly, lines with the same gradient must be parallel. So a straight line with equation $y = -3x + 4$ must be parallel to a line with equation $y = -3x - 2$ as both have a gradient of -3.

Worked examples

a) A straight line has equation $4x - 2y + 1 = 0$.

Another straight line has equation $\frac{2x-4}{y} = 1$.

Explain, giving reasons, whether the two lines are parallel to each other or not.

Rearranging the equations into gradient–intercept form gives:

$4x - 2y + 1 = 0$ $\qquad\qquad \frac{2x-4}{y} = 1$

$\qquad 2y = 4x + 1 \qquad\qquad y = 2x - 4$

$\qquad y = 2x + \frac{1}{2}$

With both equations written in gradient–intercept form it is possible to see that both lines have a gradient of 2 and are therefore parallel.

b) A straight line A has equation $y = -3x + 6$. A second line B is parallel to line A and passes through the point with coordinates $(-4, 10)$.
Calculate the equation of line B.

As line B is a straight line it must take the form $y = mx + c$. As it is parallel to line A, its gradient must be -3.
Because line B passes through the point $(-4, 10)$ these values can be substituted into the general equation of the straight line to give:

$10 = -3 \times (-4) + c$

Rearranging to find c gives: $c = -2$
The equation of line B is therefore $y = -3x - 2$.

Exercise 28.5

1. A straight line has equation $3y - 3x = 4$. Write down the equation of another straight line parallel to it.

2. A straight line has equation $y = -x + 6$. Which of the following lines is/are parallel to it?
   a) $2(y + x) = -5$
   b) $-3x - 3y + 7 = 0$
   c) $2y = -x + 12$
   d) $y + x = \frac{1}{10}$

3. Find the equation of the line parallel to $y = 4x - 1$ that passes through $(0, 0)$.

4. Find the equations of lines parallel to $y = -3x + 1$ that pass through each of the following points:
   a) $(0, 4)$
   b) $(-2, 4)$
   c) $(-\frac{5}{2}, 4)$

5. Find the equations of lines parallel to $x - 2y = 6$ that pass through each of the following points:
   a) $(-4, 1)$
   b) $(\frac{1}{2}, 0)$

● **Drawing straight-line graphs**

To draw a straight-line graph only two points need to be known. Once these have been plotted the line can be drawn between them and extended if necessary at both ends.

Worked examples a) Plot the line $y = x + 3$.

To identify two points simply choose two values of x. Substitute these into the equation and calculate their corresponding y values.

When $x = 0$, $y = 3$
When $x = 4$, $y = 7$

Therefore two of the points on the line are $(0, 3)$ and $(4, 7)$.

The straight line $y = x + 3$ is plotted below.

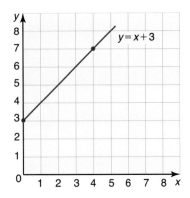

Coordinate geometry

b) Plot the line $y = -2x + 4$.

When $x = 2$, $y = 0$
When $x = -1$, $y = 6$

The coordinates of two points on the line are $(2, 0)$ and $(-1, 6)$.

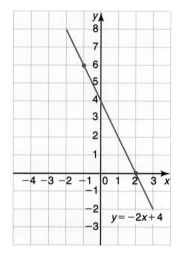

Note that, in questions of this sort, it is often easier to rearrange the equation into gradient–intercept form first.

Exercise 28.6

1. Plot the following straight lines.
   a) $y = 2x + 3$
   b) $y = x - 4$
   c) $y = 3x - 2$
   d) $y = -2x$
   e) $y = -x - 1$
   f) $-y = x + 1$
   g) $-y = 3x - 3$
   h) $2y = 4x - 2$
   i) $y - 4 = 3x$

2. Plot the following straight lines:
   a) $-2x + y = 4$
   b) $-4x + 2y = 12$
   c) $3y = 6x - 3$
   d) $2x = x + 1$
   e) $3y - 6x = 9$
   f) $2y + x = 8$
   g) $x + y + 2 = 0$
   h) $3x + 2y - 4 = 0$
   i) $4 = 4y - 2x$

3. Plot the following straight lines:
   a) $\dfrac{x+y}{2} = 1$
   b) $x + \dfrac{y}{2} = 1$
   c) $\dfrac{x}{3} + \dfrac{y}{2} = 1$
   d) $y + \dfrac{x}{2} = 3$
   e) $\dfrac{y}{5} + \dfrac{x}{3} = 0$
   f) $\dfrac{-(2x+y)}{4} = 1$
   g) $\dfrac{y-(x-y)}{3x} = -1$
   h) $\dfrac{y}{2x+3} - \dfrac{1}{2} = 0$
   i) $-2(x+y) + 4 = -y$

Graphical solution of simultaneous equations

When solving two equations simultaneously the aim is to find a solution which works for both equations. In Chapter 13 it was shown how to arrive at the solution algebraically. It is, however, possible to arrive at the same solution graphically.

Worked example i) By plotting both of the following equations on the same axes, find a common solution.

$$x + y = 4$$
$$x - y = 2$$

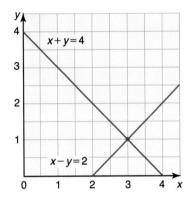

When both lines are plotted, the point at which they cross gives the common solution as it is the only point which lies on both lines.

Therefore the common solution is the point $(3, 1)$.

ii) Check the result obtained above by solving the equations algebraically.

$$x + y = 4 \quad (1)$$
$$x - y = 2 \quad (2)$$

Adding equations $(1) + (2) \to 2x = 6$
$$x = 3$$

Substituting $x = 3$ into equation (1) we have:

$$3 + y = 4$$
$$y = 1$$

Therefore the common solution occurs at $(3, 1)$.

Coordinate geometry

Exercise 28.7 Solve the simultaneous equations below:
i) by graphical means,
ii) by algebraic means.

1. a) $x + y = 5$
   $x - y = 1$
   b) $x + y = 7$
   $x - y = 3$
   c) $2x + y = 5$
   $x - y = 1$
   d) $2x + 2y = 6$
   $2x - y = 3$
   e) $x + 3y = -1$
   $x - 2y = -6$
   f) $x - y = 6$
   $x + y = 2$

2. a) $3x - 2y = 13$
   $2x + y = 4$
   b) $4x - 5y = 1$
   $2x + y = -3$
   c) $x + 5 = y$
   $2x + 3y - 5 = 0$
   d) $x = y$
   $x + y + 6 = 0$
   e) $2x + y = 4$
   $4x + 2y = 8$
   f) $y - 3x = 1$
   $y = 3x - 3$

● **Calculating the length of a line segment**

A line segment is formed when two points are joined by a straight line. To calculate the distance between two points, and therefore the length of the line segment, their coordinates need to be given. Once these are known, Pythagoras' theorem can be used to calculate the distance.

Worked example The coordinates of two points are (1, 3) and (5, 6). Draw a pair of axes, plot the given points and calculate the distance between them.

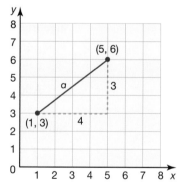

By dropping a vertical line from the point (5, 6) and drawing a horizontal line from (1, 3), a right-angled triangle is formed. The length of the hypotenuse of the triangle is the length we wish to find.

Using Pythagoras' theorem, we have:

$a^2 = 3^2 + 4^2 = 25$
$a = \sqrt{25} = 5$

The length of the line segment is 5 units.

To find the distance between two points directly from their coordinates, use the following formula:

$$d = \sqrt{(x_1 - x_2)^2 + (y_1 - y_2)^2}$$

Worked example Without plotting the points, calculate the distance between the points (1, 3) and (5, 6).

$$d = \sqrt{(1-5)^2 + (3-6)^2}$$
$$= \sqrt{(-4)^2 + (-3)^2}$$
$$= \sqrt{25} = 5$$

The distance between the two points is 5 units.

● **The midpoint of a line segment**

To find the midpoint of a line segment, use the coordinates of its end points. To find the x-coordinate of the midpoint, find the mean of the x-coordinates of the end points. Similarly, to find the y-coordinate of the midpoint, find the mean of the y-coordinates of the end points.

Worked examples a) Find the coordinates of the midpoint of the line segment AB where A is (1, 3) and B is (5, 6).

The x-coordinate of the midpoint will be $\frac{1+5}{2} = 3$

The y-coordinate of the midpoint will be $\frac{3+6}{2} = 4.5$

So the coordinates of the midpoint are (3, 4.5)

b) Find the coordinates of the midpoint of a line segment PQ where P is (−2, −5) and Q is (4, 7).

The x-coordinate of the midpoint will be $\frac{-2+4}{2} = 1$

The y-coordinate of the midpoint will be $\frac{-5+7}{2} = 1$

So the coordinates of the midpoint are (1, 1).

Exercise 28.8 1. i) Plot each of the following pairs of points.
   ii) Calculate the distance between each pair of points.
   iii) Find the coordinates of the midpoint of the line segment joining the two points.

   a) (5, 6) (1, 2)
   b) (6, 4) (3, 1)
   c) (1, 4) (5, 8)
   d) (0, 0) (4, 8)
   e) (2, 1) (4, 7)
   f) (0, 7) (−3, 1)
   g) (−3, −3) (−1, 5)
   h) (4, 2) (−4, −2)
   i) (−3, 5) (4, 5)
   j) (2, 0) (2, 6)
   k) (−4, 3) (4, 5)
   l) (3, 6) (−3, −3)

Coordinate geometry

2. Without plotting the points:
   i) calculate the distance between each of the following pairs of points
   ii) find the coordinates of the midpoint of the line segment joining the two points.

   a) (1, 4) (4, 1)
   b) (3, 6) (7, 2)
   c) (2, 6) (6, −2)
   d) (1, 2) (9, −2)
   e) (0, 3) (−3, 6)
   f) (−3, −5) (−5, −1)
   g) (−2, 6) (2, 0)
   h) (2, −3) (8, 1)
   i) (6, 1) (−6, 4)
   j) (−2, 2) (4, −4)
   k) (−5, −3) (6, −3)
   l) (3, 6) (5, −2)

● The equation of a line through two points

The equation of a straight line can be deduced once the coordinates of two points on the line are known.

Worked example Calculate the equation of the straight line passing through the points $(-3, 3)$ and $(5, 5)$.

The equation of any straight line can be written in the general form $y = mx + c$. Here we have:

$$\text{gradient} = \frac{5-3}{5-(-3)} = \frac{2}{8}$$

$$\text{gradient} = \frac{1}{4}$$

The equation of the line now takes the form $y = \frac{1}{4}x + c$.

Since the line passes through the two given points, their coordinates must satisfy the equation. So to calculate the value of 'c' the x and y coordinates of one of the points are substituted into the equation. Substituting (5, 5) into the equation gives:

$$5 = \frac{1}{4} \times 5 + c$$

$$5 = \frac{5}{4} + c$$

Therefore $c = 5 - 1\frac{1}{4} = 3\frac{3}{4}$

The equation of the straight-line passing through $(-3, 3)$ and $(5, 5)$ is:

$$y = \frac{1}{4}x + 3\frac{3}{4}$$

Exercise 28.9 Find the equation of the straight-line which passes through each of the following pairs of points:

1. a) (1, 1) (4, 7)
    b) (1, 4) (3, 10)
    c) (1, 5) (2, 7)
    d) (0, −4) (3, −1)
    e) (1, 6) (2, 10)
    f) (0, 4) (1, 3)
    g) (3, −4) (10, −18)
    h) (0, −1) (1, −4)
    i) (0, 0) (10, 5)

2. a) (−5, 3) (2, 4)
    b) (−3, −2) (4, 4)
    c) (−7, −3) (−1, 6)
    d) (2, 5) (1, −4)
    e) (−3, 4) (5, 0)
    f) (6, 4) (−7, 7)
    g) (−5, 2) (6, 2)
    h) (1, −3) (−2, 6)
    i) (6, −4) (6, 6)

● **Perpendicular lines**

The two lines shown below are perpendicular to each other.

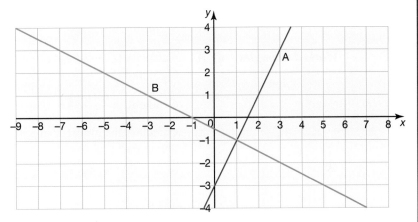

Line A has a gradient of 2.
Line B has a gradient of $-\frac{1}{2}$.

The diagram below also shows two lines perpendicular to each other.

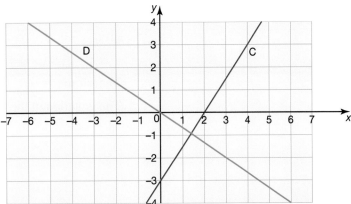

Coordinate geometry

Line C has a gradient of $\frac{3}{2}$.

Line D has a gradient of $-\frac{2}{3}$.

Notice that in both cases, the product of the two gradients is equal to -1.

In the first example $2 \times (-\frac{1}{2}) = -1$.

In the second example $\frac{3}{2} \times (-\frac{2}{3}) = -1$.

This is in fact the case for the gradients of any two perpendicular lines.

If two lines L_1 and L_2 are perpendicular to each other, the product of their gradients m_1 and m_2 is -1.

i.e. $m_1 m_2 = -1$

Therefore the gradient of one line is the negative reciprocal of the other line.

i.e. $m_1 = -\dfrac{1}{m_2}$

Worked examples
a) i) Calculate the gradient of the line joining the two points $(3, 6)$ and $(1, -6)$.

$$\text{Gradient} = \frac{6-(-6)}{3-1} = \frac{12}{2} = 6$$

ii) Calculate the gradient of a line perpendicular to the one in part i) above.

$m_1 = -\dfrac{1}{m_2}$, therefore the gradient of the perpendicular line is $-\frac{1}{6}$.

iii) The perpendicular line also passes through the point $(-1, 6)$. Calculate the equation of the perpendicular line.

The equation of the perpendicular line will take the form $y = mx + c$.

As its gradient is $-\frac{1}{6}$ and it passes though the point $(-1, 6)$, this can be substituted into the equation to give:

$$6 = -\tfrac{1}{6} \times (-1) + c$$

Therefore $c = \frac{35}{6}$.

The equation of the perpendicular line is $y = -\frac{1}{6}x + \frac{35}{6}$.

b) **i)** Show that the point $(-4, -1)$ lies on the line $y = -\frac{1}{4}x - 2$.
If the point $(-4, -1)$ lies on the line, its values of x and y will satisfy the equation. Substituting the values of x and y into the equation gives:

$$-1 = -\frac{1}{4} \times (-4) - 2$$
$$-1 = -1$$

Therefore the point lies on the line.

ii) Deduce the gradient of a line perpendicular to the one given in part i) above.

$$m_1 = -\frac{1}{m_2} \quad \text{therefore} \quad m_1 = -\frac{1}{-\frac{1}{4}} = 4$$

Therefore the gradient of the perpendicular line is 4.

iii) The perpendicular line also passes through the point $(-4, -1)$. Calculate its equation.
The equation of the perpendicular line takes the general form $y = mx + c$.
Substituting in the values of x, y and m gives:

$$-1 = 4 \times (-4) + c$$

Therefore $c = 15$.
The equation of the perpendicular line is $y = 4x + 15$.

Exercise 28.10

1. Calculate:
i) the gradient of the line joining the following pairs of points
ii) the gradient of a line perpendicular to this line
iii) the equation of the perpendicular line if it passes through the second point each time.

a) (1, 4) (4, 1)
b) (3, 6) (7, 2)
c) (2, 6) (6, −2)
d) (1, 2) (9, −2)
e) (0, 3) (−3, 6)
f) (−3, −5) (−5, −1)
g) (−2, 6) (2, 0)
h) (2, −3) (8, 1)
i) (6, 1) (−6, 4)
j) (−2, 2) (4, −4)
k) (−5, −3) (6, −3)
l) (3, 6) (5, −2)

Coordinate geometry

2. The diagram below show a square ABCD. The coordinates of A and B are given.

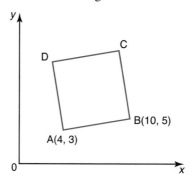

Calculate:
a) the gradient of the line AB
b) the equation of the line passing through A and B
c) the gradient of the line AD
d) the equation of the line passing through A and D
e) the equation of the line passing through B and C
f) the coordinates of C
g) the coordinates of D
h) the equation of the line passing through C and D
i) the length of the sides of the square to 1 d.p.
j) the coordinates of the midpoint of the line segment AC.

3. The diagram below shows a right-angled isosceles triangle ABC, where AB = AC.

The coordinates of A and B are given.

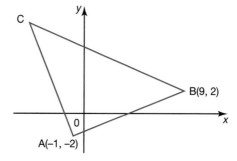

Calculate:
a) the equation of the line passing through the points A and B
b) the equation of the line passing through A and C
c) the length of the line segment BC to 1 d.p.
d) the coordinates of the midpoints of all three sides of the triangle.

Student assessment 1

1. For each of the following lines, select two points on the line and then calculate its gradient.

a)

b)

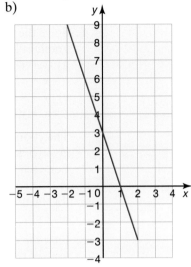

2. Find the equation of the straight line for each of the following:

a)

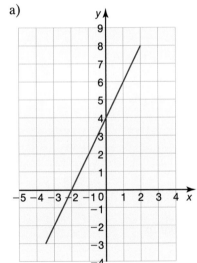

b)

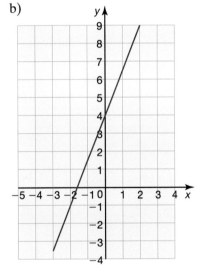

3. Calculate the gradient and y-intercept for each of the following linear equations:
   a) $y = -3x + 4$
   b) $\frac{1}{3}y - x = 2$
   c) $2x + 4y - 6 = 0$

4. Write down the equation of the line parallel to the line $y = -\frac{2}{3}x + 4$ which passes through the point $(6, 2)$.

Coordinate geometry

5. Plot the following graphs on the same pair of axes, labelling each clearly.
   a) $x = -2$
   b) $y = 3$
   c) $y = 2x$
   d) $y = -\dfrac{x}{2}$

6. Solve the following pairs of simultaneous equations graphically:
   a) $x + y = 4$
      $x - y = 0$
   b) $3x + y = 2$
      $x - y = 2$
   c) $y + 4x + 4 = 0$
      $x + y = 2$
   d) $x - y = -2$
      $3x + 2y + 6 = 0$

7. The coordinates of the end points of two line segments are given below.
   For each line segment calculate:
   i) the length
   ii) the midpoint.
   a) $(-6, -1)\ (6, 4)$
   b) $(1, 2)\ (7, 10)$

8. Find the equation of the straight line which passes through each of the following pairs of points:
   a) $(1, -1)\ (4, 8)$
   b) $(0, 7)\ (3, 1)$

9. A line L_1 passes through the points $(-2, 5)$ and $(5, 3)$.
   a) Write down the equation of the line L_1.
   Another line L_2 is perpendicular to L_1 and also passes through the point $(-2, 5)$.
   b) Write down the equation of the line L_2.

10. The diagram below show a rhombus ABCD. The coordinates of A, B and D are given.

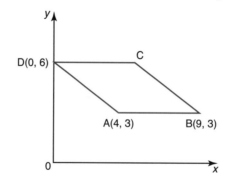

   a) Calculate:
      i) the coordinate of the point C
      ii) the equation of the line passing through A and C
      iii) the equation of the line passing through B and D.
   b) Are the diagonals of the rhombus perpendicular to each other? Justify your answer.

Student assessment 2

1. For each of the following lines, select two points on the line and then calculate its gradient.

a)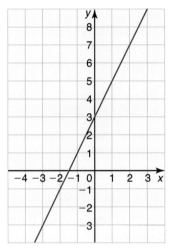

b)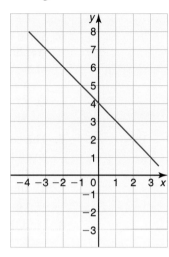

2. Find the equation of the straight line for each of the following:

a)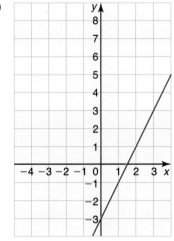

b)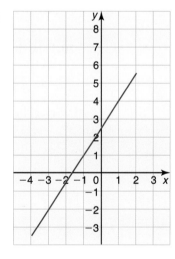

3. Calculate the gradient and *y*-intercept for each of the following linear equations:
   a) $y = \frac{1}{2}x$
   b) $-4x + y = 6$
   c) $2y - (5 - 3x) = 0$

4. Write down the equation of the line parallel to the line $y = 5x + 6$ which passes through the origin.

Coordinate geometry

5. Plot the following graphs on the same pair of axes, labelling each clearly.
   a) $x = 3$
   b) $y = -2$
   c) $y = -3x$
   d) $y = \dfrac{x}{4} + 4$

6. Solve the following pairs of simultaneous equations graphically.
   a) $x + y = 6$
      $x - y = 0$
   b) $x + 2y = 8$
      $x - y = -1$
   c) $2x - y = -5$
      $x - 3y = 0$
   d) $4x - 2y = -2$
      $3x - y + 2 = 0$

7. The coordinates of the end points of two line segments are given below. For each line segment calculate:
   i) the length
   ii) the midpoint.
   a) $(2, 6)\ (-2, 3)$
   b) $(-10, -10)\ (0, 14)$

8. Find the equation of the straight line which passes through each of the following pairs of points:
   a) $(-2, -9)\ (5, 5)$
   b) $(1, -1)\ (-1, 7)$

9. a) Write down the equation of the line L_1 that passes through the points $(3, 7)$ and $(-4, 9)$.
   b) Write down the equation of the line L_2 that is parallel to L_1 and passes through the point $(-6, -1)$.

10. The diagram below show an isosceles triangle ABC, where $AB = BC$. The coordinates of A and C are given.

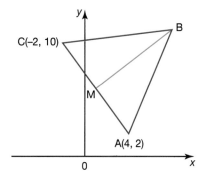

   a) The midpoint of AC is the point M. Calculate the coordinates of M.
   b) Calculate the equation of the line passing through B and M.
   c) If the y-coordinate of point B is $10\tfrac{1}{2}$, show that the x-coordinate is 7.
   d) Calculate the length of the line segment BM.

Mathematical investigations and ICT

● Plane trails

In an aircraft show, planes are made to fly with a coloured smoke trail. Depending on the formation of the planes, the trails can intersect in different ways.

In the diagram below the three smoke trails do not cross, as they are parallel.

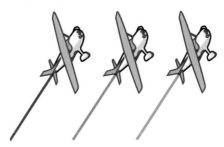

In the following diagram there are two crossing points.

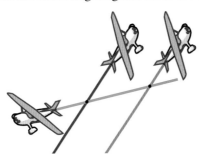

By flying differently, the three planes can produce trails that cross at three points.

1. Investigate the connection between the maximum number of crossing points and the number of planes.
2. Record the results of your investigation in an ordered table.
3. Write an algebraic rule linking the number of planes (p) and the maximum number of crossing points (n).

Coordinate geometry

● Hidden treasure

A television show sets up a puzzle for its contestants to try and solve. Some buried treasure is hidden on a 'treasure island'. The treasure is hidden in one of the 12 treasure chests shown (left). Each contestant stands by one of the treasure chests.

The treasure is hidden according to the following rule:

> It is not hidden in chest 1.
> Chest 2 is left empty for the time being.
> It is not hidden in chest 3.
> Chest 4 is left empty for the time being.
> It is not hidden in chest 5.
> The pattern of crossing out the first chest and then alternate chests is continued until only one chest is left. This will involve going round the circle several times continuing the pattern.
> The treasure is hidden in the last chest left.

The diagrams below show how the last chest is chosen:

After 1 round, chests 1, 3, 5, 7, 9 and 11 have been discounted.

After the second round, chests 2, 6 and 10 have also been discounted.

After the third round, chests 4 and 12 have also been discounted. This leaves only chest 8.
The treasure is therefore hidden in chest 8.

Unfortunately for participants, the number of contestants changes each time.

1. Investigate which treasure chest you would choose if there are:
   a) 4 contestants
   b) 5 contestants
   c) 8 contestants
   d) 9 contestants
   e) 15 contestants.

2. Investigate the winning treasure chest for other numbers of contestants and enter your results in an ordered table.
3. State any patterns you notice in your table of results.
4. Use your patterns to predict the winning chest for 31, 32 and 33 contestants.
5. Write a rule linking the winning chest x and the number of contestants n.

● **ICT activity**

A graphics calculator is able to graph inequalities and shade the appropriate region. The examples below show some screenshots taken from a graphics calculator.

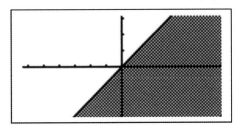 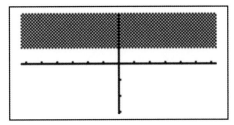

Investigate how a graphics calculator can graph linear inequalities.

Topic 6: Trigonometry

Syllabus

E6.1
Interpret and use three-figure bearings.

E6.2
Apply Pythagoras' theorem and the sine, cosine and tangent ratios for acute angles to the calculation of a side or of an angle of a right-angled triangle.
Solve trigonometrical problems in two dimensions involving angles of elevation and depression.
Extend sine and cosine values to angles between 90° and 180°.

E6.3
Solve problems using the sine and cosine rules for any triangle and the formula area of triangle $= \frac{1}{2}ab \sin C$.

E6.4
Solve simple trigonometrical problems in three dimensions including angle between a line and a plane.

Contents

Chapter 29 Bearings (E6.1)
Chapter 30 Trigonometry (E6.2)
Chapter 31 Further trigonometry (E6.3, E6.4)

The Swiss

● Leonhard Euler

Euler, like Newton, was the greatest mathematician of his generation. He studied all areas of mathematics and continued to work hard after he had gone blind.

As a young man, Euler discovered and proved:

the sum of the infinite series $\sum \left(\dfrac{1}{n^2}\right) = \dfrac{\pi^2}{6}$

i.e. $\dfrac{1}{1^2} + \dfrac{1}{2^2} + \dfrac{1}{3^2} + \ldots + \dfrac{1}{n^2} = \dfrac{\pi^2}{6}$

This brought him to the attention of other mathematicians.

Euler made discoveries in many areas of mathematics, especially calculus and trigonometry. He also developed the ideas of Newton and Leibniz.

Euler worked on graph theory and functions and was the first to prove several theorems in geometry. He studied relationships between a triangle's height, midpoint, and circumscribing and inscribing circles, and an expression for the area of a tetrahedron (a triangular pyramid) in terms of its sides.

He also worked on number theory and found the largest prime number known at the time.

Some of the most important constant symbols in mathematics, π, e and i (the square root of −1), were introduced by Euler.

Leonhard Euler (1707−1783)

● The Bernoulli family

The Bernoullis were a family of Swiss merchants who were friends of Euler. The two brothers, Johann and Jacob, were very gifted mathematicians and scientists, as were their children and grandchildren. They made discoveries in calculus, trigonometry and probability theory in mathematics. In science, they worked on astronomy, magnetism, mechanics, thermodynamics and more.

Unfortunately many members of the Bernoulli family were not pleasant people. The older members of the family were jealous of each other's successes and often stole the work of their sons and grandsons and pretended that it was their own.

29 Bearings

NB: All diagrams are not drawn to scale.

● Bearings

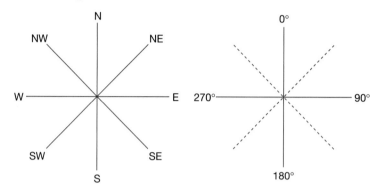

In the days when sailing ships travelled the oceans of the world, compass bearings like the ones in the diagram above were used.

As the need for more accurate direction arose, extra points were added to N, S, E, W, NE, SE, SW and NW. Midway between North and North East was North North East, and midway between North East and East was East North East, and so on. This gave sixteen points of the compass. This was later extended even further, eventually to sixty four points.

As the speed of travel increased, a new system was required. The new system was the **three figure bearing** system. North was given the bearing zero. 360° in a clockwise direction was one full rotation.

● Back bearings

The bearing of B from A is 135° and the distance from A to B is 8 cm, as shown (left). The bearing of A from B is called the **back bearing**.

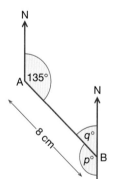

Since the two North lines are parallel:
$p = 135°$ (alternate angles), so the back bearing is $(180 + 135)°$.
That is, 315°.
(There are a number of methods of solving this type of problem.)

29　　Bearings

Worked example　　The bearing of B from A is 245°.
What is the bearing of A from B?

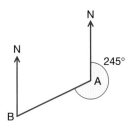

Since the two North lines are parallel:
$b = 65°$ (alternate angles), so the bearing is $(245 - 180)°$.
That is, 065°.

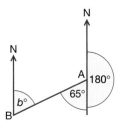

Exercise 29.1

1. Draw a diagram to show the following compass bearings and journey. Use a scale of 1 cm : 1 km. North can be taken to be a line vertically up the page.

   Start at point A. Travel a distance of 7 km on a bearing of 135° to point B. From B, travel 12 km on a bearing of 250° to point C. Measure the distance and bearing of A from C.

2. Given the following bearings of point B from point A, draw diagrams and use them to calculate the bearings of A from B.
   a) bearing 163°　　b) bearing 214°

3. Given the following bearings of point D from point C, draw diagrams and use them to calculate the bearings of C from D.
   a) bearing 300°　　b) bearing 282°

Trigonometry

Student assessment 1

1. A climber gets to the top of Mont Blanc. He can see in the distance a number of ski resorts. He uses his map to find the bearing and distance of the resorts, and records them as shown below:

    Val d'Isère 30 km   bearing 082°
    Les Arcs 40 km   bearing 135°
    La Plagne 45 km   bearing 205°
    Méribel 35 km   bearing 320°

    Choose an appropriate scale and draw a diagram to show the position of each resort. What are the distance and bearing of the following?
    a) Val d'Isère from La Plagne
    b) Méribel from Les Arcs

2. A coastal radar station picks up a distress call from a ship. It is 50 km away on a bearing of 345°. The radar station contacts a lifeboat at sea which is 20 km away on a bearing of 220°.
    Make a scale drawing and use it to find the distance and bearing of the ship from the lifeboat.

3. An aircraft is seen on radar at Milan airport. The aircraft is 210 km away from the airport on a bearing of 065°. The aircraft is diverted to Rome airport, which is 130 km away from Milan on a bearing of 215°. Use an appropriate scale and make a scale drawing to find the distance and bearing of Rome airport from the aircraft.

30 Trigonometry

NB: All diagrams are not drawn to scale.

In this chapter, unless instructed otherwise, give your answers exactly or correct to three significant figures as appropriate. Answers in degrees should be given to one decimal place.

There are three basic trigonometric ratios: sine, cosine and tangent.

Each of these relates an angle of a right-angled triangle to a ratio of the lengths of two of its sides.

The sides of the triangle have names, two of which are dependent on their position in relation to a specific angle.

The longest side (always opposite the right angle) is called the **hypotenuse**. The side opposite the angle is called the **opposite** side and the side next to the angle is called the **adjacent** side.

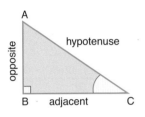

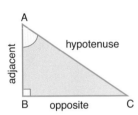

Note that, when the chosen angle is at A, the sides labelled opposite and adjacent change (left).

● **Tangent**

$$\tan C = \frac{\text{length of opposite side}}{\text{length of adjacent side}}$$

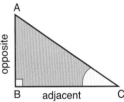

Worked examples

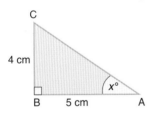

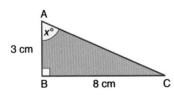

a) Calculate the size of angle BAC in each of the triangles on the left.

i) $\tan x° = \dfrac{\text{opposite}}{\text{adjacent}} = \dfrac{4}{5}$

$x = \tan^{-1}\left(\dfrac{4}{5}\right)$

$x = 38.7$ (3 s.f.)

$\angle BAC = 38.7°$ (1 d.p.)

ii) $\tan x° = \dfrac{8}{3}$

$x = \tan^{-1}\left(\dfrac{8}{3}\right)$

$x = 69.4$ (3 s.f.)

$\angle BAC = 69.4°$ (1 d.p.)

b) Calculate the length of the opposite side QR (right).

$\tan 42° = \dfrac{p}{6}$

$6 \times \tan 42° = p$

$p = 5.40$ (3 s.f.)

$QR = 5.40$ cm (3 s.f.)

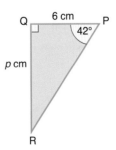

Trigonometry

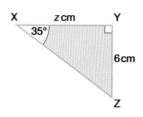

c) Calculate the length of the adjacent side XY.

$$\tan 35° = \frac{6}{z}$$

$$z \times \tan 35° = 6$$

$$z = \frac{6}{\tan 35°}$$

$$z = 8.57 \text{ (3 s.f.)}$$

$$XY = 8.57 \text{ cm (3 s.f.)}$$

Exercise 30.1 Calculate the length of the side marked x cm in each of the diagrams in questions 1 and 2.

1. a) 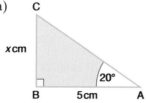   b) 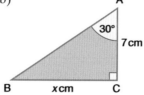   c)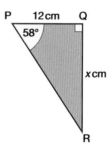

   d) 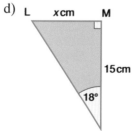   e) 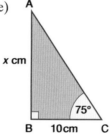   f)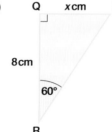

2. a) 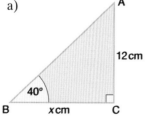   b) 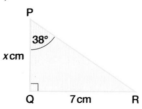   c)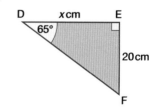

   d) 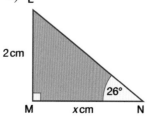 e) 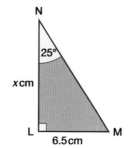 f)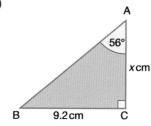

3. Calculate the size of the marked angle $x°$ in each of the following diagrams.

a)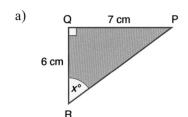
b)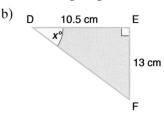
c)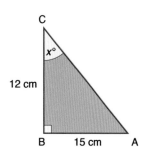

d)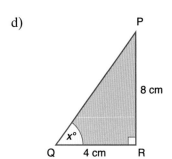
e)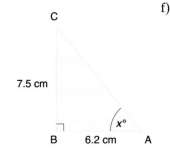
f)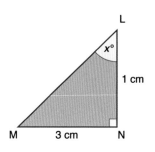

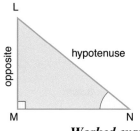

● **Sine**

$$\sin N = \frac{\text{length of opposite side}}{\text{length of hypotenuse}}$$

Worked examples a) Calculate the size of angle BAC.

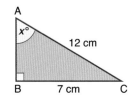

$$\sin x = \frac{\text{opposite}}{\text{hypotenuse}} = \frac{7}{12}$$

$$x = \sin^{-1}\left(\frac{7}{12}\right)$$

$x = 35.7$ (1 d.p.)
$\angle BAC = 35.7°$ (1 d.p.)

b) Calculate the length of the hypotenuse PR.

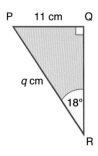

$$\sin 18° = \frac{11}{q}$$

$q \times \sin 18° = 11$

$$q = \frac{11}{\sin 18°}$$

$q = 35.6$ (3 s.f.)

PR = 35.6 cm (3 s.f.)

Trigonometry

Exercise 30.2

1. Calculate the length of the marked side in each of the following diagrams.

a)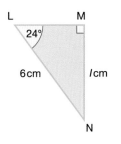
b)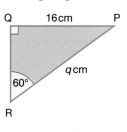
c)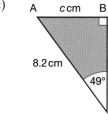

d)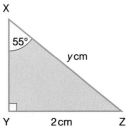
e)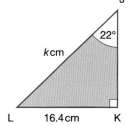
f)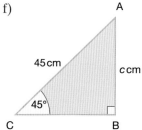

2. Calculate the size of the angle marked x in each of the following diagrams.

a)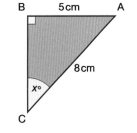
b)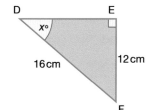
c)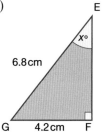

d)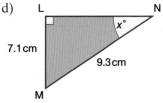
e)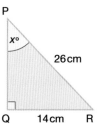
f)

● **Cosine**

$$\cos Z = \frac{\text{length of adjacent side}}{\text{length of hypotenuse}}$$

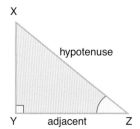

Worked examples **a)** Calculate the length XY.

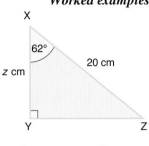

$$\cos 62° = \frac{\text{adjacent}}{\text{hypotenuse}} = \frac{z}{20}$$

$z = 20 \times \cos 62°$
$z = 9.39$ (3 s.f.)
XY = 9.39 cm (3 s.f.)

b) Calculate the size of angle ABC.

$$\cos x = \frac{5.3}{12}$$

$$x = \cos^{-1}\left(\frac{5.3}{12}\right)$$

$x = 63.8$ (1 d.p.)
$\angle ABC = 63.8°$ (1 d.p.)

Exercise 30.3 **1.** Calculate the marked side or angle in each of the following diagrams.

a)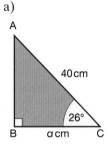

b)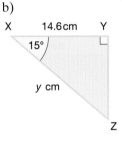

c)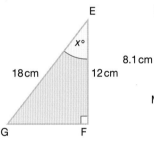

d)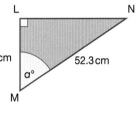

e)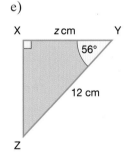

f)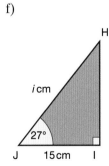

g)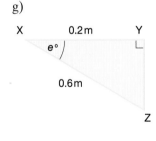

h)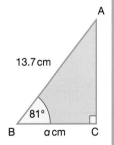

Pythagoras' theorem

Pythagoras' theorem states the relationship between the lengths of the three sides of a right-angled triangle.

Pythagoras' theorem states that:

$a^2 = b^2 + c^2$

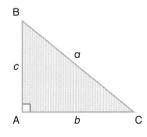

Trigonometry

Worked examples

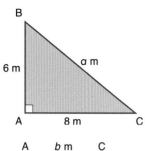

a) Calculate the length of the side BC.

Using Pythagoras:
$$a^2 = b^2 + c^2$$
$$a^2 = 8^2 + 6^2$$
$$a^2 = 64 + 36 = 100$$
$$a = \sqrt{100}$$
$$a = 10$$
$$BC = 10 \text{ m}$$

b) Calculate the length of the side AC.

Using Pythagoras:
$$a^2 = b^2 + c^2$$
$$a^2 - c^2 = b^2$$
$$b^2 = 144 - 25 = 119$$
$$b = \sqrt{119}$$
$$b = 10.9 \text{ (3 s.f.)}$$
$$AC = 10.9 \text{ m (3 s.f.)}$$

Exercise 30.4 In each of the diagrams in questions 1 and 2, use Pythagoras' theorem to calculate the length of the marked side.

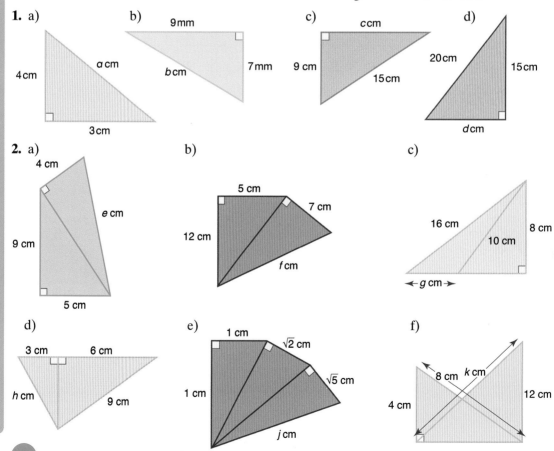

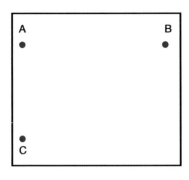

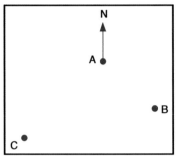

3. Villages A, B and C (left) lie on the edge of the Namib desert. Village A is 30 km due North of village C. Village B is 65 km due East of A.
   Calculate the shortest distance between villages C and B, giving your answer to the nearest 0.1 km.

4. Town X is 54 km due West of town Y. The shortest distance between town Y and town Z is 86 km. If town Z is due South of X calculate the distance between X and Z, giving your answer to the nearest kilometre.

5. Village B is on a bearing of 135° and at a distance of 40 km from village A, as shown (left). Village C is on a bearing of 225° and a distance of 62 km from village A.
   a) Show that triangle ABC is right-angled.
   b) Calculate the distance from B to C, giving your answer to the nearest 0.1 km.

6. Two boats set off from X at the same time (below). Boat A sets off on a bearing of 325° and with a velocity of 14 km/h. Boat B sets off on a bearing of 235° with a velocity of 18 km/h. Calculate the distance between the boats after they have been travelling for 2.5 hours. Give your answer to the nearest metre.

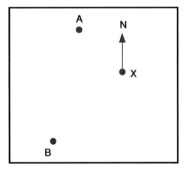

7. A boat sets off on a trip from S. It heads towards B, a point 6 km away and due North. At B it changes direction and heads towards point C, also 6 km away and due East of B. At C it changes direction once again and heads on a bearing of 135° towards D which is 13 km from C.
   a) Calculate the distance between S and C to the nearest 0.1 km.
   b) Calculate the distance the boat will have to travel if it is to return to S from D.

8. Two trees are standing on flat ground.
   The height of the smaller tree is 7 m. The distance between the top of the smaller tree and the base of the taller tree is 15 m.
   The distance between the top of the taller tree and the base of the smaller tree is 20 m.
   a) Calculate the horizontal distance between the two trees.
   b) Calculate the height of the taller tree.

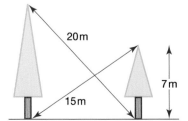

Trigonometry

Exercise 30.5

1. By using Pythagoras' theorem, trigonometry or both, calculate the marked value in each of the following diagrams. In each case give your answer to 1 d.p.

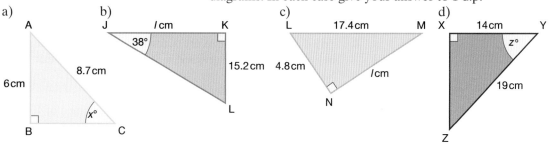

2. A sailing boat sets off from a point X and heads towards Y, a point 17 km North. At point Y it changes direction and heads towards point Z, a point 12 km away on a bearing of 090°. Once at Z the crew want to sail back to X. Calculate:
   a) the distance ZX,
   b) the bearing of X from Z.

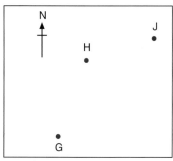

3. An aeroplane sets off from G (left) on a bearing of 024° towards H, a point 250 km away. At H it changes course and heads towards J on a bearing of 055° and a distance of 180 km away.
   a) How far is H to the North of G?
   b) How far is H to the East of G?
   c) How far is J to the North of H?
   d) How far is J to the East of H?
   e) What is the shortest distance between G and J?
   f) What is the bearing of G from J?

4. Two trees are standing on flat ground. The angle of elevation of their tops from a point X on the ground is 40°. If the horizontal distance between X and the small tree is 8 m and the distance between the tops of the two trees is 20 m, calculate:
   a) the height of the small tree,
   b) the height of the tall tree,
   c) the horizontal distance between the trees.

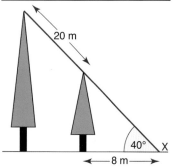

5. PQRS is a quadrilateral. The sides RS and QR are the same length. The sides QP and RS are parallel. Calculate:
   a) angle SQR,
   b) angle PSQ,
   c) length PQ,
   d) length PS,
   e) the area of PQRS.

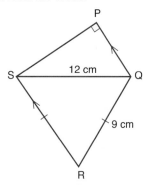

356

Angles of elevation and depression

The **angle of elevation** is the angle above the horizontal through which a line of view is raised. The **angle of depression** is the angle below the horizontal through which a line of view is lowered.

Worked examples

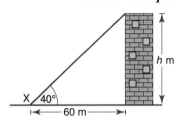

a) The base of a tower is 60 m away from a point X on the ground. If the angle of elevation of the top of the tower from X is 40° calculate the height of the tower.

Give your answer to the nearest metre.

$$\tan 40° = \frac{h}{60}$$

$$h = 60 \times \tan 40° = 50.3$$

The height is 50.3 m (3 s.f.).

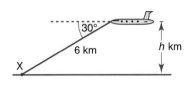

b) An aeroplane receives a signal from a point X on the ground. If the angle of depression of point X from the aeroplane is 30° calculate the height at which the plane is flying.

Give your answer to the nearest 0.1 km.

$$\sin 30° = \frac{h}{6}$$

$$h = 6 \times \sin 30° = 3$$

The height is 3 km.

Exercise 30.6

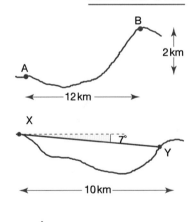

1. A and B are two villages. If the horizontal distance between them is 12 km and the vertical distance between them is 2 km calculate:
   a) the shortest distance between the two villages,
   b) the angle of elevation of B from A.

2. X and Y are two towns. If the horizontal distance between them is 10 km and the angle of depression of Y from X is 7° calculate:
   a) the shortest distance between the two towns,
   b) the vertical height between the two towns.

3. A girl standing on a hill at A, overlooking a lake, can see a small boat at a point B on the lake. If the girl is at a height of 50 m above B and at a horizontal distance of 120 m away from B, calculate:
   a) the angle of depression of the boat from the girl,
   b) the shortest distance between the girl and the boat.

Trigonometry

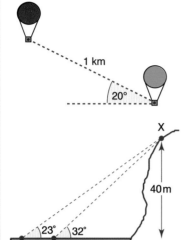

4. Two hot air balloons are 1 km apart in the air. If the angle of elevation of the higher from the lower balloon is 20°, calculate, giving your answers to the nearest metre:
   a) the vertical height between the two balloons,
   b) the horizontal distance between the two balloons.

5. A boy X can be seen by two of his friends Y and Z, who are swimming in the sea. If the angle of elevation of X from Y is 23° and from Z is 32°, and the height of X above Y and Z is 40 m calculate:
   a) the horizontal distance between X and Z,
   b) the horizontal distance between Y and Z.
   Note: XYZ is a vertical plane

6. A plane is flying at an altitude of 6 km directly over the line AB. It spots two boats A and B, on the sea. If the angles of depression of A and B from the plane are 60° and 30° respectively, calculate the horizontal distance between A and B.

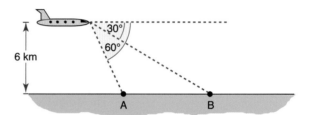

7. Two planes are flying directly above each other. A person standing at P can see both of them. The horizontal distance between the two planes and the person is 2 km. If the angles of elevation of the planes from the person are 65° and 75° calculate:
   a) the altitude at which the higher plane is flying,
   b) the vertical distance between the two planes.

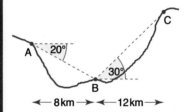

8. Three villages A, B and C can see each other across a valley. The horizontal distance between A and B is 8 km, and the horizontal distance between B and C is 12 km. The angle of depression of B from A is 20° and the angle of elevation of C from B is 30°. Calculate, giving all answers to 1 d.p.:
   a) the vertical height between A and B,
   b) the vertical height between B and C,
   c) the angle of elevation of C from A,
   d) the shortest distance between A and C.
   Note: A, B and C are in the same vertical plane.

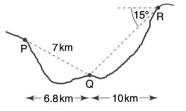

9. Using binoculars, three people P, Q and R can see each other across a valley. The horizontal distance between P and Q is 6.8 km and the horizontal distance between Q and R is 10 km. If the shortest distance between P and Q is 7 km and the angle of depression of Q from R is 15°, calculate, giving appropriate answers:
   a) the vertical height between Q and R,
   b) the vertical height between P and R,
   c) the angle of elevation of R from P,
   d) the shortest distance between P and R.
   Note: P, Q and R are in the same vertical plane.

10. Two people A and B are standing either side of a transmission mast. A is 130 m away from the mast and B is 200 m away.

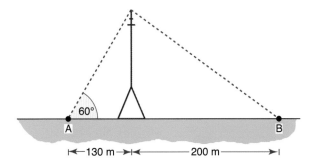

If the angle of elevation of the top of the mast from A is 60°, calculate:
a) the height of the mast to the nearest metre,
b) the angle of elevation of the top of the mast from B.

● Angles between 0° and 180°

When calculating the size of angles using trigonometry, there are often two solutions. Most calculators, however, will only give the first solution. To be able to calculate the value of the second possible solution, an understanding of the shape of trigonometrical graphs is needed.

● The sine curve

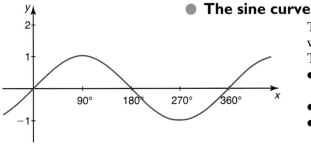

The graph of $y = \sin x$ is plotted (left), where x is the size of the angle in degrees. The graph of $y = \sin x$ has
- a period of 360° (i.e. it repeats itself every 360°),
- a maximum value of 1 (at 90°),
- a minimum value of −1 (at 270°).

Trigonometry

Worked example sin 30° = 0.5. Which other angle between 0° and 180° has a sine of 0.5?

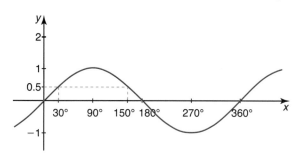

From the graph above it can be seen that sin 150° = 0.5.

sin x = sin(180° − x)

Exercise 30.7

1. Express each of the following in terms of the sine of another angle between 0° and 180°:
   a) sin 60°   b) sin 80°   c) sin 115°
   d) sin 140°   e) sin 128°   f) sin 167°

2. Express each of the following in terms of the sine of another angle between 0° and 180°:
   a) sin 35°   b) sin 50°   c) sin 30°
   d) sin 48°   e) sin 104°   f) sin 127°

3. Find the two angles between 0° and 180° which have the following sine. Give each angle to the nearest degree.
   a) 0.33   b) 0.99   c) 0.09
   d) 0.95   e) 0.22   f) 0.47

4. Find the two angles between 0° and 180° which have the following sine. Give each angle to the nearest degree.
   a) 0.94   b) 0.16   c) 0.80
   d) 0.56   e) 0.28   f) 0.33

● **The cosine curve**

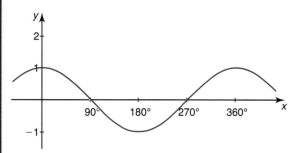

The graph of $y = \cos x$ is plotted (left), where x is the size of the angle in degrees.

As with the sine curve, the graph of $y = \cos x$ has

- a period of 360°,
- a maximum value of 1,
- a minimum value of −1.

Note cos $x°$ = − cos(180 − x)°

Worked examples **a)** cos 60° = 0.5. Which angle between 0° and 180° has a cosine of −0.5?

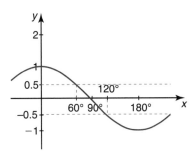

From the graph above it can be seen that cos 120° = −0.5 so cos 60° = −cos 120°.

b) The cosine of which angle between 0° and 180° is equal to the negative of cos 50°?

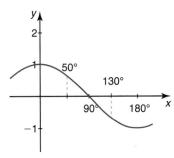

cos 130° = − cos 50°

Exercise 30.8

1. Express each of the following in terms of the cosine of another angle between 0° and 180°:
   a) cos 20°
   b) cos 85°
   c) cos 32°
   d) cos 95°
   e) cos 147°
   f) cos 106°

2. Express each of the following in terms of the cosine of another angle between 0° and 180°:
   a) cos 98°
   b) cos 144°
   c) cos 160°
   d) cos 143°
   e) cos 171°
   f) cos 123°

3. Express each of the following in terms of the cosine of another angle between 0° and 180°:
   a) −cos 100°
   b) cos 90°
   c) −cos 110°
   d) − cos 45°
   e) −cos 122°
   f) − cos 25°

4. The cosine of which acute angle has the same value as:
   a) cos 125°
   b) cos 107°
   c) −cos 120°
   d) − cos 98°
   e) −cos 92°
   f) − cos 110°?

Trigonometry

Student assessment 1

1. Calculate the length of the side marked x cm in each of the following.

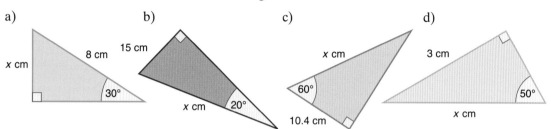

2. Calculate the size of the angle marked $\theta°$ in each of the following.

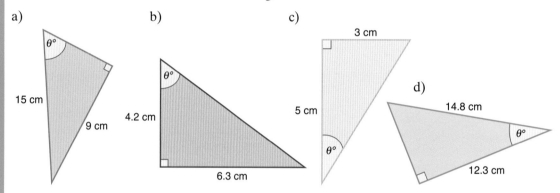

3. Calculate the length of the side marked q cm in each of the following.

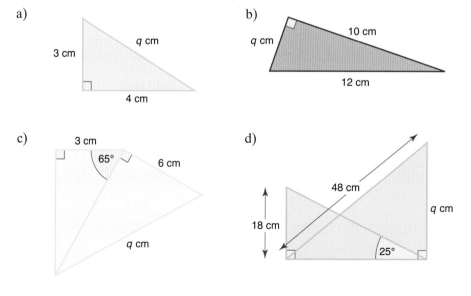

Student assessment 2

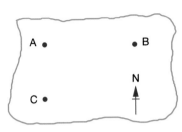

1. A map shows three towns A, B and C. Town A is due North of C. Town B is due East of A. The distance AC is 75 km and the bearing of C from B is 245°. Calculate, giving your answers to the nearest 100 m:
   a) the distance AB,
   b) the distance BC.

2. Two trees stand 16 m apart. Their tops make an angle of $\theta°$ at point A on the ground.
   a) Express $\theta°$ in terms of the height of the shorter tree and its distance x metres from point A.
   b) Express $\theta°$ in terms of the height of the taller tree and its distance from A.
   c) Form an equation in terms of x.
   d) Calculate the value of x.
   e) Calculate the value θ.

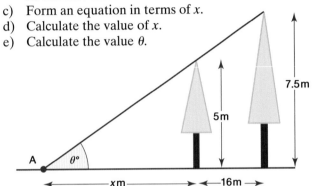

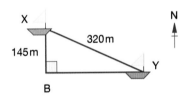

3. Two boats X and Y, sailing in a race, are shown in the diagram (left). Boat X is 145 m due North of a buoy B. Boat Y is due East of buoy B. Boats X and Y are 320 m apart. Calculate:
   a) the distance BY,
   b) the bearing of Y from X,
   c) the bearing of X from Y.

4. Two hawks P and Q are flying vertically above one another. Hawk Q is 250 m above hawk P. They both spot a snake at R.

   Using the information given, calculate:
   a) the height of P above the ground,
   b) the distance between P and R,
   c) the distance between Q and R.

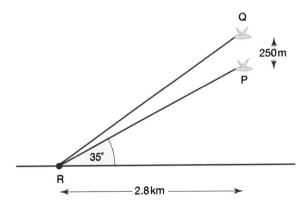

Trigonometry

Student assessment 3

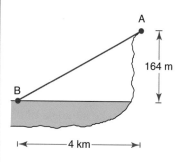

1. A boy standing on a cliff top at A can see a boat sailing in the sea at B. The vertical height of the boy above sea level is 164 m, and the horizontal distance between the boat and the boy is 4 km. Calculate:
   a) the distance AB to the nearest metre,
   b) the angle of depression of the boat from the boy.

2. Draw the graph of $y = \sin x°$ for $0° \leq x° \leq 180°$. Mark on the graph the angles $0°, 90°, 180°$, and also the maximum and minimum values of y.

3. Express each of the following in terms of another angle between 0° and 180°.
   a) sin 50°
   b) sin 150°
   c) cos 45°
   d) cos 120°

4. Find an angle between 0° and 180° which has the following cosine. Give each angle to the nearest degree.
   a) 0.79
   b) −0.28

5. An airship is travelling in a horizontal direction as shown, at a speed of 4 km/h. Its vertical height above the ground is 3.2 km. At 15 00 its horizontal distance from A is 7 km. A is under the flight path of the airship. Calculate:

   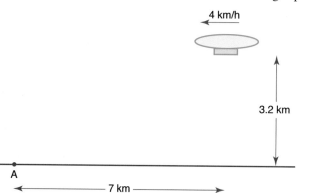

   a) the angle of elevation of the airship from A at 15 00,
   b) the angle of elevation of the airship from A at 15 30,
   c) the distance between A and the airship at 15 30,
   d) at what time, to the nearest minute, the angle of elevation of the airship from A will be 85°.

Student assessment 4

1. Draw a graph of $y = \cos \theta°$, for $0° \leq \theta° \leq 180°$. Mark on the angles $0°, 90°, 180°$, and also the maximum and minimum values of y.

2. The cosine of which other angle between 0 and 180° has the same value as
   a) cos 128°
   b) −cos 80°?

3. The Great Pyramid at Giza is 146 m high. Two people A and B are looking at the top of the pyramid. The angle of elevation of the top of the pyramid from B is 12°. The distance between A and B is 25 m.

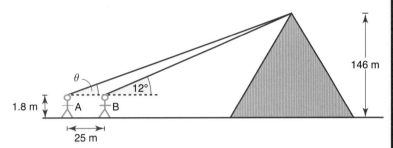

If both A and B are 1.8 m tall, calculate:
a) the distance on the ground from B to the centre of the base of the pyramid,
b) the angle of elevation θ of the top of the pyramid from A,
c) the distance between A and the top of the pyramid.
Note: A, B and the top of the pyramid are in the same vertical plane.

4. Two hot air balloons A and B are travelling in the same horizontal direction as shown in the diagram below. A is travelling at 2 m/s and B at 3 m/s. Their heights above the ground are 1.6 km and 1 km, respectively.
At midday, their horizontal distance apart is 4 km and balloon B is directly above a point X on the ground.

Calculate:
a) the angle of elevation of A from X at midday,
b) the angle of depression of B from A at midday,
c) their horizontal distance apart at 12 30,
d) the angle of elevation of B from X at 12 30,
e) the angle of elevation of A from B at 12 30,
f) how much closer A and B are at 12 30 compared with midday.

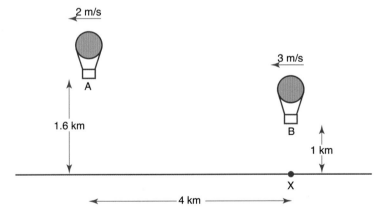

5. a) On one diagram plot the graph of $y = \sin \theta°$ and the graph of $y = \cos \theta°$, for $0° \leq \theta° \leq 180°$.
b) Use your graph to find the angles for which $\sin \theta° = \cos \theta°$.

31 Further trigonometry

● **The sine rule**

With right-angled triangles we can use the basic trigonometric ratios of sine, cosine and tangent. The **sine rule** is a relationship which can be used with non right-angled triangles.

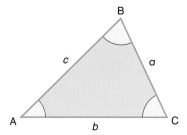

The sine rule states that:

$$\frac{a}{\sin A} = \frac{b}{\sin B} = \frac{c}{\sin C}$$

or alternatively

$$\frac{\sin A}{a} = \frac{\sin B}{b} = \frac{\sin C}{c}$$

Worked examples

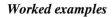

a) Calculate the length of side BC.

Using the sine rule:

$$\frac{a}{\sin A} = \frac{b}{\sin B}$$

$$\frac{a}{\sin 40°} = \frac{6}{\sin 30°}$$

$$a = \frac{6 \times \sin 40°}{\sin 30°}$$

$a = 7.7$ (1 d.p.)

BC = 7.71 cm (3 s.f.)

b) Calculate the size of angle C.

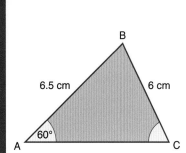

Using the sine rule:

$$\frac{\sin A}{a} = \frac{\sin C}{c}$$

$$\sin C = \frac{6.5 \times \sin 60°}{6}$$

$C = \sin^{-1}(0.94)$
$C = 69.8°$ (1 d.p.)

Further trigonometry

Exercise 31.1

1. Calculate the length of the side marked x in each of the following.

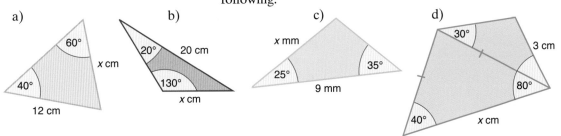

2. Calculate the size of the angle marked $\theta°$ in each of the following.

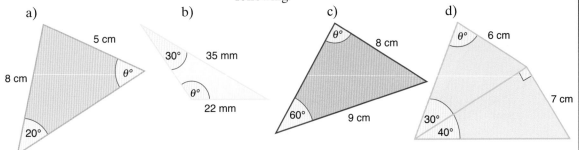

3. △ABC has the following dimensions:

   AC = 10 cm, AB = 8 cm and ∠ACB = 20°.

   a) Calculate the two possible values for ∠CBA.
   b) Sketch and label the two possible shapes for △ABC.

4. △PQR has the following dimensions:

   PQ = 6 cm, PR = 4 cm and ∠PQR = 40°.

   a) Calculate the two possible values for ∠QRP.
   b) Sketch and label the two possible shapes for △PQR.

● **The cosine rule**

The **cosine rule** is another relationship which can be used with non right-angled triangles.

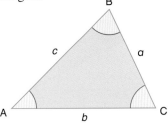

The cosine rule states that:

$$a^2 = b^2 + c^2 - 2bc \cos A$$

Trigonometry

Worked examples **a)** Calculate the length of the side BC.

Using the cosine rule:

$$a^2 = b^2 + c^2 - 2bc \cos A$$
$$a^2 = 9^2 + 7^2 - (2 \times 9 \times 7 \times \cos 50°)$$
$$= 81 + 49 - (126 \times \cos 50°) = 49.0$$
$$a = \sqrt{49.0}$$
$$a = 7.00 \text{ (3 s.f.)}$$
$$BC = 7.00 \text{ cm (3 s.f.)}$$

b) Calculate the size of angle A.

Using the cosine rule:

$$a^2 = b^2 + c^2 - 2bc \cos A.$$

Rearranging the equation gives:

$$\cos A = \frac{b^2 + c^2 - a^2}{2bc}$$

$$\cos A = \frac{15^2 + 12^2 - 20^2}{2 \times 15 \times 12} = -0.086$$

$$A = \cos^{-1}(-0.086)$$
$$A = 94.9° \text{ (1 d.p.)}$$

Exercise 31.2 **1.** Calculate the length of the side marked x in each of the following.

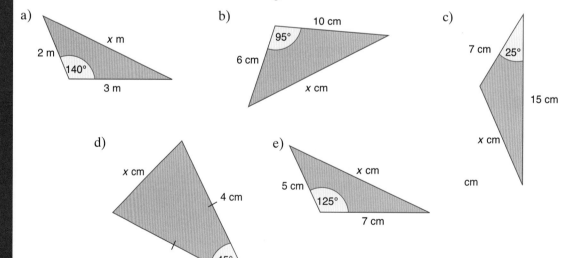

Further trigonometry

2. Calculate the angle marked $\theta°$ in each of the following.

a)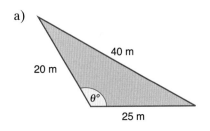

b)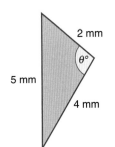

c)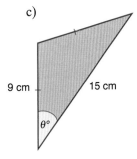

d)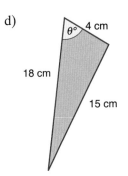

e)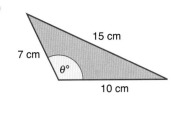

Exercise 31.3

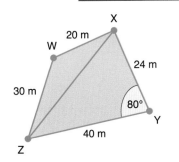

1. Four players W, X, Y and Z are on a rugby pitch. The diagram (left) shows a plan view of their relative positions. Calculate:
  a) the distance between players X and Z,
  b) $\angle ZWX$,
  c) $\angle WZX$,
  d) $\angle YZX$,
  e) the distance between players W and Y.

2. Three yachts A, B and C are racing off Cape Comorin in India. Their relative positions are shown (below).

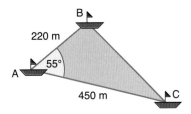

Calculate the distance between B and C to the nearest 10 m.

Trigonometry

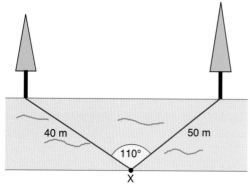

3. There are two trees standing on one side of a river bank. On the opposite side, a boy is standing at X.

 Using the information given, calculate the distance between the two trees.

● **The area of a triangle**

Area $= \frac{1}{2}bh$

Also:

$\sin C = \dfrac{h}{a}$

Rearranging:

$h = a \sin C$

Therefore

area $= \frac{1}{2}ab \sin C$

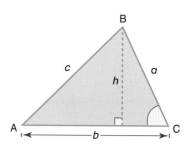

Exercise 31.4 1. Calculate the area of the following triangles.

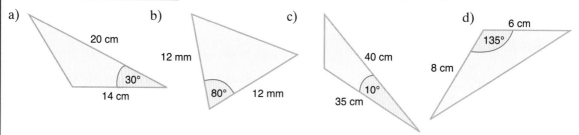

2. Calculate the value of x in each of the following.

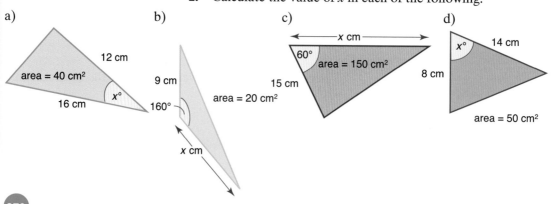

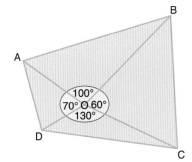

3. ABCD is a school playing field (left). The following lengths are known:

OA = 83 m,
OB = 122 m,
OC = 106 m,
OD = 78 m

Calculate the area of the school playing field to the nearest 100 m².

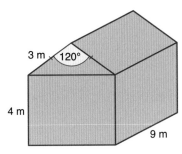

4. The roof of a garage has a slanting length of 3 m and makes an angle of 120° at its vertex (left). The height of the walls of the garage is 4 m and its depth is 9m.
Calculate:
a) the cross-sectional area of the roof,
b) the volume occupied by the whole garage.

● Trigonometry in three dimensions

Worked example The diagram (below) shows a cube of edge length 3 cm.

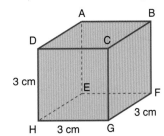

i) Calculate the length EG.

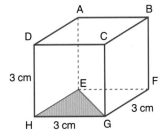 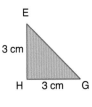

Triangle EHG (above) is right angled. Use Pythagoras' theorem to calculate the length EG.

$$EG^2 = EH^2 + HG^2$$
$$EG^2 = 3^2 + 3^2 = 18$$
$$EG = \sqrt{18} \text{ cm} = 4.24 \text{ cm (3 s.f.)}$$

Trigonometry

ii) Calculate the length AG.

Triangle AEG (below) is right angled. Use Pythagoras' theorem to calculate the length AG.

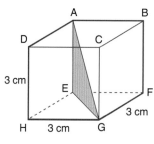 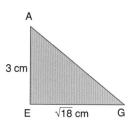

$AG^2 = AE^2 + EG^2$
$AG^2 = 3^2 + (\sqrt{18})^2$
$AG^2 = 9 + 18$
$AG = \sqrt{27}$ cm $= 5.20$ cm (3 s.f.)

iii) Calculate the angle EGA.

To calculate angle EGA we use the triangle EGA:

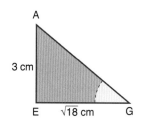

$\tan G = \dfrac{3}{\sqrt{18}}$

$G = 35.3°$ (1 d.p.)

Exercise 31.5

1. a) Calculate the length HF.
   b) Calculate the length HB.
   c) Calculate the angle BHG.

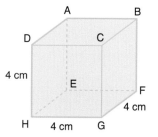

2. a) Calculate the length CA.
   b) Calculate the length CE.
   c) Calculate the angle ACE.

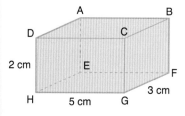

3. In the cuboid (right):
   a) Calculate the length EG.
   b) Calculate the length AG.
   c) Calculate the angle AGE.

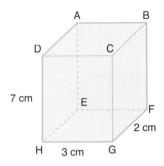

4. In the cuboid (left) calculate:
   a) the angle BCE,
   b) the angle GFH.

5. The diagram (right) shows a right pyramid where A is vertically above X.
   a) i) Calculate the length DB.
      ii) Calculate the angle DAX.
   b) i) Calculate the angle CED.
      ii) Calculate the angle DBA.

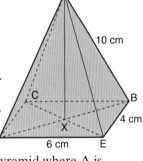

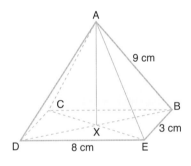

6. The diagram (left) shows a right pyramid where A is vertically above X.
   a) i) Calculate the length CE.
      ii) Calculate the angle CAX.
   b) i) Calculate the angle BDE.
      ii) Calculate the angle ADB.

7. In this cone (right) the angle YXZ = 60°. Calculate:
   a) the length XY,
   b) the length YZ,
   c) the circumference of the base.

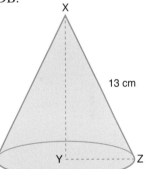

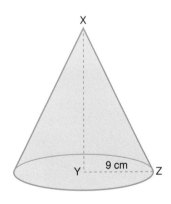

8. In this cone (left) the angle XZY = 40°. Calculate:
   a) the length XZ,
   b) the length XY.

9. One corner of this cuboid has been sliced off along the plane QTU. WU = 4 cm.
   a) Calculate the length of the three sides of the triangle QTU.
   b) Calculate the three angles P, Q and T in triangle PQT.
   c) Calculate the area of triangle PQT.

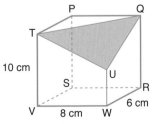

Trigonometry

● The angle between a line and a plane

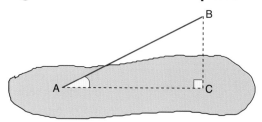

To calculate the size of the angle between the line AB and the shaded plane, drop a perpendicular from B. It meets the shaded plane at C. Then join AC.

The angle between the lines AB and AC represents the angle between the line AB and the shaded plane.

The line AC is the projection of the line AB on the shaded plane.

Worked example

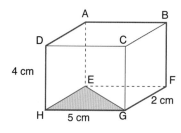

i) Calculate the length EC.

First use Pythagoras' theorem to calculate the length EG:

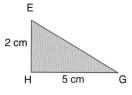

$EG^2 = EH^2 + HG^2$
$EG^2 = 2^2 + 5^2$
$EG^2 = 29$
$EG = \sqrt{29}$ cm

Now use Pythagoras' theorem to calculate CE:

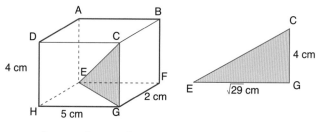

$EC^2 = EG^2 + CG^2$
$EC^2 = (\sqrt{29})^2 + 4^2$
$EC^2 = 29 + 16$
$EC = \sqrt{45}$ cm $= 6.71$ cm (3 s.f.)

Further trigonometry

ii) Calculate the angle between the line CE and the plane ADHE.

To calculate the angle between the line CE and the plane ADHE use the right-angled triangle CED and calculate the angle CED.

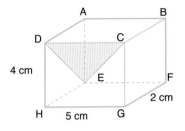
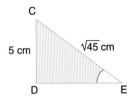

$$\sin E = \frac{CD}{CE} = \frac{5}{\sqrt{45}}$$

$$E = \sin^{-1}\left(\frac{5}{\sqrt{45}}\right) = 48.2° \text{ (1 d.p.)}$$

Exercise 31.6

1. Name the projection of each line onto the given plane (right):
a) TR onto RSWV
b) TR onto PQUT
c) SU onto PQRS
d) SU onto TUVW
e) PV onto QRVU
f) PV onto RSWV

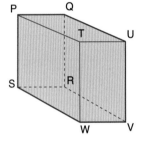

2. Name the projection of each line onto the given plane (left):
a) KM onto IJNM
b) KM onto JKON
c) KM onto HIML
d) IO onto HLOK
e) IO onto JKON
f) IO onto LMNO

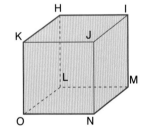

3. Name the angle between the given line and plane (left):
a) PT and PQRS
b) PU and PQRS
c) SV and PSWT
d) RT and TUVW
e) SU and QRVU
f) PV and PSWT

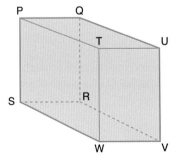

4. a) Calculate the length BH (right).
b) Calculate the angle between the line BH and the plane EFGH.

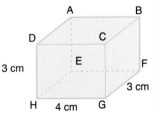

Trigonometry

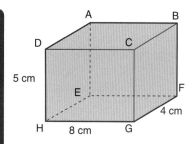

5. a) Calculate the length AG.
    b) Calculate the angle between the line AG and the plane EFGH.
    c) Calculate the angle between the line AG and the plane ADHE.

6. The diagram (right) shows a right pyramid where A is vertically above X.
    a) Calculate the length BD.
    b) Calculate the angle between AB and CBED.

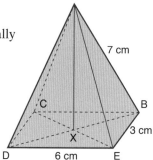

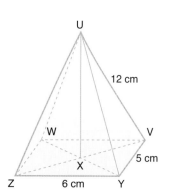

7. The diagram (left) shows a right pyramid where U is vertically above X.
    a) Calculate the length WY.
    b) Calculate the length UX.
    c) Calculate the angle between UX and UZY.

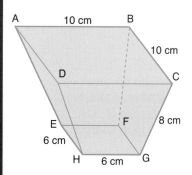

8. ABCD and EFGH are square faces lying parallel to each other.
    Calculate:
    a) the length DB,
    b) the length HF,
    c) the vertical height of the object,
    d) the angle DH makes with the plane ABCD.

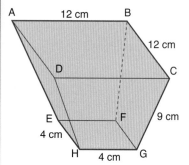

9. ABCD and EFGH are square faces lying parallel to each other.
    Calculate:
    a) the length AC,
    b) the length EG,
    c) the vertical height of the object,
    d) the angle CG makes with the plane EFGH.

Student assessment 1

1. Calculate the size of the obtuse angle marked $\theta°$ in the triangle (right).

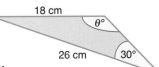

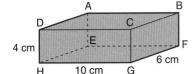

2. For the cuboid (left), calculate:
   a) the length EG,
   b) the length EC,
   c) ∠BEC.

3. For the quadrilateral (right), calculate:
   a) the length JL,
   b) ∠KJL,
   c) the length JM,
   d) the area of JKLM.

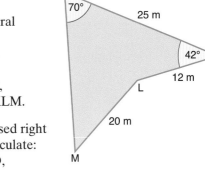

4. For the square-based right pyramid (left), calculate:
   a) the length BD,
   b) ∠ABD,
   c) the area of △ABD,
   d) the vertical height of the pyramid.

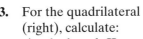

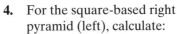

Student assessment 2

1. Using the triangular prism (left), calculate:
   a) the length AD,
   b) the length AC,
   c) the angle AC makes with the plane CDEF,
   d) the angle AC makes with the plane ABFE.

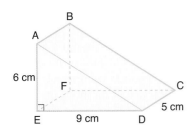

2. For the triangle (left), calculate:
   a) the length PS,
   b) ∠QRS,
   c) the length SR.

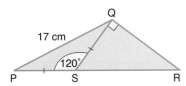

3. The cuboid (right) has one of its corners removed to leave a flat triangle BDC. Calculate:
   a) length DC,
   b) length BC,
   c) length DB,
   d) ∠CBD,
   e) the area of △BDC,
   f) the angle AC makes with the plane AEHD.

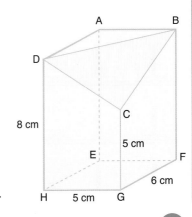

Topic 6: Mathematical investigations and ICT

● Numbered balls

The balls below start with the number 25 and then subsequent numbered balls are added according to a rule. The process stops when ball number 1 is added.

1. Express in words the rule for generating the sequence of numbered balls.
2. What is the longest sequence of balls starting with a number less than 100?
3. Is there a strategy for generating a long sequence?
4. Use your rule to state the longest sequence of balls starting with a number less than 1000.
5. Extend the investigation by having a different term-to-term rule.

● Towers of Hanoi

This investigation is based on an old Vietnamese legend. The legend is as follows:

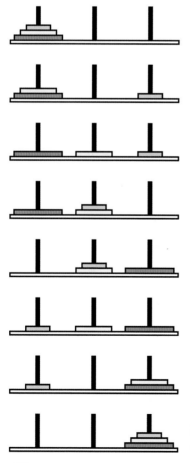

At the beginning of time a temple was created by the Gods. Inside the temple stood three giant rods. On one of these rods, 64 gold discs, all of different diameters, were stacked in descending order of size, i.e the largest at the bottom rising to the smallest at the top. Priests at the temple were responsible for moving the discs onto the remaining two rods until all 64 discs were stacked in the same order on one of the other rods. When this task was completed, time would cease and the world would come to an end.

The discs however could only be moved according to certain rules. These were:

- Only one disc could be moved at a time.
- A disc could only be placed on top of a larger one.

The diagram (left) shows the smallest number of moves required to transfer three discs from the rod on the left to the rod on the right.
  With three discs, the smallest number of moves is seven.

1. What is the smallest number of moves needed for 2 discs?
2. What is the smallest number of moves needed for 4 discs?
3. Investigate the smallest number of moves needed to move different numbers of discs.

4. Display the results of your investigation in an ordered table.
5. Describe any patterns you see in your results.
6. Predict, from your results, the smallest number of moves needed to move 10 discs.
7. Determine a formula for the smallest number of moves for n discs.
8. Assume the priests have been transferring the discs at the rate of one per second and assume the Earth is approximately 4.54 billion years old (4.54×10^9 years).
   According to the legend, is the world coming to an end soon? Justify your answer with relevant calculations.

● ICT activity

In this activity you will need to use a graphics calculator to investigate the relationship between different trigonometric ratios.

Note that this activity goes beyond the syllabus by considering angles greater than 180°.

1. a) Using the calculator, plot the graph of $y = \sin x$ for $0° \leq x \leq 360°$.
   The graph should look similar to the one shown below:

   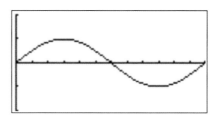

   b) Using the equation solving facility evaluate the following:
      i) sin 70°
      ii) sin 125°
      iii) sin 300°
   c) Referring to the graph explain why $\sin x = 0.7$ has two solutions between 0° and 360°.
   d) Use the graph to solve the equation $\sin x = 0.5$.

2. a) On the same axes as before plot $y = \cos x$.
   b) How many solutions are there to the equation $\sin x = \cos x$ between 0° and 360°?
   c) What is the solution to the equation $\sin x = \cos x$ between 180° and 270°?

3. By plotting appropriate graphs solve the following for $0° \leq x \leq 360°$.
   a) $\sin x = \tan x$
   b) $\cos x = \tan x$

Topic 7: Matrices and transformations

Syllabus

E7.1

Describe a translation by using a vector represented by e.g. $\binom{x}{y}$, $\overrightarrow{AB}$ or **a**.
Add and subtract vectors.
Multiply a vector by a scalar.

E7.2

Reflect simple plane figures in horizontal or vertical lines.
Rotate simple plane figures about the origin, vertices or midpoints of edges of the figures, through multiples of 90°.
Construct given translations and enlargements of simple plane figures.
Recognise and describe reflections, rotations, translations and enlargements.

E7.3

Calculate the magnitude of a vector $\binom{x}{y}$ as $\sqrt{x^2 + y^2}$.
Represent vectors by directed line segments.
Use the sum and difference of two vectors to express given vectors in terms of two coplanar vectors.
Use position vectors.

E7.4

Display information in the form of a matrix of any order.
Calculate the sum and product (where appropriate) of two matrices.
Calculate the product of a matrix and a scalar quantity.
Use the algebra of 2 × 2 matrices including the zero and identity 2 × 2 matrices.
Calculate the determinant |**A**| and inverse $\mathbf{A}^{-1}$ of a non-singular matrix **A**.

E7.5

Use the following transformations of the plane: reflection (M); rotation (R); translation (T); enlargement (E); and their combinations.
Identify and give precise descriptions of transformations connecting given figures.
Describe transformations using co-ordinates and matrices (singular matrices are excluded).

Contents

Chapter 32   Vectors (E7.1, E7.3)
Chapter 33   Matrices (E7.4)
Chapter 34   Transformations (E7.2, E7.5)

The Italians

Leonardo Pisano (known today as Fibonacci) introduced new methods of arithmetic to Europe, from the Hindus, Persians and Arabs. He discovered the sequence 1, 1, 2, 3, 5, 8, 13, … which is now called the Fibonacci sequence, and some of its occurrences in nature. He also brought the decimal system, algebra and the 'lattice' method of multiplication to Europe. Fibonacci has been called the 'most talented mathematician of the middle ages'. Many books say that he brought Islamic mathematics to Europe, but in Fibonacci's own introduction to *Liber Abaci*, he credits the Hindus.

Fibonacci (1170–1250)

The Renaissance began in Italy. Art, architecture, music and the sciences flourished. Girolamo Cardano (1501–1576) wrote his great mathematical book *Ars Magna* (Great Art) in which he showed, among much algebra that was new, calculations involving the solutions to cubic equations. He wrote this book, the first algebra book in Latin, to great acclaim. However, although he continued to study mathematics, no other work of his was ever published.

32 Vectors

● **Translations**

A **translation** (a sliding movement) can be described using column vectors. A column vector describes the movement of the object in both the x direction and the y direction.

Worked example i) Describe the translation from A to B in the diagram (left) in terms of a column vector.

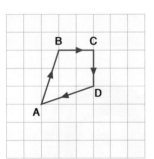

$$\overrightarrow{AB} = \begin{pmatrix} 1 \\ 3 \end{pmatrix}$$

i.e. 1 unit in the x direction, 3 units in the y direction

ii) Describe $\overrightarrow{BC}, \overrightarrow{CD}$ and $\overrightarrow{DA}$ in terms of column vectors.

$$\overrightarrow{BC} = \begin{pmatrix} 2 \\ 0 \end{pmatrix} \qquad \overrightarrow{CD} = \begin{pmatrix} 0 \\ -2 \end{pmatrix} \qquad \overrightarrow{DA} = \begin{pmatrix} -3 \\ -1 \end{pmatrix}$$

Translations can also be named by a single letter. The direction of the arrow indicates the direction of the translation.

Worked example Define **a** and **b** in the diagram (left) using column vectors.

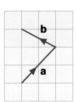

$$\mathbf{a} = \begin{pmatrix} 2 \\ 2 \end{pmatrix} \qquad \qquad \mathbf{b} = \begin{pmatrix} -2 \\ 1 \end{pmatrix}$$

Note: When you represent vectors by single letters, i.e. **a**, in handwritten work, you should write them as a̲.

If $\mathbf{a} = \begin{pmatrix} 2 \\ 5 \end{pmatrix}$ and $\mathbf{b} = \begin{pmatrix} -3 \\ -2 \end{pmatrix}$, they can be represented diagrammatically as shown (below).

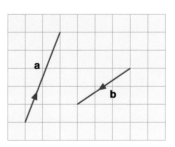

The diagrammatic representation of −**a** and −**b** is shown below.

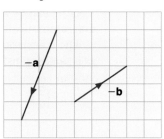

It can be seen from the diagram above that $-\mathbf{a} = \begin{pmatrix} -2 \\ -5 \end{pmatrix}$ and $-\mathbf{b} = \begin{pmatrix} 3 \\ 2 \end{pmatrix}$.

Exercise 32.1 In Q.1 and 2 describe each translation using a column vector.

1. a) $\overrightarrow{AB}$
   b) $\overrightarrow{BC}$
   c) $\overrightarrow{CD}$
   d) $\overrightarrow{DE}$
   e) $\overrightarrow{EA}$
   f) $\overrightarrow{AE}$
   g) $\overrightarrow{DA}$
   h) $\overrightarrow{CA}$
   i) $\overrightarrow{DB}$

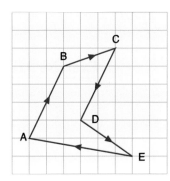

2. a) **a**
   b) **b**
   c) **c**
   d) **d**
   e) **e**
   f) −**b**
   g) −**c**
   h) −**d**
   i) −**a**

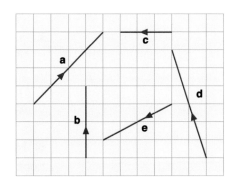

3. Draw and label the following vectors on a square grid:

   a) $\mathbf{a} = \begin{pmatrix} 2 \\ 4 \end{pmatrix}$
   b) $\mathbf{b} = \begin{pmatrix} -3 \\ 6 \end{pmatrix}$
   c) $\mathbf{c} = \begin{pmatrix} 3 \\ -5 \end{pmatrix}$
   d) $\mathbf{d} = \begin{pmatrix} -4 \\ -3 \end{pmatrix}$
   e) $\mathbf{e} = \begin{pmatrix} 0 \\ -6 \end{pmatrix}$
   f) $\mathbf{f} = \begin{pmatrix} -5 \\ 0 \end{pmatrix}$
   g) $\mathbf{g} = -\mathbf{c}$
   h) $\mathbf{h} = -\mathbf{b}$
   i) $\mathbf{i} = -\mathbf{f}$

Matrices and transformations

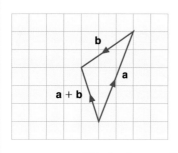

● Addition and subtraction of vectors

Vectors can be added together and represented diagrammatically as shown (left).

The translation represented by **a** followed by **b** can be written as a single transformation **a** + **b**:

i.e. $\begin{pmatrix} 2 \\ 5 \end{pmatrix} + \begin{pmatrix} -3 \\ -2 \end{pmatrix} = \begin{pmatrix} -1 \\ 3 \end{pmatrix}$

Worked example

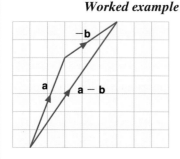

$\mathbf{a} = \begin{pmatrix} 2 \\ 5 \end{pmatrix} \quad \mathbf{b} = \begin{pmatrix} -3 \\ -2 \end{pmatrix}$

i) Draw a diagram to represent **a** − **b**, where
   a − **b** = (**a**) + (−**b**).

ii) Calculate the vector represented by **a** − **b**.

$\begin{pmatrix} 2 \\ 5 \end{pmatrix} - \begin{pmatrix} -3 \\ -2 \end{pmatrix} = \begin{pmatrix} 5 \\ 7 \end{pmatrix}$

Exercise 32.2 In the following questions,

$\mathbf{a} = \begin{pmatrix} 3 \\ 4 \end{pmatrix} \quad \mathbf{b} = \begin{pmatrix} -2 \\ 1 \end{pmatrix} \quad \mathbf{c} = \begin{pmatrix} -4 \\ -3 \end{pmatrix} \quad \mathbf{d} = \begin{pmatrix} 3 \\ -2 \end{pmatrix}.$

1. Draw vector diagrams to represent the following:
   a) **a** + **b** b) **b** + **a** c) **a** + **d**
   d) **d** + **a** e) **b** + **c** f) **c** + **b**

2. What conclusions can you draw from your answers to question 1 above?

3. Draw vector diagrams to represent the following:
   a) **b** − **c** b) **d** − **a** c) −**a** − **c**
   d) **a** + **c** − **b** e) **d** − **c** − **b** f) −**c** + **b** + **d**

4. Represent each of the vectors in question 3 by a single column vector.

● Multiplying a vector by a scalar

Look at the two vectors in the diagram.

$\mathbf{a} = \begin{pmatrix} 1 \\ 2 \end{pmatrix} \quad 2\mathbf{a} = 2\begin{pmatrix} 1 \\ 2 \end{pmatrix} = \begin{pmatrix} 2 \\ 4 \end{pmatrix}$

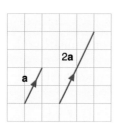

Worked example If $\mathbf{a} = \begin{pmatrix} 2 \\ -4 \end{pmatrix}$, express the vectors **b**, **c**, **d** and **e** in terms of **a**.

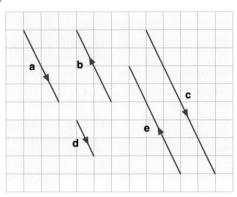

$\mathbf{b} = -\mathbf{a} \qquad \mathbf{c} = 2\mathbf{a} \qquad \mathbf{d} = \tfrac{1}{2}\mathbf{a} \qquad \mathbf{e} = -\tfrac{3}{2}\mathbf{a}$

Exercise 32.3 1. $\mathbf{a} = \begin{pmatrix} 1 \\ 4 \end{pmatrix} \quad \mathbf{b} = \begin{pmatrix} -4 \\ -2 \end{pmatrix} \quad \mathbf{c} = \begin{pmatrix} -4 \\ 6 \end{pmatrix}$

Express the following vectors in terms of either **a**, **b** or **c**.

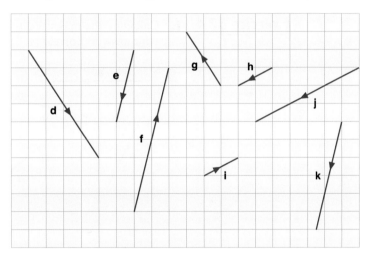

2. $\mathbf{a} = \begin{pmatrix} 2 \\ 3 \end{pmatrix} \quad \mathbf{b} = \begin{pmatrix} -4 \\ -1 \end{pmatrix} \quad \mathbf{c} = \begin{pmatrix} -2 \\ 4 \end{pmatrix}$

Represent each of the following as a single column vector:
a) $2\mathbf{a}$ b) $3\mathbf{b}$ c) $-\mathbf{c}$ d) $\mathbf{a} + \mathbf{b}$ e) $\mathbf{b} - \mathbf{c}$
f) $3\mathbf{c} - \mathbf{a}$ g) $2\mathbf{b} - \mathbf{a}$ h) $\tfrac{1}{2}(\mathbf{a} - \mathbf{b})$ i) $2\mathbf{a} - 3\mathbf{c}$

3. $\mathbf{a} = \begin{pmatrix} -2 \\ 3 \end{pmatrix} \quad \mathbf{b} = \begin{pmatrix} 0 \\ -3 \end{pmatrix} \quad \mathbf{c} = \begin{pmatrix} 4 \\ -1 \end{pmatrix}$

Express each of the following vectors in terms of **a**, **b** and **c**:
a) $\begin{pmatrix} -4 \\ 6 \end{pmatrix}$ b) $\begin{pmatrix} 0 \\ 3 \end{pmatrix}$ c) $\begin{pmatrix} 4 \\ -4 \end{pmatrix}$
d) $\begin{pmatrix} -2 \\ 6 \end{pmatrix}$ e) $\begin{pmatrix} 8 \\ -2 \end{pmatrix}$ f) $\begin{pmatrix} 10 \\ -5 \end{pmatrix}$

Matrices and transformations

● The magnitude of a vector

The **magnitude** or size of a vector is represented by its length, i.e. the longer the length, the greater the magnitude. The magnitude of a vector **a** or $\overrightarrow{AB}$ is denoted by |**a**| or $|\overrightarrow{AB}|$ respectively and is calculated using Pythagoras' theorem.

Worked examples $\mathbf{a} = \begin{pmatrix} 3 \\ 4 \end{pmatrix}$ $\overrightarrow{BC} = \begin{pmatrix} -6 \\ 8 \end{pmatrix}$

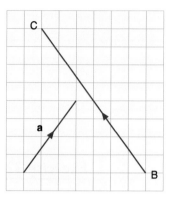

a) Represent both of the above vectors diagramatically.

b) i) Calculate |**a**|.
$$|\mathbf{a}| = \sqrt{(3^2 + 4^2)}$$
$$= \sqrt{25} = 5$$

   ii) Calculate $|\overrightarrow{BC}|$.
$$|\overrightarrow{BC}| = \sqrt{(-6)^2 + 8^2}$$
$$= \sqrt{100} = 10$$

Exercise 32.4

1. Calculate the magnitude of the vectors shown below. Give your answers correct to 1 d.p.

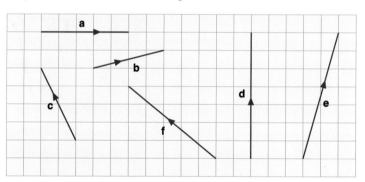

2. Calculate the magnitude of the following vectors, giving your answers to 1 d.p.

   a) $\overrightarrow{AB} = \begin{pmatrix} 0 \\ 4 \end{pmatrix}$
   b) $\overrightarrow{BC} = \begin{pmatrix} 2 \\ 5 \end{pmatrix}$
   c) $\overrightarrow{CD} = \begin{pmatrix} -4 \\ -6 \end{pmatrix}$
   d) $\overrightarrow{DE} = \begin{pmatrix} -5 \\ 12 \end{pmatrix}$
   e) $2\overrightarrow{AB}$
   f) $-\overrightarrow{CD}$

3. $\mathbf{a} = \begin{pmatrix} 4 \\ -3 \end{pmatrix}$ $\mathbf{b} = \begin{pmatrix} -5 \\ 7 \end{pmatrix}$ $\mathbf{c} = \begin{pmatrix} -1 \\ -8 \end{pmatrix}$

   Calculate the magnitude of the following, giving your answers to 1 d.p.
   a) **a** + **b**
   b) 2**a** − **b**
   c) **b** − **c**
   d) 2**c** + 3**b**
   e) 2**b** − 3**a**
   f) **a** + 2**b** − **c**

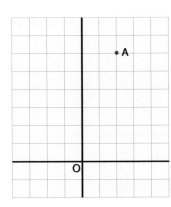

● Position vectors

Sometimes a vector is fixed in position relative to a specific point. In the diagram (left), the position vector of A relative to O is $\binom{2}{6}$.

Exercise 32.5

1. Give the position vectors of A, B, C, D, E, F, G and H relative to O in the diagram (below).

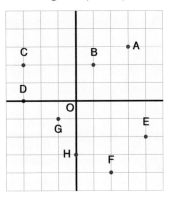

● Vector geometry

In general vectors are not fixed in position. If a vector **a** has a specific magnitude and direction, then any other vector with the same magnitude and direction as **a** can also be labelled **a**.

If $\mathbf{a} = \binom{3}{2}$ then all the vectors shown in the diagram (left) can also be labelled **a**, as they all have the same magnitude and direction.

This property of vectors can be used to solve problems in vector geometry.

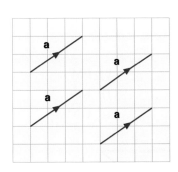

Worked example

i) Name a vector equal to $\overrightarrow{AD}$.
$\overrightarrow{BC} = \overrightarrow{AD}$

ii) Write $\overrightarrow{BD}$ in terms of $\overrightarrow{BE}$.
$\overrightarrow{BD} = 2\overrightarrow{BE}$

iii) Express $\overrightarrow{CD}$ in terms of $\overrightarrow{AB}$.
$\overrightarrow{CD} = \overrightarrow{BA} = -\overrightarrow{AB}$

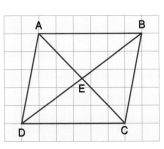

Matrices and transformations

Exercise 32.6

1. If $\vec{AG} = \mathbf{a}$ and $\vec{AE} = \mathbf{b}$, express the following in terms of $\mathbf{a}$ and $\mathbf{b}$:
   a) $\vec{EI}$
   b) $\vec{HC}$
   c) $\vec{FC}$
   d) $\vec{DE}$
   e) $\vec{GH}$
   f) $\vec{CD}$
   g) $\vec{AI}$
   h) $\vec{GE}$
   i) $\vec{FD}$

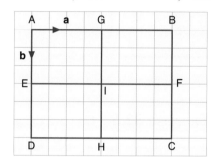

2. If $\vec{LP} = \mathbf{a}$ and $\vec{LR} = \mathbf{b}$, express the following in terms of $\mathbf{a}$ and $\mathbf{b}$:
   a) $\vec{LM}$
   b) $\vec{PQ}$
   c) $\vec{PR}$
   d) $\vec{MQ}$
   e) $\vec{MP}$
   f) $\vec{NP}$

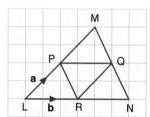

3. ABCDEF is a regular hexagon.
   If $\vec{GA} = \mathbf{a}$ and $\vec{GB} = \mathbf{b}$, express the following in terms of $\mathbf{a}$ and $\mathbf{b}$:
   a) $\vec{AD}$
   b) $\vec{FE}$
   c) $\vec{DC}$
   d) $\vec{AB}$
   e) $\vec{FC}$
   f) $\vec{EC}$
   g) $\vec{BE}$
   h) $\vec{FD}$
   i) $\vec{AE}$

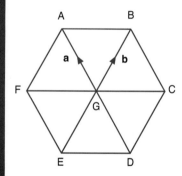

4. If $\vec{AB} = \mathbf{a}$ and $\vec{AG} = \mathbf{b}$, express the following in terms of $\mathbf{a}$ and $\mathbf{b}$:
   a) $\vec{AF}$
   b) $\vec{AM}$
   c) $\vec{FM}$
   d) $\vec{FO}$
   e) $\vec{EI}$
   f) $\vec{KF}$
   g) $\vec{CN}$
   h) $\vec{AN}$
   i) $\vec{DN}$

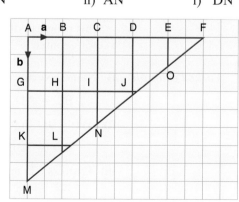

Exercise 32.7

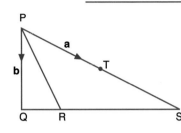

1. T is the midpoint of the line PS and R divides the line QS in the ratio 1 : 3.
   $\vec{PT}$ = **a** and $\vec{PQ}$ = **b**.
   a) Express each of the following in terms of **a** and **b**:
      i) $\vec{PS}$
      ii) $\vec{QS}$
      iii) $\vec{PR}$
   b) Show that $\vec{RT} = \frac{1}{4}(2\mathbf{a} - 3\mathbf{b})$.

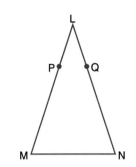

2. $\vec{PM} = 3\vec{LP}$ and $\vec{QN} = 3\vec{LQ}$

   Prove that:
   a) the line PQ is parallel to the line MN,
   b) the line MN is four times the length of the line PQ.

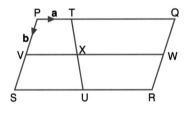

3. PQRS is a parallelogram. The point T divides the line PQ in the ratio 1 : 3, and U, V and W are the midpoints of SR, PS and QR respectively.
   $\vec{PT}$ = **a** and $\vec{PV}$ = **b**.
   a) Express each of the following in terms of **a** and **b**:
      i) $\vec{PQ}$ ii) $\vec{SU}$
      iii) $\vec{PU}$ iv) $\vec{VX}$
   b) Show that $\vec{XR} = \frac{1}{2}(5\mathbf{a} + 2\mathbf{b})$.

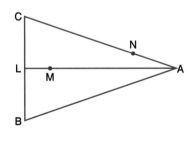

4. ABC is an isosceles triangle. L is the midpoint of BC. M divides the line LA in the ratio 1 : 5, and N divides AC in the ratio 2 : 5.
   a) $\vec{BC}$ = **p** and $\vec{BA}$ = **q**. Express the following in terms of **p** and **q**:
      i) $\vec{LA}$ ii) $\vec{AN}$
   b) Show that $\vec{MN} = \frac{1}{84}(46\mathbf{q} - 11\mathbf{p})$.

Matrices and transformations

Student assessment 1

1. Using the diagram (below), describe the following translations using column vectors.
   a) $\overrightarrow{AB}$ b) $\overrightarrow{DA}$ c) $\overrightarrow{CA}$

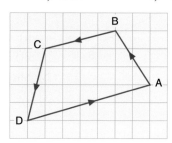

2. Describe each of the translations shown (left) using column vectors.

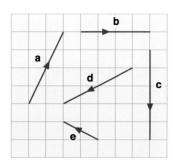

3. Using the vectors in question 2, draw diagrams to represent:
   a) $\mathbf{a} + \mathbf{b}$ b) $\mathbf{e} - \mathbf{d}$ c) $\mathbf{c} - \mathbf{e}$ d) $2\mathbf{e} + \mathbf{b}$

4. In the following, $\mathbf{a} = \begin{pmatrix} 2 \\ 6 \end{pmatrix}$ $\mathbf{b} = \begin{pmatrix} -3 \\ -1 \end{pmatrix}$ $\mathbf{c} = \begin{pmatrix} -2 \\ 4 \end{pmatrix}$.
   Calculate:
   a) $\mathbf{a} + \mathbf{b}$ b) $\mathbf{c} - \mathbf{b}$ c) $2\mathbf{a} + \mathbf{b}$ d) $3\mathbf{c} - 2\mathbf{b}$

Student assessment 2

1. Using the diagram (below), describe the following translations using column vectors.
   a) $\overrightarrow{AB}$ b) $\overrightarrow{DA}$ c) $\overrightarrow{CA}$

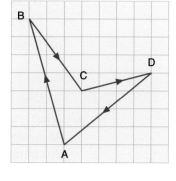

2. Describe each of the translations shown (left) using column vectors.

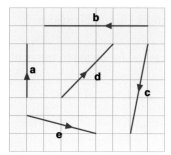

3. Using the vectors in question 2, draw diagrams to represent:
   a) $\mathbf{a} + \mathbf{e}$ b) $\mathbf{c} - \mathbf{d}$ c) $-\mathbf{c} - \mathbf{e}$ d) $-\mathbf{b} + 2\mathbf{a}$

4. In the following,

$$a = \begin{pmatrix} 3 \\ -5 \end{pmatrix} \quad b = \begin{pmatrix} 0 \\ 4 \end{pmatrix} \quad c = \begin{pmatrix} -4 \\ 6 \end{pmatrix}.$$

Calculate:
a) $a - c$ b) $b - a$ c) $2a + b$ d) $3c - 2a$

Student assessment 3

1. a) Calculate the magnitude of the vector $\overrightarrow{AB}$ shown in the diagram (left).
   b) Calculate the magnitude of the following vectors:

$$a = \begin{pmatrix} 2 \\ 9 \end{pmatrix} \quad b = \begin{pmatrix} -7 \\ -4 \end{pmatrix} \quad c = \begin{pmatrix} -5 \\ 12 \end{pmatrix}$$

2. $p = \begin{pmatrix} 3 \\ 2 \end{pmatrix} \quad q = \begin{pmatrix} -4 \\ 1 \end{pmatrix} \quad r = \begin{pmatrix} 3 \\ -4 \end{pmatrix}$

   Calculate the magnitude of the following, giving your answers to 3 s.f.
   a) $3p - 2q$ b) $\frac{1}{2}r + q$

3. Give the position vectors of A, B, C, D and E relative to O for the diagram below.

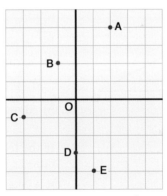

4. a) Name another vector equal to $\overrightarrow{DE}$ in the diagram (left).
   b) Express $\overrightarrow{DF}$ in terms of $\overrightarrow{BC}$.
   c) Express $\overrightarrow{CF}$ in terms of $\overrightarrow{DE}$.

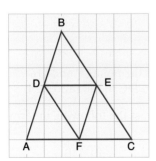

Matrices and transformations

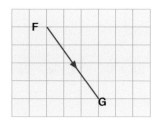

Student assessment 4

1. a) Calculate the magnitude of the vector $\overrightarrow{FG}$ shown in the diagram (left).
   b) Calculate the magnitude of each of the following vectors:
   $$\mathbf{a} = \begin{pmatrix} 1 \\ 6 \end{pmatrix} \quad \mathbf{b} = \begin{pmatrix} 12 \\ -3 \end{pmatrix} \quad \mathbf{c} = \begin{pmatrix} -10 \\ 10 \end{pmatrix}$$

2. $\mathbf{p} = \begin{pmatrix} -3 \\ 5 \end{pmatrix} \quad \mathbf{q} = \begin{pmatrix} -4 \\ -4 \end{pmatrix} \quad \mathbf{r} = \begin{pmatrix} 8 \\ -2 \end{pmatrix}$

   Calculate the magnitude of each of the following, giving your answers to 1 d.p.
   a) $4\mathbf{p} - \mathbf{r}$
   b) $\tfrac{3}{2}\mathbf{q} - \mathbf{p}$

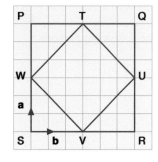

3. If $\overrightarrow{SW} = \mathbf{a}$ and $\overrightarrow{SV} = \mathbf{b}$ in the diagram (left), express each of the following in terms of $\mathbf{a}$ and $\mathbf{b}$:
   a) $\overrightarrow{SP}$
   b) $\overrightarrow{QT}$
   c) $\overrightarrow{TU}$

Student assessment 5

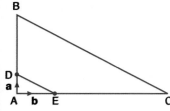

1. In the triangle PQR, the point S divides the line PQ in the ratio 1 : 3, and T divides the line RQ in the ratio 3 : 2. $\overrightarrow{PR} = \mathbf{a}$ and $\overrightarrow{PQ} = \mathbf{b}$.
   a) Express the following in terms of $\mathbf{a}$ and $\mathbf{b}$:
      i) $\overrightarrow{PS}$
      ii) $\overrightarrow{SR}$
      iii) $\overrightarrow{TQ}$
   b) Show that $\overrightarrow{ST} = \tfrac{1}{20}(8\mathbf{a} + 7\mathbf{b})$.

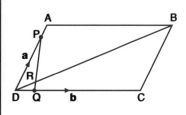

2. In the triangle ABC, the point D divides the line AB in the ratio 1 : 3, and E divides the line AC also in the ratio 1 : 3. If $\overrightarrow{AD} = \mathbf{a}$ and $\overrightarrow{AE} = \mathbf{b}$ prove that:
   a) $\overrightarrow{BC} = 4\overrightarrow{DE}$,
   b) BCED is a trapezium.

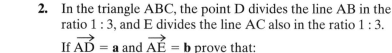

3. The parallelogram ABCD shows the points P and Q dividing each of the lines AD and DC in the ratio 1 : 4.
   a) If $\overrightarrow{DA} = \mathbf{a}$ and $\overrightarrow{DC} = \mathbf{b}$ express the following in terms of $\mathbf{a}$ and $\mathbf{b}$:
      i) $\overrightarrow{AC}$
      ii) $\overrightarrow{CB}$
      iii) $\overrightarrow{DB}$
   b) i) Find the ratio in which R divides DB.
      ii) Find the ratio in which R divides PQ.

Student assessment 6

1. ABCDEFGH is a regular octagon.
   $\vec{AB} = \mathbf{a}$ and $\vec{AH} = \mathbf{b}$. Express the following vectors in terms of **a** and **b**:
   a) $\vec{FE}$ b) $\vec{ED}$ c) $\vec{BG}$

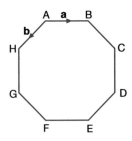

2. In the triangle ABC, $\vec{AB} = \mathbf{a}$ and $\vec{AD} = \mathbf{b}$. D divides the side AC in the ratio 1 : 4 and E is the midpoint of BC. Express the following in terms of **a** and **b**:
   a) $\vec{AC}$ b) $\vec{BC}$ c) $\vec{DE}$

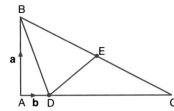

3. In the square PQRS, T is the midpoint of the side PQ and U is the midpoint of the side SR. $\vec{PQ} = \mathbf{a}$ and $\vec{PS} = \mathbf{b}$.
   a) Express the following in terms of **a** and **b**:
      i) $\vec{PT}$ ii) $\vec{QS}$
   b) Calculate the ratio $\vec{PV} : \vec{PU}$.

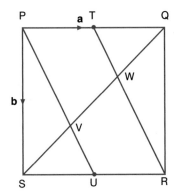

33 Matrices

A matrix represents another way of writing information. Here the information is written as a rectangular array. For example, two pupils Lea and Pablo sit a maths exam, a science exam and an English exam. Lea scores 73%, 67% and 81% respectively, whilst Pablo scores 64%, 82% and 48% respectively. This can be written as

$$\begin{pmatrix} 73 & 67 & 81 \\ 64 & 82 & 48 \end{pmatrix}$$

A matrix can take any size.

$$\mathbf{A} = \begin{pmatrix} 4 & 1 & 3 \\ 2 & 6 & 4 \\ 9 & 1 & 0 \end{pmatrix} \qquad \mathbf{B} = \begin{pmatrix} 7 & 1 & 4 & 5 \\ 6 & 8 & 2 & 1 \end{pmatrix}$$

$$\mathbf{C} = (9 \quad 6 \quad 2) \qquad \mathbf{D} = \begin{pmatrix} 4 \\ 0 \\ 6 \end{pmatrix} \qquad \mathbf{E} = \begin{pmatrix} 1 & 0 \\ 0 & 1 \end{pmatrix}$$

A size of a matrix is known as its **order** and is denoted by the number of rows times the number of columns. Therefore the order of matrix **A** is 3 × 3, whilst the order of matrix **B** is 2 × 4. Each of the numbers in the matrix is called an **element**. A 2 × 4 matrix consists of 8 elements.

Matrix **A** and matrix **E** above are called **square matrices** as they have the same number of rows and columns. Matrix **C** is called a **row matrix** as it consists of only one row, and matrix **D** is called a **column matrix** as it consists of only one column.

Therefore, for any matrix of order $m \times n$:

- m is the number of rows,
- n is the number of columns,
- if $m = n$, it is a square matrix,
- if $m = 1$, it is a row matrix,
- if $n = 1$, it is a column matrix.

Exercise 33.1

1. Give the order of the following matrices:

   a) $\mathbf{P} = \begin{pmatrix} 9 & 0 & 3 \\ 4 & 6 & 2 \end{pmatrix}$

   b) $\mathbf{Q} = \begin{pmatrix} 8 & 6 & 5 & -6 \\ 7 & 2 & 4 & 0 \end{pmatrix}$

   c) $\mathbf{R} = \begin{pmatrix} 1 & 12 \\ 4 & 0 \\ 9 & 10 \\ 2 & -6 \end{pmatrix}$

   d) $\mathbf{S} = \begin{pmatrix} 8 & 2 & 1 & 4 & -3 \\ 6 & 7 & 9 & 3 & 12 \\ 8 & 5 & 1 & 6 & 1 \\ 7 & 3 & 2 & 8 & 9 \end{pmatrix}$

e) $\mathbf{T} = \begin{pmatrix} 4 \\ 0 \\ -9 \\ 8 \\ 7 \end{pmatrix}$

f) $\mathbf{F} = (6 \ 6 \ 8 \ 4 \ 2)$

2. Write matrices of the following orders:
   a) 3×2 b) 2×3 c) 4×1
   d) 1×4 e) 4×4 f) 2×2

3. A small factory produces televisions and videos. In 2008 it manufactured 6500 televisions and 900 videos. In 2009 it made 7200 televisions and 1100 videos, and in 2010 it made 7300 televisions and 1040 videos. Write this information in a 3×2 matrix.

4. A shop selling beds records the number and type of beds it sells over a three-week period. In the first week it sells three cots, four single beds, two double beds and one king-size bed. In the second week it sells only six single beds and two double beds. In the third week it sells one cot, three single beds and two king-size beds. Write this information as a 3×4 matrix.

5. A shoe shop sells shoes for girls, boys, ladies and gentlemen. One Saturday it sells eight pairs of girls' shoes, six pairs of boys' shoes, nine pairs of ladies' shoes and three pairs of men's shoes. Write this information as a row matrix.

6. Four students sit two tests. Carlos achieved 37% in the first test and 49% in the second. Cristina achieved 74% in the first test and 58% in the second. Ali got 76% in the first test and 62% in the second. Helena got 89% in the first test and 56% in the second. Write this information in a 4×2 matrix.

7. The pie charts below show the nationalities of students at three different schools A, B and C.

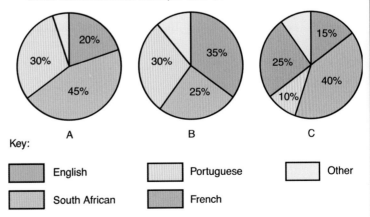

Write this information as a matrix.

8. The graph below shows the number of units sold by a computer manufacturer for each quarter of the years 2008, 2009 and 2010.
   Represent this information as a matrix.

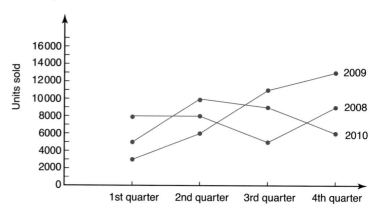

9. The percentage of pupils achieving each of the pass grades in a maths exam for the years 2008, 2009 and 2010 is shown below.

   2008: A 6%, B 12%, C 43%, D 18%, E 6%, F 9%, G 6%

   2009: A 9%, B 15%, C 28%, D 18%, E 12%, F 12%, G 6%

   2010: A 12%, B 19%, C 30%, D 12%, E 9%, F 9%, G 9%

   Represent this information as a matrix.

10. Collect some data of your own, either from newspapers, books or from your own surveys, and write it as a matrix.

Matrices

● Addition and subtraction of matrices

Worked examples **a)** A music store sells music through its website in three formats: records, CDs and mp3s. It also sells the following types of music: classical, rock/pop and hip-hop. It records the number of each type and format sold on two consecutive Saturdays. The results of this are presented in the matrices below:

$$
\begin{array}{c}
\text{1st Saturday} \\
\begin{array}{c} \text{Record} \quad \text{CD} \quad \text{mp3} \end{array} \\
\begin{array}{c} \text{Classical} \\ \text{Rock/pop} \\ \text{Hip-hop} \end{array}
\begin{pmatrix} 28 & 14 & 46 \\ 56 & 91 & 15 \\ 17 & 5 & 7 \end{pmatrix}
\end{array}
\qquad
\begin{array}{c}
\text{2nd Saturday} \\
\begin{array}{c} \text{Record} \quad \text{CD} \quad \text{mp3} \end{array} \\
\begin{array}{c} \text{Classical} \\ \text{Rock/pop} \\ \text{Hip-hop} \end{array}
\begin{pmatrix} 24 & 10 & 51 \\ 35 & 82 & 24 \\ 15 & 8 & 6 \end{pmatrix}
\end{array}
$$

i) Calculate the total number of records sold on the 1st Saturday.

$$\text{Records} = 28 + 56 + 17 = 101$$

ii) Calculate the total sales of hip-hop on the 2nd Saturday.

$$\text{Hip-hop} = 15 + 8 + 6 = 29$$

iii) Calculate the total sales of each type of music and format. Express your answer as a matrix.

$$\begin{pmatrix} 28 & 14 & 46 \\ 56 & 91 & 15 \\ 17 & 5 & 7 \end{pmatrix} + \begin{pmatrix} 24 & 10 & 51 \\ 35 & 82 & 24 \\ 15 & 8 & 6 \end{pmatrix} = \begin{pmatrix} 52 & 24 & 97 \\ 91 & 173 & 39 \\ 32 & 13 & 13 \end{pmatrix}$$

To add matrices together they must be of the same order. Corresponding elements are then added together.

b) A shop sells shoes for both adults and children. Matrix **A** below shows how many of each type it has in stock. Matrix **B** shows the number of each type it sells in a particular week.

$$
\begin{array}{c}
\mathbf{A} \\
\begin{array}{c} \text{Male} \quad \text{Female} \end{array} \\
\begin{array}{c} \text{Child} \\ \text{Adult} \end{array}
\begin{pmatrix} 65 & 42 \\ 111 & 154 \end{pmatrix}
\end{array}
\qquad
\begin{array}{c}
\mathbf{B} \\
\begin{array}{c} \text{Male} \quad \text{Female} \end{array} \\
\begin{array}{c} \text{Child} \\ \text{Adult} \end{array}
\begin{pmatrix} 15 & 21 \\ 19 & 28 \end{pmatrix}
\end{array}
$$

Matrices and transformations

i) Calculate how many shoes the shop has in stock at the start of the week.

$$\text{Total stock} = 65 + 42 + 111 + 154$$
$$= 372$$

ii) Calculate the number of shoes the shop sells over the week to females.

$$\text{Total female sales} = 21 + 28$$
$$= 49$$

iii) Calculate the number of each type of shoe still in stock at the end of the week. Give your answer as a matrix.

$$\begin{pmatrix} 65 & 42 \\ 111 & 154 \end{pmatrix} - \begin{pmatrix} 15 & 21 \\ 19 & 28 \end{pmatrix} = \begin{pmatrix} 50 & 21 \\ 92 & 126 \end{pmatrix}$$

To subtract matrices from each other, they also need to be of the same order. Corresponding elements are then subtracted from each other.

Exercise 33.2

1. Add the following matrices:

   a) $\begin{pmatrix} 8 & 2 \\ 7 & 9 \end{pmatrix} + \begin{pmatrix} 6 & 6 \\ 0 & 7 \end{pmatrix}$

   b) $\begin{pmatrix} 0 & 1 & 3 \\ 12 & 15 & 9 \end{pmatrix} + \begin{pmatrix} 9 & 2 & 21 \\ 7 & 20 & -4 \end{pmatrix}$

   c) $\begin{pmatrix} 9 \\ 3 \\ 12 \end{pmatrix} + \begin{pmatrix} 15 \\ 0 \\ 6 \end{pmatrix}$

   d) $\begin{pmatrix} -14 & 7 & 8 \\ 0 & -5 & 12 \\ 1 & 3 & 6 \end{pmatrix} + \begin{pmatrix} 4 & 6 & 9 \\ 2 & -8 & 6 \\ 4 & 7 & 5 \end{pmatrix}$

   e) $(-1 \quad -1 \quad 1) + (-2 \quad 2 \quad -2)$

   f) $\begin{pmatrix} 6 & -1 \\ 9 & -8 \\ 4 & 6 \end{pmatrix} + \begin{pmatrix} 8 & -8 \\ 6 & 3 \\ -9 & 4 \end{pmatrix}$

2. Subtract the following matrices:

   a) $\begin{pmatrix} 8 & 6 \\ 7 & 5 \end{pmatrix} - \begin{pmatrix} 3 & 2 \\ 6 & 5 \end{pmatrix}$

   b) $\begin{pmatrix} 9 & 12 & 8 \\ 15 & 7 & 2 \end{pmatrix} - \begin{pmatrix} 7 & 10 & 6 \\ 13 & 5 & 0 \end{pmatrix}$

   c) $\begin{pmatrix} 15 \\ 6 \\ 8 \end{pmatrix} - \begin{pmatrix} 20 \\ 6 \\ 4 \end{pmatrix}$

   d) $\begin{pmatrix} 8 & -9 & 7 \\ 6 & 12 & 10 \\ -5 & 6 & 4 \end{pmatrix} - \begin{pmatrix} 4 & 5 & 8 \\ 4 & 12 & 6 \\ 4 & 7 & 9 \end{pmatrix}$

e) $\begin{pmatrix} -6 & 14 \\ 9 & -12 \end{pmatrix} - \begin{pmatrix} -3 & -4 \\ 10 & 0 \end{pmatrix}$

f) $\begin{pmatrix} 6 & -8 \\ -15 & 4 \\ 9 & -6 \end{pmatrix} - \begin{pmatrix} 9 & -4 \\ 6 & -9 \\ 12 & -8 \end{pmatrix}$

3. Four matrices are given below:

$$\mathbf{A} = \begin{pmatrix} 2 & 7 \\ 6 & 3 \\ 1 & 4 \end{pmatrix} \qquad \mathbf{B} = \begin{pmatrix} 7 & 4 \\ 9 & 2 \\ 8 & 4 \end{pmatrix}$$

$$\mathbf{C} = \begin{pmatrix} -3 & -2 \\ 0 & 3 \\ 6 & -9 \end{pmatrix} \qquad \mathbf{D} = \begin{pmatrix} 8 & -4 \\ -9 & 11 \\ 2 & 0 \end{pmatrix}$$

Calculate the following:
a) **A + B**   b) **B + A**   c) **A + C**
d) **D + A**   e) **B − C**   f) **D − C**
g) **A − C**   h) **A + D − B**   i) **B + D − A**

4. Three teams compete in a two-day athletics competition. On day 1, team A won two gold medals, three silver medals and one bronze medal. Team B won three gold, one silver and four bronze medals. Team C won three gold medals, four silver medals and three bronze medals. On day 2, team A won five gold medals, one silver medal and one bronze medal. Team B won one gold, four silver and three bronze medals and team C won one gold, two silver and three bronze medals.
a) Write this information down in two matrices.
b) What was the total number of races over the two days?
c) Write down a matrix to represent the total number of each medal won by each team for the whole competition.

5. A shop selling clothes keeps a record of its stock. Matrix **A** below shows the number and type of shirts and dresses it has in stock at the start of the week. Matrix **B** shows the number and type of shirts and dresses it has at the end of the week.

$$\mathbf{A} \qquad\qquad \mathbf{B}$$

$$\begin{array}{c} \text{Child} \\ \text{Adult} \end{array} \begin{pmatrix} \overset{\text{Shirts}}{265} & \overset{\text{Dresses}}{312} \\ 140 & 132 \end{pmatrix} \qquad \begin{array}{c} \text{Child} \\ \text{Adult} \end{array} \begin{pmatrix} \overset{\text{Shirts}}{189} & \overset{\text{Dresses}}{204} \\ 121 & 68 \end{pmatrix}$$

a) Calculate the total number of dresses at the start of the week.
b) Write the matrix which shows the number of each type sold over the week.
c) What is the total number of shirts and dresses sold over the whole week?

Matrices and transformations

● Multiplying matrices by a scalar quantity

Worked example Two children A and B record the number of hours of television and DVDs they watch over the period of one week. This information is represented in the matrix below.

$$\begin{array}{c} \\ \text{TV} \\ \text{DVDs} \end{array} \begin{array}{cc} \text{A} & \text{B} \\ \begin{pmatrix} 8 & 6 \\ 4 & 2 \end{pmatrix} \end{array}$$

i) The following week, both children watch twice as many hours of TV and DVDs as the first week. Write a second matrix to show the number of hours of TV and DVDs they watch in the second week.

$$2\begin{pmatrix} 8 & 6 \\ 4 & 2 \end{pmatrix} = \begin{pmatrix} 16 & 12 \\ 8 & 4 \end{pmatrix}$$

ii) In the third week they watch half as many hours of TV and DVDs as they did in the first week. Write another matrix to show the number of hours of TV and DVDs they watched in the third week.

$$\frac{1}{2}\begin{pmatrix} 8 & 6 \\ 4 & 2 \end{pmatrix} = \begin{pmatrix} 4 & 3 \\ 2 & 1 \end{pmatrix}$$

When multiplying a matrix by a scalar quantity, each element in the matrix is multiplied by that quantity.

Exercise 33.3 Evaluate the following:

1. a) $2\begin{pmatrix} 4 & 6 \\ 7 & 3 \end{pmatrix}$ b) $3\begin{pmatrix} 7 & 2 \\ 1 & 0 \end{pmatrix}$

    c) $1\begin{pmatrix} 9 & 3 \\ 0 & 6 \end{pmatrix}$ d) $5\begin{pmatrix} 2 & 1 \\ 0 & 4 \end{pmatrix}$

    e) $4\begin{pmatrix} 7 & 5 \\ 4 & 8 \end{pmatrix}$ f) $3.5\begin{pmatrix} 6 & 12 \\ 4 & 2 \end{pmatrix}$

2. a) $\frac{1}{2}\begin{pmatrix} 8 & 2 \\ 0 & 4 \end{pmatrix}$ b) $\frac{1}{3}\begin{pmatrix} 12 & 9 \\ 6 & 3 \end{pmatrix}$

    c) $\frac{3}{4}\begin{pmatrix} 4 & 8 \\ 1 & 2 \end{pmatrix}$ d) $\frac{2}{5}\begin{pmatrix} 15 & 10 \\ 2 & 5 \end{pmatrix}$

    e) $\frac{5}{8}\begin{pmatrix} 4 & 12 \\ 12 & 16 \end{pmatrix}$ f) $\frac{3}{7}\begin{pmatrix} 28 & 7 \\ 7 & 14 \end{pmatrix}$

● Multiplying a matrix by another matrix

Worked examples a) Paula and Gregori have a choice of shopping at one of two supermarkets, X and Y. The first matrix below shows the type and quantity of certain foods they both wish to buy and the second matrix shows the cost of the items (in cents) at each of the supermarkets.

$$\begin{array}{c} \text{Cereal packets} \\ \text{Loaves of bread} \\ \text{Potatoes (kg)} \end{array}$$

$$\begin{array}{c}\text{Paula}\\ \text{Gregori}\end{array}\begin{pmatrix} 2 & 4 & 5 \\ 1 & 7 & 3 \end{pmatrix} \quad \begin{array}{cc} X & Y \end{array} \\ \begin{pmatrix} 120 & 110 \\ 55 & 60 \\ 35 & 30 \end{pmatrix} \begin{array}{l}\text{Cereal packets}\\ \text{Loaves of bread}\\ \text{Potatoes (kg)}\end{array}$$

Calculate their shopping bill for these items at each supermarket and decide where they should buy their food.

The shopping bill for each can be calculated by multiplying the two matrices together.

Paula at X $= (2 \times 120) + (4 \times 55) + (5 \times 35) = 635$
Paula at Y $= (2 \times 110) + (4 \times 60) + (5 \times 30) = 610$
Gregori at X $= (1 \times 120) + (7 \times 55) + (3 \times 35) = 610$
Gregori at Y $= (1 \times 110) + (7 \times 60) + (3 \times 30) = 620$

This can also be written as a matrix:

$$\begin{array}{c}\text{Paula}\\ \text{Gregori}\end{array}\begin{array}{cc} X & Y \end{array} \\ \begin{pmatrix} 635 & 610 \\ 610 & 620 \end{pmatrix}$$

Paula should therefore shop at Y and Gregori at X.

Multiplying matrices together involves multiplying the elements in the rows of the first matrix by the elements in the columns of the second matrix.

b) $\begin{pmatrix} 6 & 3 \\ 2 & 4 \\ 0 & 1 \end{pmatrix} \times \begin{pmatrix} 2 & 4 \\ 1 & 3 \end{pmatrix}$

$$\begin{pmatrix} (6 \times 2) + (3 \times 1) & (6 \times 4) + (3 \times 3) \\ (2 \times 2) + (4 \times 1) & (2 \times 4) + (4 \times 3) \\ (0 \times 2) + (1 \times 1) & (0 \times 4) + (1 \times 3) \end{pmatrix} = \begin{pmatrix} 15 & 33 \\ 8 & 20 \\ 1 & 3 \end{pmatrix}$$

Note: Not all matrices can be multiplied together. For matrices to be multiplied together, the number of columns in the first must be equal to the number of rows in the second. This can be seen clearly if their orders are considered.

e.g.

$$\begin{pmatrix} 4 & 3 & 2 \\ 1 & 2 & 7 \end{pmatrix}\begin{pmatrix} 4 & 2 & 8 \\ 6 & 1 & 7 \\ 7 & 2 & 0 \end{pmatrix} = \begin{pmatrix} 48 & 15 & 53 \\ 65 & 18 & 22 \end{pmatrix}$$

Order: $2 \times \mathbf{3} \quad \mathbf{3} \times 3 \ = 2 \times 3$

For multiplication to be possible, the two middle numbers (in bold) must be the same. The result will be a matrix of the order of the outer two numbers (i.e. 2×3).

Matrices and transformations

Exercise 33.4 Multiply the following pairs of matrices (remember row by column):

1. a) $\begin{pmatrix} 2 & 8 \\ 6 & 4 \end{pmatrix}\begin{pmatrix} 1 & 3 \\ 2 & 9 \end{pmatrix}$
   b) $\begin{pmatrix} 3 & 6 \\ 4 & 0 \end{pmatrix}\begin{pmatrix} 0 & 6 \\ 2 & 9 \end{pmatrix}$

2. a) $\begin{pmatrix} 1 & 6 \\ 3 & 4 \end{pmatrix}\begin{pmatrix} 4 & 6 & 0 \\ 8 & 1 & 3 \end{pmatrix}$
   b) $\begin{pmatrix} 1 & 4 \\ 2 & 1 \\ 3 & 0 \end{pmatrix}\begin{pmatrix} 2 & 2 \\ 4 & 4 \end{pmatrix}$

3. a) $\begin{pmatrix} 4 & -1 & 2 \\ 3 & 6 & -4 \end{pmatrix}\begin{pmatrix} 3 & 1 & 0 \\ -2 & 4 & 6 \\ 7 & 2 & -1 \end{pmatrix}$

   b) $(8 \;\; -2 \;\; -3 \;\; 0)\begin{pmatrix} 2 & 2 \\ -1 & 6 \\ -3 & 7 \\ 2 & -4 \end{pmatrix}$

4. a) $\begin{pmatrix} -2 \\ 1 \\ 4 \\ -6 \end{pmatrix}(-3 \;\; -2 \;\; 1)$

   b) $\begin{pmatrix} 2 & -6 \\ 8 & 1 \\ -1 & -3 \\ -4 & -9 \end{pmatrix}\begin{pmatrix} 3 & 2 & 1 \\ -4 & -3 & -6 \end{pmatrix}$

Exercise 33.5 In the following, calculate $\mathbf{V} \times \mathbf{W}$ and where possible $\mathbf{W} \times \mathbf{V}$.

1. $\mathbf{V} = \begin{pmatrix} -3 & 2 \\ 0 & 4 \end{pmatrix}$ $\mathbf{W} = \begin{pmatrix} 4 & -1 \\ 6 & 2 \end{pmatrix}$

2. $\mathbf{V} = \begin{pmatrix} 4 & 6 & 1 \\ 3 & 2 & -3 \end{pmatrix}$ $\mathbf{W} = \begin{pmatrix} 2 & 1 \\ -4 & -3 \\ 6 & 5 \end{pmatrix}$

3. $\mathbf{V} = (2 \;\; -5 \;\; 9 \;\; 2)$ $\mathbf{W} = \begin{pmatrix} 2 \\ 0 \\ -3 \\ 6 \end{pmatrix}$

4. $\mathbf{V} = \begin{pmatrix} 3 & 2 \\ -4 & -2 \end{pmatrix}$ $\mathbf{W} = \begin{pmatrix} -1 & -3 & 5 \\ 5 & 8 & -3 \end{pmatrix}$

5. $\mathbf{V} = \begin{pmatrix} -3 & -2 & 5 & 8 \\ -1 & 4 & -3 & 6 \end{pmatrix}$ $\mathbf{W} = \begin{pmatrix} 3 & -3 & 6 \\ 2 & -2 & 1 \\ -4 & 4 & 2 \\ 0 & 1 & 0 \end{pmatrix}$

In general if matrix **A** is multiplied by matrix **B** then this is not the same as matrix **B** multiplied by matrix **A**.

i.e. $\mathbf{A} \times \mathbf{B} \neq \mathbf{B} \times \mathbf{A}$

However there are some exceptions to this.

The identity matrix

The matrix $\begin{pmatrix} 1 & 0 \\ 0 & 1 \end{pmatrix}$ is known as the **identity matrix** of order 2.
The identity matrix is always represented by **I**.

Exercise 33.6 In the following calculate, where possible, **AI** and **IA**.

1. $\mathbf{A} = \begin{pmatrix} 2 & 1 \\ 3 & 2 \end{pmatrix}$
2. $\mathbf{A} = \begin{pmatrix} -2 & -4 \\ 3 & 6 \end{pmatrix}$
3. $\mathbf{A} = \begin{pmatrix} 4 & 8 \\ -2 & 4 \end{pmatrix}$
4. $\mathbf{A} = \begin{pmatrix} 3 & 2 \\ 1 & 6 \\ -2 & 5 \end{pmatrix}$
5. $\mathbf{A} = (-5 \quad -6)$
6. $\mathbf{A} = \begin{pmatrix} 4 & -3 \\ 5 & -6 \\ 3 & 2 \\ 1 & 4 \end{pmatrix}$

7. What conclusions can you make about **AI** in each case?
8. What conclusions can you make about **AI** and **IA** where **A** is a 2 × 2 matrix?

The zero matrix

A matrix in which all the elements are zero is called a zero matrix. Multiplying a matrix by a zero matrix gives a zero matrix.

i.e. $\begin{pmatrix} 4 & 2 \\ -3 & 0 \end{pmatrix} \begin{pmatrix} 0 & 0 \\ 0 & 0 \end{pmatrix} = \begin{pmatrix} 0 & 0 \\ 0 & 0 \end{pmatrix}$

The determinant of a 2 × 2 matrix

Worked examples

a) Find the determinant |**R**| of matrix **R** if $\mathbf{R} = \begin{pmatrix} 4 & 2 \\ 3 & 5 \end{pmatrix}$.

The product of the elements in the leading diagonal = 4 × 5 = 20.
The product of the elements in the secondary diagonal = 2 × 3 = 6.

|**R**| = 20 − 6
     = 14

b) If $\mathbf{S} = \begin{pmatrix} 3 & -4 \\ 2 & 1 \end{pmatrix}$, calculate |**S**|.

The product of the elements in the leading diagonal = 3 × 1 = 3.
The product of the elements in the secondary diagonal = −4 × 2 = −8.

|**S**| = 3 − (−8)
     = 11

Matrices and transformations

Exercise 33.7 Calculate the determinant of each of the matrices in questions 1–3.

1. a) $\begin{pmatrix} 6 & 3 \\ 5 & 3 \end{pmatrix}$ b) $\begin{pmatrix} 5 & 7 \\ 6 & 9 \end{pmatrix}$

   c) $\begin{pmatrix} 6 & 10 \\ 5 & 9 \end{pmatrix}$ d) $\begin{pmatrix} 7 & 9 \\ 6 & 8 \end{pmatrix}$

2. a) $\begin{pmatrix} 4 & 9 \\ 4 & 8 \end{pmatrix}$ b) $\begin{pmatrix} 3 & 7 \\ 4 & 6 \end{pmatrix}$

   c) $\begin{pmatrix} 2 & 8 \\ 3 & 7 \end{pmatrix}$ d) $\begin{pmatrix} 5 & 4 \\ 9 & 7 \end{pmatrix}$

3. a) $\begin{pmatrix} -3 & -5 \\ 6 & 4 \end{pmatrix}$ b) $\begin{pmatrix} 5 & 6 \\ -4 & -2 \end{pmatrix}$

   c) $\begin{pmatrix} 8 & 6 \\ -3 & -9 \end{pmatrix}$ d) $\begin{pmatrix} 1 & 4 \\ -1 & 0 \end{pmatrix}$

4. Write two matrices with a determinant of 5.
5. Write two matrices with a determinant of 0.
6. Write two matrices with a determinant of −7.
7. $\mathbf{A} = \begin{pmatrix} 4 & 2 \\ 3 & 8 \end{pmatrix}$ $\mathbf{B} = \begin{pmatrix} -3 & 2 \\ -4 & 4 \end{pmatrix}$ $\mathbf{C} = \begin{pmatrix} -5 & 3 \\ -2 & -6 \end{pmatrix}$

   Calculate:
   a) |**A** + **B**| b) |**A** − **C**| c) |**C** − **B**|
   d) |**AC**| e) |**BA**| f) |2**BC**|
   g) |3**A** − 2**B**| h) |2**CB**| i) |**B** + **C** − **A**|

● **The inverse of a matrix**

Consider the two matrices $\begin{pmatrix} 2 & 3 \\ 3 & 5 \end{pmatrix}$ and $\begin{pmatrix} 5 & -3 \\ -3 & 2 \end{pmatrix}$.

$$\begin{pmatrix} 2 & 3 \\ 3 & 5 \end{pmatrix} \begin{pmatrix} 5 & -3 \\ -3 & 2 \end{pmatrix} = \begin{pmatrix} 1 & 0 \\ 0 & 1 \end{pmatrix}$$

The product of these two matrices gives the identity matrix.

If $\mathbf{A} = \begin{pmatrix} 2 & 3 \\ 3 & 5 \end{pmatrix}$ then $\begin{pmatrix} 5 & -3 \\ -3 & 2 \end{pmatrix}$ is known as the **inverse** of **A**

and is written as $\mathbf{A}^{-1}$.

Finding the inverse of a matrix can be done in two ways: by simultaneous equations, and by use of a formula.

Worked example Find the inverse of $\begin{pmatrix} 6 & 8 \\ 2 & 3 \end{pmatrix}$.

$$\begin{pmatrix} 6 & 8 \\ 2 & 3 \end{pmatrix}\begin{pmatrix} w & y \\ x & z \end{pmatrix} = \begin{pmatrix} 1 & 0 \\ 0 & 1 \end{pmatrix}$$

Simultaneous equations

$6w + 8x = 1$ (1)
$2w + 3x = 0$ (2)

Multiplying eq.(2) by 3 and subtracting it from eq.(1) gives:

$6w + 8x = 1$
$6w + 9x = 0$
$\overline{\quad -x = 1}$ therefore $x = -1$

Substituting $x = -1$ into eq.(1) gives:

$6w + 8(-1) = 1$
$6w - 8 = 1$
$6w = 9$ therefore $w = 1.5$

$6y + 8z = 0$ (3)
$2y + 3z = 1$ (4)

Multiplying eq.(4) by 3 and subtracting it from eq.(3) gives:

$6y + 8z = 0$
$6y + 9z = 3$
$\overline{\quad -z = -3}$ therefore $z = 3$

Substituting $z = 3$ into eq.(4) gives:

$2y + 9 = 1$
$2y = -8$ therefore $y = -4$

The inverse of $\begin{pmatrix} 6 & 8 \\ 2 & 3 \end{pmatrix} = \begin{pmatrix} 1.5 & -4 \\ -1 & 3 \end{pmatrix}$.

Use of a formula

If $\mathbf{A} = \begin{pmatrix} w & y \\ x & z \end{pmatrix}$, $\mathbf{A}^{-1} = \dfrac{1}{wz - xy}\begin{pmatrix} z & -y \\ -x & w \end{pmatrix}$

Note: $wz - xy = |\mathbf{A}|$ so $\dfrac{1}{wz - xy} = \dfrac{1}{|\mathbf{A}|}$

Therefore if $\mathbf{A} = \begin{pmatrix} 6 & 8 \\ 2 & 3 \end{pmatrix}$

$$\mathbf{A}^{-1} = \dfrac{1}{18 - 16}\begin{pmatrix} 3 & -8 \\ -2 & 6 \end{pmatrix}$$

$$= \dfrac{1}{2}\begin{pmatrix} 3 & -8 \\ -2 & 6 \end{pmatrix} = \begin{pmatrix} 1.5 & -4 \\ -1 & 3 \end{pmatrix}$$

Matrices and transformations

Exercise 33.8

1. Using simultaneous equations find the inverse of each of the following matrices:

   a) $\begin{pmatrix} 9 & 5 \\ 7 & 4 \end{pmatrix}$
   b) $\begin{pmatrix} 10 & 7 \\ 7 & 5 \end{pmatrix}$
   c) $\begin{pmatrix} 5 & 5 \\ 4 & 5 \end{pmatrix}$
   d) $\begin{pmatrix} 6 & -9 \\ 3 & -4 \end{pmatrix}$
   e) $\begin{pmatrix} -5 & -4 \\ 10 & 9 \end{pmatrix}$
   f) $\begin{pmatrix} -3 & 2 \\ 6 & -4 \end{pmatrix}$

2. Using the formula find the inverse of the matrices in question 1, if possible.

3. Explain why (f) in question 1 above has no inverse.

4. Which of the following four matrices have no inverse?

   $A = \begin{pmatrix} 6 & 8 \\ 3 & 4 \end{pmatrix}$
   $B = \begin{pmatrix} 6 & 9 \\ 4 & 6 \end{pmatrix}$
   $C = \begin{pmatrix} 6 & 15 \\ -2 & 5 \end{pmatrix}$
   $D = \begin{pmatrix} -4 & -2 \\ 8 & 4 \end{pmatrix}$

5. $L = \begin{pmatrix} 3 & 4 \\ 7 & 9 \end{pmatrix}$
   $M = \begin{pmatrix} 2 & 3 \\ 3 & 6 \end{pmatrix}$
   $N = \begin{pmatrix} -4 & -3 \\ 8 & 5 \end{pmatrix}$
   $O = \begin{pmatrix} 4 & 3 \\ -8 & -5 \end{pmatrix}$

   Calculate the following:
   a) L^{-1}
   b) O^{-1}
   c) $(LO)^{-1}$
   d) $(M + N)^{-1}$
   e) $(NM)^{-1}$
   f) $(MO + LN)^{-1}$

Student assessment I

1. Calculate the following:

   a) $\begin{pmatrix} 2 & 4 \\ -3 & 5 \end{pmatrix} + \begin{pmatrix} 2 & -5 \\ -4 & 3 \end{pmatrix}$
   b) $\begin{pmatrix} 6 & 8 \\ 4 & -6 \\ 9 & -1 \end{pmatrix} + \begin{pmatrix} 0 & 3 \\ -5 & -6 \\ 1 & -4 \end{pmatrix}$
   c) $\begin{pmatrix} -5 & -3 \\ 7 & 2 \end{pmatrix} - \begin{pmatrix} 4 & 3 \\ 8 & 1 \end{pmatrix}$
   d) $\begin{pmatrix} 6 & -9 \\ 4 & -3 \\ 0 & 7 \end{pmatrix} - \begin{pmatrix} 1 & -5 \\ 3 & -8 \\ -1 & 2 \end{pmatrix}$
   e) $2\begin{pmatrix} 3 & 8 & -4 \\ 1 & -6 & 7 \end{pmatrix}$
   f) $\frac{1}{4}\begin{pmatrix} 4 & -8 \\ 0 & -6 \end{pmatrix}$

2. Multiply the following matrices:

   a) $\begin{pmatrix} 8 & -1 \\ 3 & 6 \\ -4 & 7 \end{pmatrix} \begin{pmatrix} 3 & -2 \\ 6 & 1 \end{pmatrix}$

   b) $(1 \quad -4 \quad 6) \begin{pmatrix} 3 & 4 \\ 8 & 0 \\ -9 & -3 \end{pmatrix}$

3. $\mathbf{A} = \begin{pmatrix} 4 & 9 \\ 3 & 7 \end{pmatrix} \quad \mathbf{B} = \begin{pmatrix} -6 & 7 \\ -5 & 6 \end{pmatrix}$

   Calculate: a) $|\mathbf{A}|$ b) $|\mathbf{A} + \mathbf{B}|$ c) $|\mathbf{B} - \mathbf{A}|$ d) $|2\mathbf{BA}|$

4. $\mathbf{X} = \begin{pmatrix} 7 & 4 \\ 3 & 2 \end{pmatrix} \quad \mathbf{Y} = \begin{pmatrix} 8 & 7 \\ 9 & 8 \end{pmatrix}$

   Calculate: a) $\mathbf{X}^{-1}$ b) $\mathbf{Y}^{-1}$ c) $(\mathbf{X} + \mathbf{Y})^{-1}$ d) $(\mathbf{Y} - \mathbf{X})^{-1}$

Student assessment 2

1. Calculate the following:

   a) $\begin{pmatrix} 3 & 6 \\ -2 & 0 \end{pmatrix} + \begin{pmatrix} 0 & -7 \\ -6 & 2 \end{pmatrix}$

   b) $\begin{pmatrix} 7 & 1 \\ 8 & -9 \\ 4 & -2 \end{pmatrix} + \begin{pmatrix} 4 & 6 \\ -2 & -3 \\ 4 & -3 \end{pmatrix}$

   c) $\begin{pmatrix} -7 & -2 \\ 2 & 1 \end{pmatrix} - \begin{pmatrix} 3 & 5 \\ 6 & 0 \end{pmatrix}$

   d) $\begin{pmatrix} 5 & -6 \\ 1 & -2 \\ 4 & 3 \end{pmatrix} - \begin{pmatrix} 2 & -7 \\ 5 & -3 \\ -2 & 6 \end{pmatrix}$

   e) $3 \begin{pmatrix} 1 & 4 & -1 \\ 2 & -4 & 5 \end{pmatrix}$

   f) $\frac{1}{3} \begin{pmatrix} 3 & -6 \\ 0 & -12 \end{pmatrix}$

2. Multiply the following matrices:

   a) $\begin{pmatrix} 1 & -2 \\ 4 & 2 \\ -5 & 3 \end{pmatrix} \begin{pmatrix} 9 & -1 \\ 3 & 7 \end{pmatrix}$

   b) $(2 \quad -3 \quad 5) \begin{pmatrix} 6 & 3 \\ 2 & 1 \\ -5 & -7 \end{pmatrix}$

3. $\mathbf{A} = \begin{pmatrix} 2 & 5 \\ 3 & 8 \end{pmatrix} \quad \mathbf{B} = \begin{pmatrix} -7 & 10 \\ -3 & 4 \end{pmatrix}$

   Calculate: a) $|\mathbf{A}|$ b) $|\mathbf{A} - \mathbf{B}|$ c) $|\mathbf{B} - \mathbf{A}|$ d) $|3\mathbf{AB}|$

4. $\mathbf{X} = \begin{pmatrix} 9 & 8 \\ 10 & 9 \end{pmatrix} \quad \mathbf{Y} = \begin{pmatrix} 8 & 6 \\ 9 & 7 \end{pmatrix}$

   Calculate: a) $\mathbf{X}^{-1}$ b) $\mathbf{Y}^{-1}$ c) $(\mathbf{X} - \mathbf{Y})^{-1}$ d) $(\mathbf{YX})^{-1}$

34 Transformations

An object undergoing a transformation changes in either position or shape. In its simplest form this change can occur as a result of either a **reflection**, **rotation**, **translation** or **enlargement**. If an object undergoes a transformation, then its new position or shape is known as the **image**.

● Reflection

If an object is reflected it undergoes a 'flip' movement about a dashed (broken) line known as the **mirror line**, as shown in the diagram.

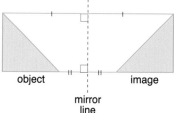

A point on the object and its equivalent point on the image are equidistant from the mirror line. This distance is measured at right angles to the mirror line. The line joining the point to its image is perpendicular to the mirror line.

The position of the mirror line is essential when describing a reflection. At times its equation as well as its position will be required.

Worked examples

a) Find the equation of the mirror line in the reflection given in the diagram (left).

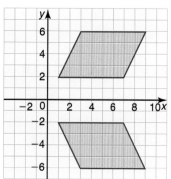

Here the mirror line is the *x*-axis. The equation of the mirror line is therefore $y = 0$.

b) A reflection is shown below.

i) Draw the position of the mirror line.

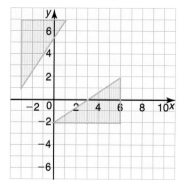 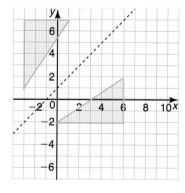

ii) Give the equation of the mirror line.

Equation of mirror line: $y = x + 1$.

Exercise 34.1 Copy each of the following diagrams, then:
a) draw the position of the mirror line(s),
b) give the equation of the mirror line(s).

1.
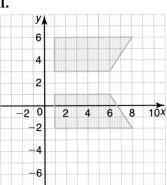

2.
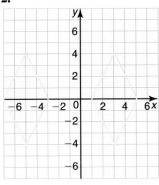

3.
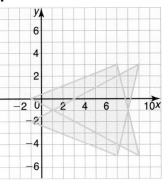

4.
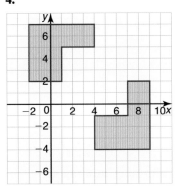

5.
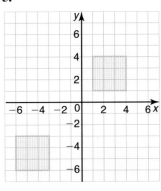

6.
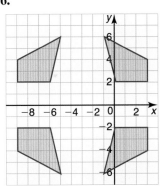

7.
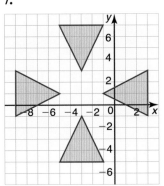

8.
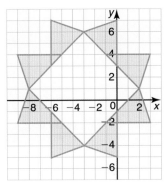

Matrices and transformations

Exercise 34.2 In questions 1 and 2, copy each diagram four times and reflect the object in each of the lines given.

1. a) $x = 2$
   b) $y = 0$
   c) $y = x$
   d) $y = -x$

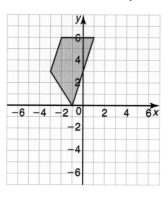

2. a) $x = -1$
   b) $y = -x - 1$
   c) $y = x + 2$
   d) $x = 0$

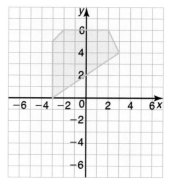

3. Copy the diagram (right), and reflect the triangles in the following lines:

   $x = 1$ and $y = -3$.

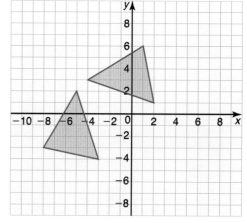

● **Rotation**

If an object is rotated it undergoes a 'turning' movement about a specific point known as the **centre of rotation**. When describing a rotation it is necessary to identify not only the position of the centre of rotation, but also the angle and direction of the turn, as shown in the diagram.

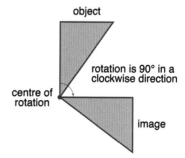

Transformations

Exercise 34.3 In the following, the object and centre of rotation have both been given. Copy each diagram and draw the object's image under the stated rotation about the marked point.

1.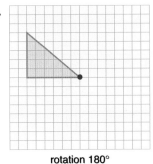
rotation 180°

2.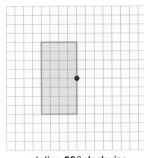
rotation 90° clockwise

3.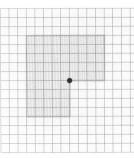
rotation 180°

4.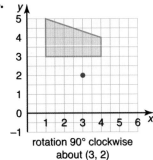
rotation 90° clockwise about (3, 2)

5.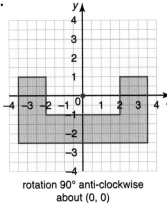
rotation 90° anti-clockwise about (0, 0)

6.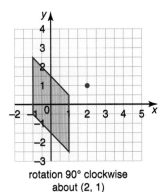
rotation 90° clockwise about (2, 1)

Exercise 34.4 In the following, the object (unshaded) and image (shaded) have been drawn. Copy each diagram.
a) Mark the centre of rotation.
b) Calculate the angle and direction of rotation.

1.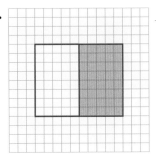

2.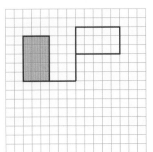

3.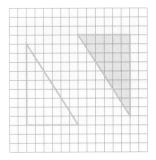

Matrices and transformations

4.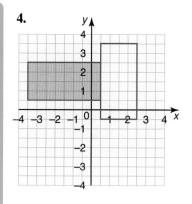

5.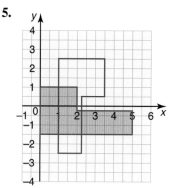

6.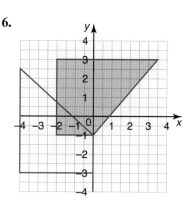

As discussed earlier, to describe a rotation, three pieces of information need to be given. These are the centre of rotation, the angle of rotation and the direction of rotation.

Finding the centre and angle of rotation

Worked example Consider the triangle ABC and its new position A'B'C' after being rotated.

    i) Find the centre of rotation.

        The centre of rotation is found in the following way:

- Join a point on the object to its corresponding point on the image, e.g. AA'.
- Find the perpendicular bisector of this line.
- Repeat this for another pair of points, e.g. BB'.

Where the two perpendicular bisectors meet gives the centre of rotation O.

    ii) Find the angle and direction of rotation.

        Angle of rotation = 90° in a clockwise direction

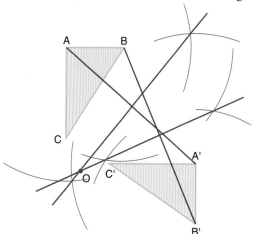

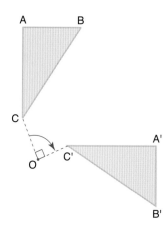

Exercise 34.5 For each of questions 1–4, draw two identical shapes in approximately the same positions as shown. For each pair, assuming the left or upper shape is the initial object, find:
a) the centre of rotation,
b) the angle and direction of rotation.
Check the accuracy of your results using tracing paper.

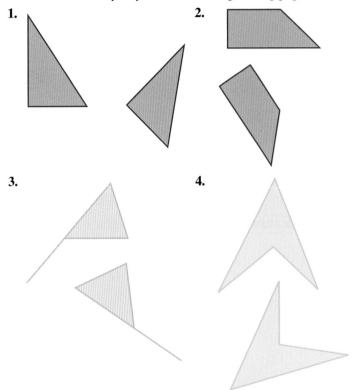

1.
2.
3.
4.

5. Draw a shape of your choice then draw its image after undergoing a rotation of 60° in a clockwise direction. Mark the centre of rotation.

6. Draw a shape of your choice then draw its image after undergoing a rotation of 240°. Mark on the centre of rotation.

Matrices and transformations

● **Translation**

If an object is translated, it undergoes a 'straight sliding' movement. When describing a translation it is necessary to give the translation vector. As no rotation is involved, each point on the object moves in the same way to its corresponding point on the image, e.g.

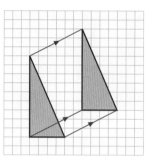 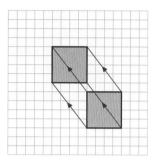

$$\text{Vector} = \begin{pmatrix} 6 \\ 3 \end{pmatrix} \qquad \text{Vector} = \begin{pmatrix} -4 \\ 5 \end{pmatrix}$$

Exercise 34.6 In the following diagrams, object A has been translated to each of images B and C. Give the translation vectors in each case.

1.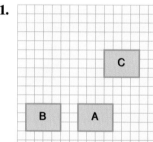

2.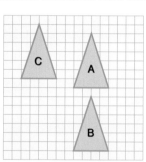

3.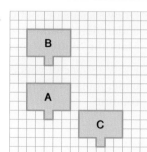

4.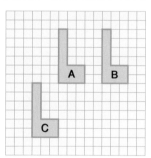

Transformations

Exercise 34.7 Copy each of the following diagrams and draw the object. Translate the object by the vector given in each case and draw the image in its position. (Note that a bigger grid than the one shown may be needed.)

1.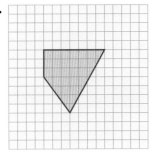

   Vector = $\begin{pmatrix} 3 \\ 5 \end{pmatrix}$

2.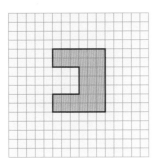

   Vector = $\begin{pmatrix} 5 \\ -4 \end{pmatrix}$

3.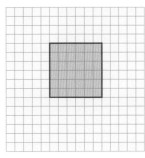

   Vector = $\begin{pmatrix} -4 \\ 6 \end{pmatrix}$

4.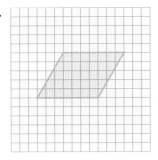

   Vector = $\begin{pmatrix} -2 \\ -5 \end{pmatrix}$

5.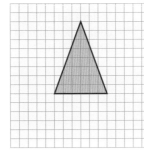

   Vector = $\begin{pmatrix} -6 \\ 0 \end{pmatrix}$

6.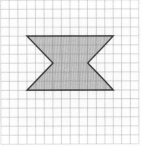

   Vector = $\begin{pmatrix} 0 \\ -1 \end{pmatrix}$

● Enlargement

If an object is enlarged, the result is an image which is mathematically similar to the object but of a different size. The image can be either larger or smaller than the original object. When describing an enlargement two pieces of information need to be given, the position of the **centre of enlargement** and the **scale factor of enlargement**.

Worked examples **a)** In the diagram below, triangle ABC is enlarged to form triangle A'B'C'.

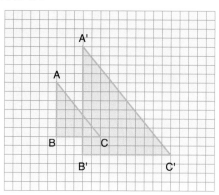

i) Find the centre of enlargement.

The centre of enlargement is found by joining corresponding points on the object and image with a straight line. These lines are then extended until they meet. The point at which they meet is the centre of enlargement O.

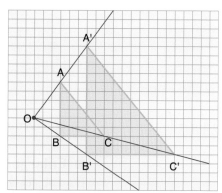

ii) Calculate the scale factor of enlargement.

The scale factor of enlargement can be calculated in one of two ways. From the diagram above it can be seen that the distance OA' is twice the distance OA. Similarly OC' and OB' are both twice OC and OB respectively, hence the scale factor of enlargement is 2.

Alternatively the scale factor can be found by considering the ratio of the length of a side on the image to the length of the corresponding side on the object. i.e.

$$\frac{A'B'}{AB} = \frac{12}{6} = 2$$

Hence the scale factor of enlargement is 2.

b) In the diagram below, the rectangle ABCD undergoes a transformation to form rectangle A'B'C'D'.

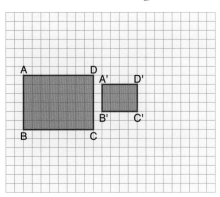

i) Find the centre of enlargement.

By joining corresponding points on both the object and the image the centre of enlargement is found at O.

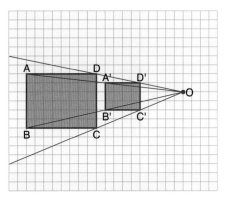

ii) Calculate the scale factor of enlargement.

$$\text{The scale factor of enlargement} = \frac{A'B'}{AB} = \frac{3}{6} = \frac{1}{2}$$

Note: If the scale factor of enlargement is greater than 1, then the image is larger than the object. If the scale factor lies between 0 and 1, then the resulting image is smaller than the object. In these cases, although the image is smaller than the object, the transformation is still known as an enlargement.

Matrices and transformations

Exercise 34.8 Copy the following diagrams and find:
a) the centre of enlargement,
b) the scale factor of enlargement.

1.
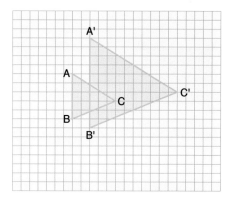

2.
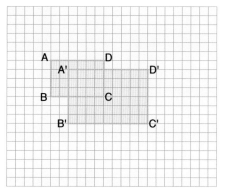

3.
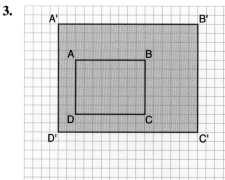

4.
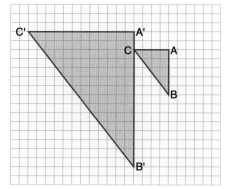

5.
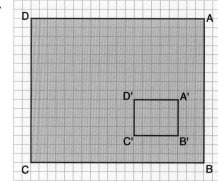

Exercise 34.9

Copy the following diagrams and enlarge the objects by the scale factor given and from the centre of enlargement shown. Grids larger than those shown may be needed.

1.

scale factor 2

2.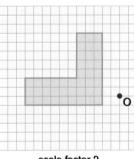

scale factor 2

3.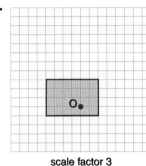

scale factor 3

4.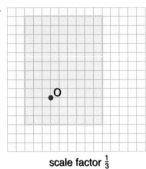

scale factor $\frac{1}{3}$

● Negative englargement

The diagram below shows an example of **negative enlargement**.

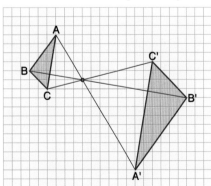

scale factor of enlargement is −2

With negative enlargement each point and its image are on opposite sides of the centre of enlargement. The scale factor of enlargement is calculated in the same way, remembering, however, to write a '−' sign before the number.

Matrices and transformations

Exercise 34.10

1. Copy the following diagram and then calculate the scale factor of enlargement and show the position of the centre of enlargement.

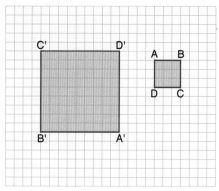

2. The scale factor of enlargement and centre of enlargement are both given. Copy and complete the diagram.

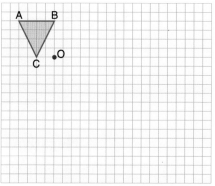

scale factor of enlargement is −2.5

3. The scale factor of enlargement and centre of enlargement are both given. Copy and complete the diagram.

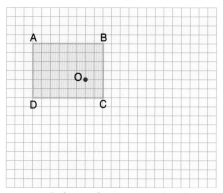

scale factor of enlargement is −2

4. Copy the following diagram and then calculate the scale factor of enlargement and show the position of the centre of enlargement.

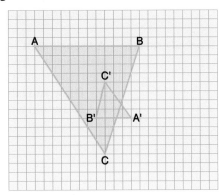

5. An object and part of its image under enlargement are given in the diagram below. Copy the diagram and complete the image. Also find the centre of enlargement and calculate the scale factor of enlargement.

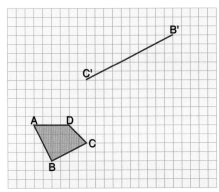

6. In the diagram below, part of an object in the shape of a quadrilateral and its image under enlargement are drawn. Copy and complete the diagram. Also find the centre of enlargement and calculate the scale factor of enlargement.

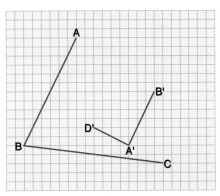

Matrices and transformations

● Combinations of transformations

An object need not be subjected to just one type of transformation. It can undergo a succession of different transformations.

Worked example A triangle ABC maps onto A'B'C' after an enlargement of scale factor 3 from the centre of enlargement (0, 7). A'B'C' is then mapped onto A"B"C" by a reflection in the line $x = 1$.
i) Draw and label the image A'B'C'.
ii) Draw and label the image A"B"C".

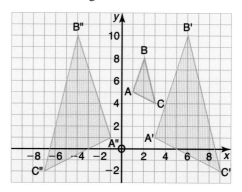

Exercise 34.11 In each of the following questions, copy the diagram. After each transformation, draw the image on the same grid and label it clearly.

1. The square ABCD is mapped onto A'B'C'D' by a reflection in the line $y = 3$. A'B'C'D' then maps onto A"B"C"D" as a result of a 90° rotation in a clockwise direction about the point $(-2, 5)$.

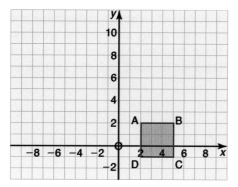

2. The rectangle ABCD is mapped onto A'B'C'D' by an enlargement of scale factor -2 with its centre at (0, 5). A'B'C'D' then maps onto A"B"C"D" as a result of a reflection in the line $y = -x + 7$.

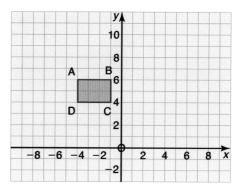

Transformations and matrices

A transformation can be represented by a matrix.

Worked example
i) Express the vertices of the trapezium PQRS in the form of a matrix.

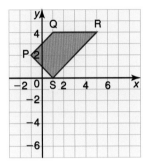

$$\begin{matrix} P & Q & R & S \end{matrix}$$
$$\begin{pmatrix} -1 & 1 & 5 & 1 \\ 2 & 4 & 4 & 0 \end{pmatrix}$$

ii) Find the coordinates of the vertices of its image if PQRS undergoes a transformation by the matrix $\begin{pmatrix} 1 & 0 \\ 0 & -1 \end{pmatrix}$.

$$\begin{pmatrix} 1 & 0 \\ 0 & -1 \end{pmatrix} \begin{pmatrix} -1 & 1 & 5 & 1 \\ 2 & 4 & 4 & 0 \end{pmatrix} = \overset{\begin{matrix} P' & Q' & R' & S' \end{matrix}}{\begin{pmatrix} -1 & 1 & 5 & 1 \\ -2 & -4 & -4 & 0 \end{pmatrix}}$$

iii) Plot the object PQRS and its image P'Q'R'S'.

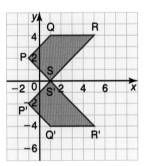

Matrices and transformations

Exercise 34.12 In questions 1–6, transform the object shown in the diagram below by the matrix given. Draw a diagram for each transformation, and plot both the object and its image on the same grid.

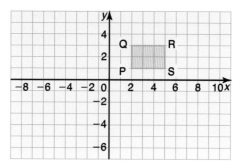

1. Transformation matrix is $\begin{pmatrix} 0 & 1 \\ 1 & 0 \end{pmatrix}$.

2. Transformation matrix is $\begin{pmatrix} 0 & -1 \\ -1 & 0 \end{pmatrix}$.

3. Transformation matrix is $\begin{pmatrix} 0 & 1 \\ -1 & 0 \end{pmatrix}$.

4. Transformation matrix is $\begin{pmatrix} 0 & -1 \\ 1 & 0 \end{pmatrix}$.

5. Transformation matrix is $\begin{pmatrix} 2 & 0 \\ 0 & 2 \end{pmatrix}$.

6. Transformation matrix is $\begin{pmatrix} -2 & 0 \\ 0 & -2 \end{pmatrix}$.

7. Describe in geometrical terms each of the transformations in questions 1–6.

8. a) Draw the image of triangle XYZ under the transformation matrix $\begin{pmatrix} -3 & 0 \\ 0 & -3 \end{pmatrix}$.
   Label it X'Y'Z'.

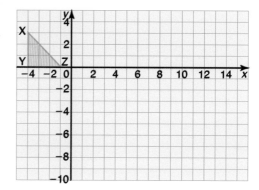

b) Calculate the area of triangle XYZ.
c) Calculate the area of triangle X'Y'Z'.
d) Calculate the area scale factor.
e) Calculate the determinant of the transformation matrix.

9. a) Draw the image of rectangle PQRS under the transformation matrix $\begin{pmatrix} 1.5 & 0 \\ 0 & 1.5 \end{pmatrix}$.

   Label it P'Q'R'S'.

   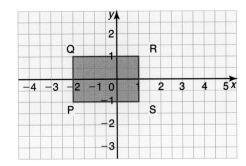

   b) Calculate the area of PQRS.
   c) Calculate the area of P'Q'R'S'.
   d) Calculate the area scale factor.
   e) Calculate the determinant of the transformation matrix.

10. a) Draw the image of ABC under the transformation matrix $\begin{pmatrix} 0 & -2.5 \\ 2.5 & 0 \end{pmatrix}$.

    Label it A'B'C'.

    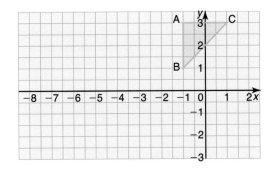

    b) Calculate the area of ABC.
    c) Calculate the area of A'B'C'.
    d) Calculate the area scale factor.
    e) Calculate the determinant of the transformation matrix.

Transformations and inverse matrices

If an object is transformed by a matrix **A**, then the inverse matrix $\mathbf{A}^{-1}$ gives the inverse transformation, i.e. it maps the image back onto the object.

Worked example i) The matrix $\begin{pmatrix} 0 & -2 \\ 2 & 0 \end{pmatrix}$ maps $\triangle PQR$ onto $\triangle P'Q'R'$.

Draw the image $P'Q'R'$.

$$\begin{pmatrix} 0 & -2 \\ 2 & 0 \end{pmatrix} \overset{P \quad Q \quad R}{\begin{pmatrix} -2 & 1 & 0 \\ 2 & 2 & -1 \end{pmatrix}} = \overset{P' \quad Q' \quad R'}{\begin{pmatrix} -4 & -4 & 2 \\ -4 & 2 & 0 \end{pmatrix}}$$

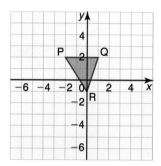 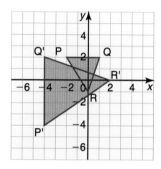

ii) Calculate the area of $\triangle P'Q'R'$.

Area of $\triangle PQR = 4.5$ units^2

The determinant of the transformation matrix is 4. As the determinant is numerically equal to the area factor the area factor is also 4.

Area of $\triangle P'Q'R' = 4 \times 4.5 = 18$ units^2

iii) Calculate the matrix which maps $\triangle P'Q'R'$ back onto $\triangle PQR$.

This is calculated by finding the inverse of the transformation matrix.

$$\begin{pmatrix} 0 & -2 \\ 2 & 0 \end{pmatrix}^{-1} = \frac{1}{0 \times 0 - 2 \times (-2)} \begin{pmatrix} 0 & 2 \\ -2 & 0 \end{pmatrix}$$

$$= \frac{1}{4} \begin{pmatrix} 0 & 2 \\ -2 & 0 \end{pmatrix} = \begin{pmatrix} 0 & 0.5 \\ 0.5 & 0 \end{pmatrix}$$

Exercise 34.13

1. a) On a grid, draw a quadrilateral of your choice and label its vertices P, Q, R and S.
   b) Draw its image under the transformation of the matrix $\begin{pmatrix} 0 & -2 \\ -2 & 0 \end{pmatrix}$ and label the vertices $P'Q'R'S'$.
   c) What matrix maps $P'Q'R'S'$ onto PQRS?

2. a) On a grid, draw a triangle of your choice and label its vertices A, B and C.
    b) Draw its image under the transformation of the matrix $\begin{pmatrix} -1 & 0 \\ 0 & -1 \end{pmatrix}$ and label the vertices A'B'C'.
    c) What matrix maps △A'B'C' onto △ABC?

3. a) On a grid, draw a kite of your choice and label its vertices A, B, C and D.
    b) Draw its image under the transformation of the matrix $\begin{pmatrix} 0 & 1.5 \\ -1.5 & 0 \end{pmatrix}$ and label the vertices A'B'C'D'.
    c) What matrix maps A'B'C'D' onto ABCD?

4. a) On a grid, draw a square of your choice and label its vertices W, X, Y and Z.
    b) Draw its image under the transformation of the matrix $\begin{pmatrix} -0.5 & 0 \\ 0 & 0.5 \end{pmatrix}$ and label the vertices W'X'Y'Z'.
    c) What matrix maps W'X'Y'Z' onto WXYZ?

5. a) On a grid, draw an isosceles triangle of your choice and label its vertices L, M and N.
    b) Draw its image under the transformation of the matrix $\begin{pmatrix} 2.5 & 0 \\ 0 & -2.5 \end{pmatrix}$ and label the vertices L'M'N'.
    c) What matrix maps △L'M'N' onto △LMN?

6. a) On a grid, draw a square of your choice and label its vertices A, B, C and D.
    b) Draw its image under the transformation of the matrix $\begin{pmatrix} -1 & \sqrt{3} \\ -\sqrt{3} & -1 \end{pmatrix}$ and label its vertices A'B'C'D'.
    c) What matrix maps A'B'C'D' onto ABCD?

● **Combinations of transformations**

An object can be transformed by a series of transformation matrices. These matrices can be replaced by a single matrix which maps the original object onto the final image.

Matrices and transformations

Worked example $A = \begin{pmatrix} -1 & 0 \\ 0 & -1 \end{pmatrix}$ $B = \begin{pmatrix} 2 & 0 \\ 0 & 2 \end{pmatrix}$

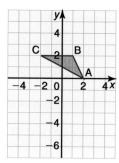

i) The triangle ABC is mapped onto A'B'C' under the transformation of matrix **A**.
Draw and label the position of A'B'C'.

$$\begin{pmatrix} -1 & 0 \\ 0 & -1 \end{pmatrix} \overset{A\ B\ C}{\begin{pmatrix} 2 & 1 & -2 \\ 0 & 2 & 2 \end{pmatrix}} = \overset{A'\ B'\ C'}{\begin{pmatrix} -2 & -1 & 2 \\ 0 & -2 & -2 \end{pmatrix}}$$

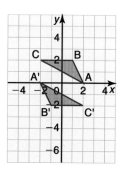

ii) A'B'C' is mapped onto A"B"C" under the transformation of matrix **B**.

Draw and label the position of A"B"C".

$$\begin{pmatrix} 2 & 0 \\ 0 & 2 \end{pmatrix} \overset{A'\ B'\ C'}{\begin{pmatrix} -2 & -1 & 2 \\ 0 & -2 & -2 \end{pmatrix}} = \overset{A"\ B"\ C"}{\begin{pmatrix} -4 & -2 & 4 \\ 0 & -4 & -4 \end{pmatrix}}$$

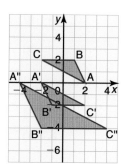

iii) Find the matrix which maps ABC onto A"B"C".

ABC undergoes a transformation firstly by matrix **A** and then by matrix **B**. The matrix which maps ABC directly onto A"B"C" is given by **BA**, i.e.

$$\begin{pmatrix} 2 & 0 \\ 0 & 2 \end{pmatrix} \begin{pmatrix} -1 & 0 \\ 0 & -1 \end{pmatrix} = \begin{pmatrix} -2 & 0 \\ 0 & -2 \end{pmatrix}$$

iv) Find the matrix which maps A"B"C" onto ABC.

This is the inverse of the matrix which maps ABC onto A"B"C", i.e.

$$\begin{pmatrix} -2 & 0 \\ 0 & -2 \end{pmatrix}^{-1} = \begin{pmatrix} -0.5 & 0 \\ 0 & -0.5 \end{pmatrix}$$

Exercise 34.14

1. $A = \begin{pmatrix} -1 & 0 \\ 0 & 1 \end{pmatrix}$ $B = \begin{pmatrix} \frac{\sqrt{2}}{2} & -\frac{\sqrt{2}}{2} \\ \frac{\sqrt{2}}{2} & \frac{\sqrt{2}}{2} \end{pmatrix}$

   Triangle XYZ is mapped onto X'Y'Z' under the transformation of matrix **A**. Triangle X'Y'Z' is subsequently mapped onto triangle X"Y"Z" under the transformation of matrix **B**.

   a) Copy the diagram below and plot the position of X'Y'Z'.

   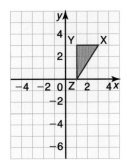

   b) Describe the transformation given by matrix **A**.
   c) Plot the position of X"Y"Z".
   d) Describe the transformation given by matrix **B**.
   e) Find the matrix which maps XYZ directly onto X"Y"Z".
   f) Find the matrix which maps X"Y"Z" directly onto XYZ.

2. $A = \begin{pmatrix} 2 & 0 \\ 0 & 2 \end{pmatrix}$ $B = \begin{pmatrix} 0 & -1 \\ -1 & 0 \end{pmatrix}$

   Quadrilateral JKLM is mapped onto J'K'L'M' under the transformation of matrix **A**. J'K'L'M' is subsequently mapped onto J"K"L"M" under the transformation of matrix **B**.

   a) Copy the diagram below and plot the position of J'K'L'M'.

   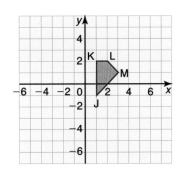

b) Describe the transformation given by matrix A.
c) Plot the position of J"K"L"M".
d) Describe the transformation given by matrix B.
e) Find the matrix which maps JKLM directly onto J"K"L"M".
f) Find the matrix which maps J"K"L"M" directly onto JKLM.

3. The vertices of a triangle ABC are given by the coordinates A(3, 2), B(0, 4) and C(5, 3). Triangle ABC is mapped onto triangle A'B'C', the coordinates of its vertices being

$A'\left(-3, \frac{9}{2}\right) B'\left(-6, 0\right)$ and $C'\left(-\frac{9}{2}, \frac{15}{2}\right)$.

a) Find the matrix which maps ABC onto A'B'C'.
b) Find the matrix which maps A'B'C' onto ABC.

4. The coordinates of the vertices of a square ABCD after two transformations are as follows:

P(3, −1)	Q(4, −2)	R(3, −3)	S(2, −2)
P'(−3, 1)	Q'(−4, 2)	R'(−3, 3)	S'(−2, 2)
P"(−1, 3)	Q"(−2, 4)	R"(−3, 3)	S"(−2, 2)

a) Find the matrix which maps PQRS onto P'Q'R'S'.
b) Find the matrix which maps P'Q'R'S' onto P"Q"R"S".
c) Find the single matrix which maps PQRS onto P"Q"R"S".

With the aid of a diagram if necessary:

d) Describe the transformation which maps PQRS onto P'Q'R'S'.
e) Describe the transformation which maps P'Q'R'S' onto P"Q"R"S".
f) Describe the single transformation which maps PQRS onto P"Q"R"S".

Student assessment 1

1. Copy the diagram below, which shows an object and its reflected image.
   a) Draw on your diagram the position of the mirror line.
   b) Find the equation of the mirror line.

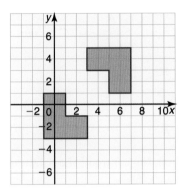

2. The triangle ABC is mapped onto triangle A'B'C' by a rotation (below).
   a) Find the coordinates of the centre of rotation.
   b) Give the angle and direction of rotation.

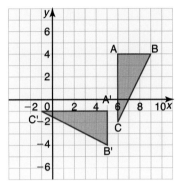

3. Write down the column vector of the translation which maps:
   a) rectangle A to rectangle B,
   b) rectangle B to rectangle C.

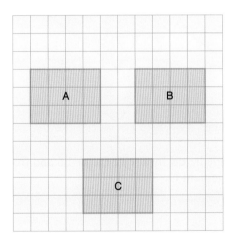

4. Enlarge the triangle below by a scale factor 2 and from the centre of enlargement O.

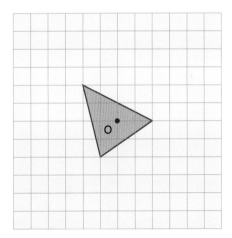

Matrices and transformations

Student assessment 2

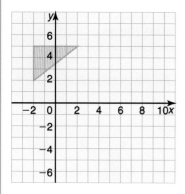

1. Copy the diagram (left).
   a) Draw in the mirror line with equation $y = x - 1$.
   b) Reflect the object in the mirror line.

2. Draw a triangle ABC and a triangle A'B'C' in positions similar to those shown (below).

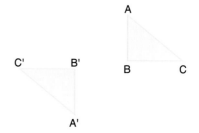

   a) Find, by construction, the centre of the rotation which maps $\triangle ABC$ onto $\triangle A'B'C'$.
   b) Calculate the angle and direction of the rotation.

3. Write down the column vector of the translation which maps:
   a) triangle A to triangle B,
   b) triangle B to triangle C.

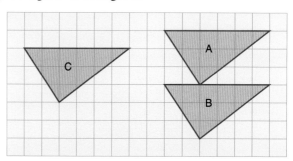

4. Enlarge the rectangle below by a scale factor 1.5 and from the centre of enlargement O.

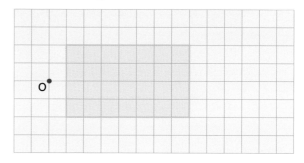

Student assessment 3

1. An object ABCD and its image A'B'C'D' are shown below.
   a) Find the position of the centre of enlargement.
   b) Calculate the scale factor of enlargement.

   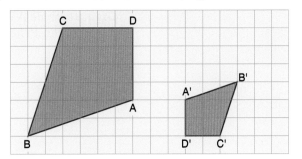

2. The square ABCD is mapped onto A'B'C'D'. A'B'C'D' is subsequently mapped onto A"B"C"D".

   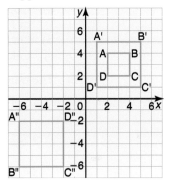

   a) Describe in full the transformation which maps ABCD onto A'B'C'D'.
   b) Describe in full the transformation which maps A'B'C'D' onto A"B"C"D".

3. Square ABCD is mapped onto square A'B'C'D'. Square A'B'C'D' is subsequently mapped onto square A"B"C"D".

   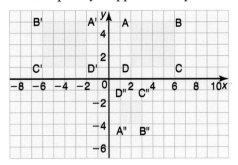

   a) Describe fully the transformation which maps ABCD onto A'B'C'D'.
   b) Describe fully the transformation which maps A'B'C'D' onto A"B"C"D".

4. The triangle JKL (below) is mapped onto triangle J'K'L' by the matrix $\begin{pmatrix} 0 & -1 \\ -1 & 0 \end{pmatrix}$.

△J'K'L' is subsequently mapped onto △J"K"L" by the matrix $\begin{pmatrix} 0 & 1 \\ -1 & 0 \end{pmatrix}$.

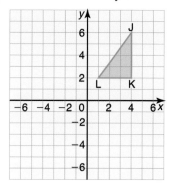

a) Copy the grid (above), and draw and label the position of J'K'L'.
b) On the same axes, draw and label the position of J"K"L".
c) Calculate the matrix which maps JKL directly onto J"K"L".

5. The quadrilateral PQRS is mapped onto P'Q'R'S' by the matrix $\begin{pmatrix} 1.5 & 0 \\ 0 & 1.5 \end{pmatrix}$.

P'Q'R'S' is mapped onto P"Q"R"S" by the matrix $\begin{pmatrix} -1 & 0 \\ 0 & -1 \end{pmatrix}$.

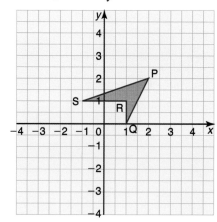

a) Copy the grid (above), and plot and label the position of P'Q'R'S'.
b) On the same axes, plot and label the position of P"Q"R"S".
c) Find the matrix which would map PQRS directly onto P"Q"R"S".
d) Find the matrix which would map P"Q"R"S" directly onto PQRS.

Student assessment 4

1. An object WXYZ and its image W'X'Y'Z' are shown below.
   a) Find the position of the centre of enlargement.
   b) Calculate the scale factor of enlargement.

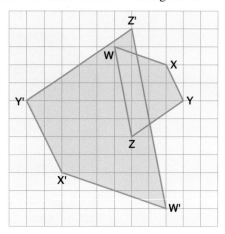

2. The triangle ABC is mapped onto A'B'C'. A'B'C' is subsequently mapped onto A"B"C".
   a) Describe in full the transformation which maps ABC onto A'B'C'.
   b) Describe in full the transformation which maps A'B'C' onto A"B"C".

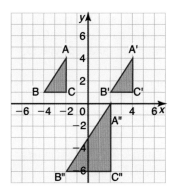

3. The triangle PQR undergoes a transformation by the matrix $\begin{pmatrix} -0.4 & 0 \\ 0 & -0.4 \end{pmatrix}$.

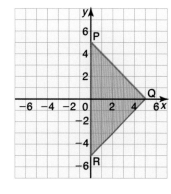

   a) Draw the image of △PQR after the transformation and label its vertices P'Q'R'.
   b) Calculate the area scale factor from △PQR to △P'Q'R'.
   c) Calculate the determinant of the transformation matrix.

Topic 7: Mathematical investigations and ICT

● A painted cube

A $3 \times 3 \times 3$ cm cube is painted on the outside as shown in the left-hand diagram below:

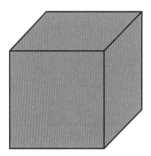
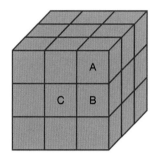

The large cube is then cut up into 27 smaller cubes, each $1\,\text{cm} \times 1\,\text{cm} \times 1\,\text{cm}$ as shown on the right.

$1 \times 1 \times 1$ cm cubes with 3 painted faces are labelled type A.
$1 \times 1 \times 1$ cm cubes with 2 painted faces are labelled type B.
$1 \times 1 \times 1$ cm cubes with 1 face painted are labelled type C.
$1 \times 1 \times 1$ cm cubes with no faces painted are labelled type D.

1. a) How many of the 27 cubes are type A?
   b) How many of the 27 cubes are type B?
   c) How many of the 27 cubes are type C?
   d) How many of the 27 cubes are type D?

2. Consider a $4 \times 4 \times 4$ cm cube cut into $1 \times 1 \times 1$ cm cubes. How many of the cubes are type A, B, C and D?

3. How many type A, B, C and D cubes are there when a $10 \times 10 \times 10$ cm cube is cut into $1 \times 1 \times 1$ cm cubes?

4. Generalise for the number of type A, B, C and D cubes in an $n \times n \times n$ cube.

5. Generalise for the number of type A, B, C and D cubes in a cuboid l cm long, w cm wide and h cm high.

● Triangle count

The diagram below shows an isosceles triangle with a vertical line drawn from its apex to its base.

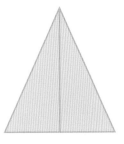

There is a total of 3 triangles in this diagram.

If a horizontal line is drawn across the triangle, it will look as shown:

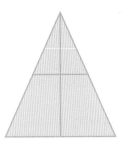

There is a total of 6 triangles in this diagram.

When one more horizontal line is added, the number of triangles increases further:

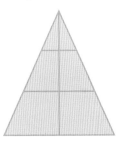

1. Calculate the total number of triangles in the diagram above with the two inner horizontal lines.
2. Investigate the relationship between the total number of triangles (t) and the number of inner horizontal lines (h). Enter your results in an ordered table.
3. Write an algebraic rule linking the total number of triangles and the number of inner horizontal lines.

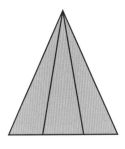

The triangle (left) has two lines drawn from the apex to the base.

There is a total of six triangles in this diagram.

If a horizontal line is drawn through this triangle, the number of triangles increases as shown:

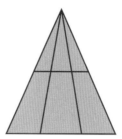

4. Calculate the total number of triangles in the diagram above with two lines from the vertex and one inner horizontal line.
5. Investigate the relationship between the total number of triangles (t) and the number of inner horizontal lines (h) when two lines are drawn from the apex. Enter your results in an ordered table.
6. Write an algebraic rule linking the total number of triangles and the number of inner horizontal lines.

● **ICT activity 1**

Using Autograph or another appropriate software package, prepare a help sheet for your revision that demonstrates the addition, subtraction and multiplication of vectors. An example is shown below:

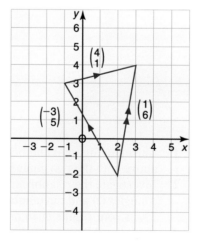

Vector addition:

$$\begin{pmatrix} -3 \\ 5 \end{pmatrix} + \begin{pmatrix} 4 \\ 1 \end{pmatrix} = \begin{pmatrix} 1 \\ 6 \end{pmatrix}$$

Topic 7

Mathematical investigations and ICT

● ICT activity 2

Using Autograph or other appropriate software, investigate the effect of different transformation matrices on an object.

1. Draw an object such as a rectangle or other basic shape.

2. Apply a 2 × 2 matrix to the object in the form $\begin{pmatrix} a & b \\ c & d \end{pmatrix}$

3. By changing the values of a, b, c, and d to $\begin{pmatrix} 0 & 1 \\ 1 & 0 \end{pmatrix}$ describe the transformation on the original object.

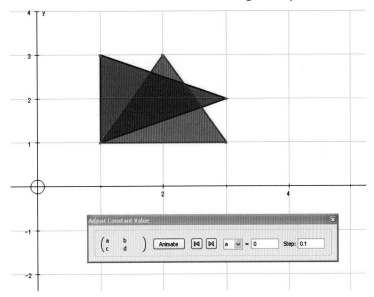

4. Apply each of the following transformation matrices in turn. Describe in geometrical terms each of the transformations:

   a) $\begin{pmatrix} 0 & -1 \\ -1 & 0 \end{pmatrix}$

   b) $\begin{pmatrix} 0 & 1 \\ -1 & 0 \end{pmatrix}$

   c) $\begin{pmatrix} 0 & -1 \\ 1 & 0 \end{pmatrix}$

   d) $\begin{pmatrix} 2 & 0 \\ 0 & 2 \end{pmatrix}$

   e) $\begin{pmatrix} -2 & 0 \\ 0 & -2 \end{pmatrix}$

5. Prepare a small display of your findings in questions 3 and 4 above.

Topic 8: Probability

Syllabus

E8.1
Calculate the probability of a single event as either a fraction, decimal or percentage.

E8.2
Understand and use the probability scale from 0 to 1.

E8.3
Understand that the probability of an event occurring = 1 − the probability of the event not occurring.

E8.4
Understand relative frequency as an estimate of probability.

E8.5
Calculate the probability of simple combined events, using possibility diagrams and tree diagrams where appropriate.

Contents

Chapter 35   Probability (E8.1, E8.2, E8.3, E8.4)
Chapter 36   Further probability (E8.5)

◯ Order and chaos

Blaise Pascal and Pierre de Fermat (known for his last theorem) corresponded about problems connected to games of chance.

Although Newton and Galileo had had some thoughts on the subject, this is accepted as the beginning of the study of what is now called probability. Later, in 1657, Christiaan Huygens wrote the first book on the subject entitled *The Value of all Chances in Games of Fortune*.

In 1821 Carl Friedrich Gauss (1777–1855) worked on normal distribution.

At the start of the nineteenth century, the French mathematician Pierre Simon de Laplace was convinced of the existence of a Newtonian universe. In other words, if you knew the position and velocities of all the particles in the universe, you would be able to predict the future because their movement would be predetermined by scientific laws. However, quantum mechanics has since shown that this is not true. Chaos theory is at the centre of understanding these limits.

Blaise Pascal (1623–1662)

35 Probability

Probability is the study of chance, or the likelihood of an event happening. However, because probability is based on chance, what theory predicts does not necessarily happen in practice.

A **favourable outcome** refers to the event in question actually happening. The **total number of possible outcomes** refers to all the different types of outcome one can get in a particular situation. In general:

$$\text{Probability of an event} = \frac{\text{number of favourable outcomes}}{\text{total number of equally likely outcomes}}$$

If the probability = 0, the event is impossible.
If the probability = 1, the event is certain to happen.

If an event can either happen or not happen then:

Probability of the event not occurring
$$= 1 - \text{the probability of the event occurring.}$$

Worked examples a) An ordinary, fair dice is rolled. Calculate the probability of getting a six.

  Number of favourable outcomes = 1   (i.e. getting a 6)
  Total number of possible outcomes = 6
      (i.e. getting a 1, 2, 3, 4, 5 or 6)
  Probability of getting a six = $\frac{1}{6}$
  Probability of not getting a six = $1 - \frac{1}{6} = \frac{5}{6}$

 b) An ordinary, fair dice is rolled. Calculate the probability of getting an even number.

  Number of favourable outcomes = 3
      (i.e. getting a 2, 4 or 6)
  Total number of possible outcomes = 6
      (i.e. getting a 1, 2, 3, 4, 5 or 6)
  Probability of getting an even number = $\frac{3}{6} = \frac{1}{2}$

 c) Thirty students are asked to choose their favourite subject out of Maths, English and Art. The results are shown in the table below:

	Maths	English	Art
Girls	7	4	5
Boys	5	3	6

  A student is chosen at random.

i) What is the probability that it is a girl?
Total number of girls is 16.
Probability of choosing a girl is $\frac{16}{30} = \frac{8}{15}$.

ii) What is the probability that it is a boy whose favourite subject is Art?
Number of boys whose favourite subject is Art is 6.
Probability is therefore $\frac{6}{30} = \frac{1}{5}$.

iii) What is the probability of **not** choosing a girl whose favourite subject is English?
There are two ways of approaching this:
Method 1:
Total number of students who are not girls whose favourite subject is English is $7 + 5 + 5 + 3 + 6 = 26$.
Therefore probability is $\frac{26}{30} = \frac{13}{15}$.
Method 2:
Total number of girls whose favourite subject is English is 4.
Probability of choosing a girl whose favourite subject is English is $\frac{4}{30}$.
Therefore the probability of **not** choosing a girl whose favourite subject is English is:
$$1 - \frac{4}{30} = \frac{26}{30} = \frac{13}{15}$$

The likelihood of an event such as 'you will play sport tomorrow' will vary from person to person. Therefore, the probability of the event is not constant. However, the probability of some events, such as the result of throwing dice, spinning a coin or dealing cards, can be found by experiment or calculation.

A probability scale goes from 0 to 1.

Exercise 35.1

1. Copy the probability scale above.
   Mark on the probability scale the probability that
   a) a day chosen at random is a Saturday,
   b) a coin will show tails when spun,
   c) the sun will rise tomorrow,
   d) a woman will run a marathon in two hours,
   e) the next car you see will be silver.

2. Express your answers to question 1 as fractions, decimals and percentages.

Probability

Exercise 35.2

1. Calculate the theoretical probability, when rolling an ordinary, fair dice, of getting each of the following:
   a) a score of 1
   b) a score of 2, 3, 4, 5 or 6
   c) an odd number
   d) a score less than 6
   e) a score of 7
   f) a score less than 7

2. a) Calculate the probability of:
      i) being born on a Wednesday,
      ii) not being born on a Wednesday.
   b) Explain the result of adding the answers to a) i) and ii) together.

3. 250 balls are numbered from 1 to 250 and placed in a box. A ball is picked at random. Find the probability of picking a ball with:
   a) the number 1
   b) an even number
   c) a three-digit number
   d) a number less than 300

4. In a class there are 25 girls and 15 boys. The teacher takes in all of their books in a random order. Calculate the probability that the teacher will:
   a) mark a book belonging to a girl first,
   b) mark a book belonging to a boy first.

5. Tiles, each lettered with one different letter of the alphabet, are put into a bag. If one tile is taken out at random, calculate the probability that it is:
   a) an A or P
   b) a vowel
   c) a consonant
   d) an X, Y or Z
   e) a letter in your first name.

6. A boy was late for school 5 times in the previous 30 school days. If tomorrow is a school day, calculate the probability that he will arrive late.

7. a) Three red, 10 white, 5 blue and 2 green counters are put into a bag. If one is picked at random, calculate the probability that it is:
      i) a green counter
      ii) a blue counter.
   b) If the first counter taken out is green and it is not put back into the bag, calculate the probability that the second counter picked is:
      i) a green counter
      ii) a red counter.

8. A circular spinner has the numbers 0 to 36 equally spaced around its edge. Assuming that it is unbiased, calculate the probability on spinning it of getting:
   a) the number 5
   b) not 5
   c) an odd number
   d) zero
   e) a number greater than 15
   f) a multiple of 3
   g) a multiple of 3 or 5
   h) a prime number.

9. The letters R, C and A can be combined in several different ways.
    a) Write the letters in as many different orders as possible.
    If a computer writes these three letters at random, calculate the probability that:
    b) the letters will be written in alphabetical order,
    c) the letter R is written before both the letters A and C,
    d) the letter C is written after the letter A,
    e) the computer will spell the word CART if the letter T is added.

10. A normal pack of playing cards contains 52 cards. These are made up of four suits (hearts, diamonds, clubs and spades). Each suit consists of 13 cards. These are labelled Ace, 2, 3, 4, 5, 6, 7, 8, 9, 10, Jack, Queen and King. The hearts and diamonds are red; the clubs and spades are black.

    If a card is picked at random from a normal pack of cards, calculate the probability of picking:
    a) a heart   b) not a heart
    c) a 4   d) a red King
    e) a Jack, Queen or King   f) the Ace of spades
    g) an even numbered card   h) a 7 or a club.

Exercise 35.3

1. A student conducts a survey on the types of vehicle that pass his house. The results are shown below.

Vehicle type	Car	Lorry	Van	Bicycle	Motorbike	Other
Frequency	28	6	20	48	32	6

    a) How many vehicles passed the student's house?
    b) A vehicle is chosen at random from the results. Calculate the probability that it is:
       i) a car
       ii) a lorry
       iii) not a van.

2. In a class, data is collected about whether each student is right-handed or left-handed. The results are shown below.

	Left-handed	Right-handed
Boys	2	12
Girls	3	15

a) How many students are in the class?
b) A student is chosen at random. Calculate the probability that the student is:
i) a girl
ii) left-handed
iii) a right-handed boy
iv) not a right-handed boy.

3. A library keeps a record of the books that are borrowed during one day. The results are shown in the chart below.

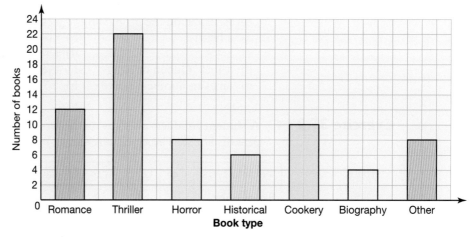

a) How many books were borrowed that day?
b) A book is chosen at random from the ones borrowed. Calculate the probability that it is:
i) a thriller
ii) a horror or a romance
iii) not a horror or romance
iv) not a biography.

● Relative frequency

A football referee always used a special coin to toss for ends. He noticed that out of the last twenty matches the coin had come down heads far more often than tails. He wanted to know if the coin was fair, that is, if it was as likely to come down heads as tails.

He decided to do a simple experiment by spinning the coin lots of times. His results are shown below:

Number of trials	Number of heads	Relative frequency
100	40	0.4
200	90	0.45
300	142	0.47…
400	210	0.525
500	260	0.52
600	290	0.48…
700	345	0.49…
800	404	0.505
900	451	0.50…
1000	499	0.499

The relative frequency $= \dfrac{\text{number of successful trials}}{\text{total number of trials}}$

In the 'long run', that is after a large number of trials, did the coin appear to be fair?

Notice that the greater the number of trials the better the estimated probability or relative frequency is likely to be. The key idea is that increasing the number of trials gives a better estimate of the probability and the closer the result obtained by experiment will be to that obtained by calculation.

Worked examples   **a)** There is a group of 250 people in a hall. A girl calculates that the probability of randomly picking someone that she knows from the group is 0.032. Calculate the number of people in the group that the girl knows.

$$\text{Probability} = \dfrac{\text{number of favourable results } (F)}{\text{number of possible results}}$$

$$0.032 = \dfrac{F}{250}$$

$250 \times 0.032 = F$ so $8 = F$

The girl knows 8 people in the group.

Probability

b) A boy enters 8 dogs into a dog show competition. His father knows how many dogs have been entered into the competition, and tells his son that they have a probability of 0.016 of winning the first prize (assuming all the dogs have an equal chance). How many dogs were entered into the competition?

$$\text{Probability} = \frac{\text{number of favourable results}}{\text{number of possible results }(T)}$$

$$0.016 = \frac{8}{T}$$

$$T = \frac{8}{0.016} = 500$$

So 500 dogs were entered into the competition.

Exercise 35.4

1. A boy calculates that he has a probability of 0.004 of winning the first prize in a photography competition if the selection is made at random. If 500 photographs are entered into the competition, how many photographs did the boy enter?

2. The probability of getting any particular number on a spinner game is given as 0.04. How many numbers are there on the spinner?

3. A bag contains 7 red counters, 5 blue, 3 green and 1 yellow. If one counter is drawn, what is the probability that it is:
   a) yellow b) red c) blue or green
   d) red, blue or green e) not blue?

4. A boy collects marbles. He has the following colours in a bag: 28 red, 14 blue, 25 yellow, 17 green and 6 purple. If he draws one marble from the bag, what is the probability that it is:
   a) red b) blue c) yellow or blue
   d) purple e) not purple?

5. The probability of a boy drawing a marble of one of the following colours from another bag of marbles is:

   blue 0.25 red 0.2 yellow 0.15 green 0.35 white 0.05

   If there are 49 green marbles, how many of each other colour does he have in his bag?

6. There are six red sweets in a bag. If the probability of randomly picking a red sweet is 0.02, calculate the number of sweets in the bag?

7. The probability of getting a bad egg in a batch of 400 is 0.035. How many bad eggs are there likely to be in a batch?

8. A sports arena has 25 000 seats, some of which are VIP seats. For a charity event all the seats are allocated randomly. The probability of getting a VIP seat is 0.008. How many VIP seats are there?

9. The probability of Juan's favourite football team winning 4−0 is 0.05. How many times are they likely to win by this score in a season of 40 matches?

Student assessment 1

1. What is the probability of throwing the following numbers with a fair dice?
   a) a 2
   b) not a 2
   c) less than 5
   d) a 7

2. If you have a normal pack of 52 cards, what is the probability of drawing:
   a) a diamond
   b) a 6
   c) a black card
   d) a picture card
   e) a card less than 5?

3. 250 coins, one of which is gold, are placed in a bag. What is the probability of getting the gold coin if I take, without looking, the following numbers of coin?
   a) 1
   b) 5
   c) 20
   d) 75
   e) 250

4. A bag contains 11 blue, 8 red, 6 white, 5 green and 10 yellow counters. If one counter is taken from the bag, what is the probability that it is:
   a) blue
   b) green
   c) yellow
   d) not red?

5. The probability of drawing a red, blue or green marble from a bag containing 320 marbles is:

   red 0.5   blue 0.3   green 0.2

   How many marbles of each colour are there?

6. In a small town there are a number of sports clubs. The clubs have 750 members in total. The table below shows the types of sports club and the number of members each has.

	Tennis	Football	Golf	Hockey	Athletics
Men	30	110	40	15	10
Women	15	25	20	45	30
Boys	10	200	5	10	40
Girls	20	35	0	30	60

A sports club member is chosen at random from the town.

Calculate the probability that the member is:
a) a man
b) a girl
c) a woman who does athletics
d) a boy who plays football
e) not a boy who plays football
f) not a golf player
g) a male who plays hockey.

7. A dice is thought to be biased. In order to test it, a boy rolls it 12 times and gets the following results:

Number	1	2	3	4	5	6
Frequency	2	2	2	2	2	2

A girl decides to test the same dice and rolls it 60 times. The table below shows her results:

Number	1	2	3	4	5	6
Frequency	3	3	47	3	2	2

a) Which results are likely to be more reliable? Justify your answer.
b) What conclusion can you make about whether the dice is biased?

Student assessment 2

1. An octagonal spinner has the numbers 1 to 8 on it as shown (below).

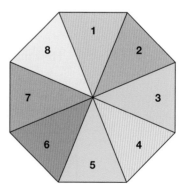

What is the probability of spinning:
a) a 7
b) not a 7
c) a factor of 12
d) a 9?

2. A game requires the use of all the playing cards in a normal pack from 6 to King inclusive.
   a) How many cards are used in the game?
   b) What is the probability of drawing:
      i) a 6   ii) a picture   iii) a club
      iv) a prime number   v) an 8 or a spade?

3. 180 students in a school are offered a chance to attend a football match for free. If the students are chosen at random, what is the chance of being picked to go if the following numbers of tickets are available?
   a) 1   b) 9   c) 15
   d) 40   e) 180

4. A bag contains 11 white, 9 blue, 7 green and 5 red counters. What is the probability that a single counter drawn will be:
   a) blue   b) red or green   c) not white?

5. The probability of drawing a red, blue or green marble from a bag containing 320 marbles is:

   red 0.4   blue 0.25   green 0.35

   If there are no other colours in the bag, how many marbles of each colour are there?

6. Students in a class conduct a survey to see how many friends they have on Facebook. The results were grouped and are shown in the pie chart below.

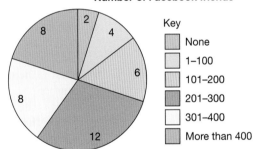

Number of Facebook friends

A student is chosen at random. What is the probability that he/she:
   a) has 101−200 Facebook friends
   b) uses Facebook
   c) has more than 200 Facebook friends?

7. a) If I enter a competition and have a 0.00002 probability of winning, how many people entered the competition?
   b) What assumption do you have to make in order to answer part a)?

36 Further probability

● **Combined events**

Combined events look at the probability of two or more events.

Worked example   i) Two coins are tossed. Show in a **two-way table** all the possible outcomes.

	Coin 1 Head	Coin 1 Tail
Coin 2 Head	HH	TH
Coin 2 Tail	HT	TT

ii) Calculate the probability of getting two heads.

All four outcomes are equally likely: therefore, the probability of getting HH is $\frac{1}{4}$.

iii) Calculate the probability of getting a head and a tail in any order.

The probability of getting a head and a tail in any order, i.e. HT or TH, is $\frac{2}{4} = \frac{1}{2}$.

Exercise 36.1

1. a) Two fair tetrahedral dice are rolled. If each is numbered 1–4, draw a two-way table to show all the possible outcomes.
   b) What is the probability that both dice show the same number?
   c) What is the probability that the number on one dice is double the number on the other?
   d) What is the probability that the sum of both numbers is prime?

2. Two fair dice are rolled. Copy and complete the diagram (left) to show all the possible combinations.
   What is the probability of getting:
   a) a double 3,   b) any double,
   c) a total score of 11,   d) a total score of 7,
   e) an even number on both dice,
   f) an even number on at least one dice,
   g) a total of 6 or a double,
   h) scores which differ by 3,
   i) a total which is either a multiple of 2 or 5?

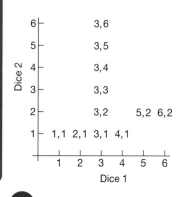

Tree diagrams

When more than two combined events are being considered then two-way tables cannot be used and therefore another method of representing information diagrammatically is needed. Tree diagrams are a good way of doing this.

Worked example
i) If a coin is tossed three times, show all the possible outcomes on a tree diagram, writing each of the probabilities at the side of the branches.

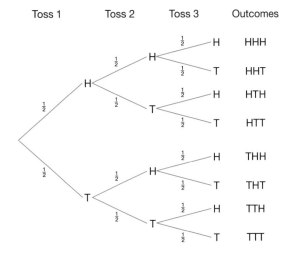

ii) What is the probability of getting three heads?

There are eight equally likely outcomes, therefore the probability of getting HHH is $\frac{1}{8}$.

iii) What is the probability of getting two heads and one tail in any order?

The successful outcomes are HHT, HTH, THH. Therefore the probability is $\frac{3}{8}$.

iv) What is the probability of getting at least one head?

This refers to any outcome with either one, two or three heads, i.e. all of them *except* TTT. Therefore the probability is $\frac{7}{8}$.

v) What is the probability of getting no heads?

The only successful outcome for this event is TTT. Therefore the probability is $\frac{1}{8}$.

Probability

Exercise 36.2

1. a) A computer uses the numbers 1, 2 or 3 at random to make three-digit numbers. Assuming that a number can be repeated, show on a tree diagram all the possible combinations that the computer can print.
   b) Calculate the probability of getting:
      i) the number 131,
      ii) an even number,
      iii) a multiple of 11,
      iv) a multiple of 3,
      v) a multiple of 2 or 3.

2. a) A cat has four kittens. Draw a tree diagram to show all the possible combinations of males and females. [assume P (male) = P (female)]
   b) Calculate the probability of getting:
      i) all female,
      ii) two females and two males,
      iii) at least one female,
      iv) more females than males.

3. a) A netball team plays three matches. In each match the team is equally likely to win, lose or draw. Draw a tree diagram to show all the possible outcomes over the three matches.
   b) Calculate the probability that the team:
      i) wins all three matches,
      ii) wins more times than loses,
      iii) loses at least one match,
      iv) either draws or loses all three matches.
   c) Explain why it is not very realistic to assume that the outcomes are equally likely in this case.

4. A spinner is split into quarters as shown.

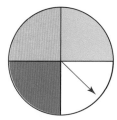

   a) If it is spun twice, draw a probability tree showing all the possible outcomes.
   b) Calculate the probability of getting:
      i) two dark blues,
      ii) two blues of either shade,
      iii) a pink and a white in any order.

36 Further probability

In each of the cases considered so far, all of the outcomes have been assumed to be equally likely. However, this need not be the case.

Worked example   In winter, the probability that it rains on any one day is $\frac{5}{7}$.

i) Using a tree diagram show all the possible combinations for two consecutive days.

ii) Write each of the probabilities by the sides of the branches.

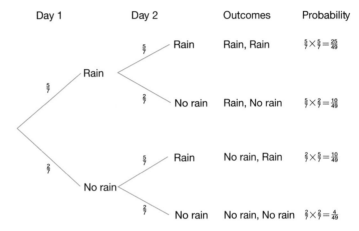

Note how the probability of each outcome is arrived at by multiplying the probabilities of the branches.

iii) Calculate the probability that it will rain on both days.

$$P(R, R) = \frac{5}{7} \times \frac{5}{7} = \frac{25}{49}$$

iv) Calculate the probability that it will rain on the first but not the second day.

$$P(R, NR) = \frac{5}{7} \times \frac{2}{7} = \frac{10}{49}$$

v) Calculate the probability that it will rain on at least one day.

The outcomes which satisfy this event are (R, R) (R, NR) and (NR, R).
Therefore the probability is $\frac{25}{49} + \frac{10}{49} + \frac{10}{49} = \frac{45}{49}$.

Probability

Exercise 36.3

1. A particular board game involves players rolling a dice. However, before a player can start, he or she needs to roll a 6.
   a) Copy and complete the tree diagram below showing all the possible combinations for the first two rolls of the dice.

   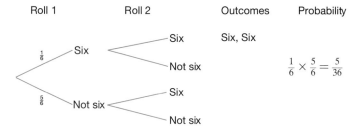

   b) Calculate the probability of the following:
      i) getting a six on the first roll,
      ii) starting within the first two rolls,
      iii) starting on the second roll,
      iv) not starting within the first three rolls,
      v) starting within the first three rolls.
   c) If you add the answers to b) iv) and v) what do you notice? Explain.

2. In Italy $\frac{3}{5}$ of the cars are foreign made. By drawing a tree diagram and writing the probabilities next to each of the branches, calculate the following probabilities:
   a) the next two cars to pass a particular spot are both Italian,
   b) two of the next three cars are foreign,
   c) at least one of the next three cars is Italian.

3. The probability that a morning bus arrives on time is 65%.
   a) Draw a tree diagram showing all the possible outcomes for three consecutive mornings.
   b) Label your tree diagram and use it to calculate the probability that:
      i) the bus is on time on all three mornings,
      ii) the bus is late the first two mornings,
      iii) the bus is on time two out of the three mornings,
      iv) the bus is on time at least twice.

4. A normal pack of 52 cards is shuffled and three cards are picked at random. Draw a tree diagram to help calculate the probability of picking:
   a) two clubs first,
   b) three clubs,
   c) no clubs,
   d) at least one club.

5. A bowl of fruit contains one kiwifruit, one banana, two mangos and two lychees. Two pieces of fruit are chosen at random and eaten.
   a) Draw a probability tree showing all the possible combinations of the two pieces of fruit.

b) Use your tree diagram to calculate the probability that:
  i) both the pieces of fruit eaten are mangos,
  ii) a kiwifruit and a banana are eaten,
  iii) at least one lychee is eaten.

6. Light bulbs are packaged in cartons of three. 10% of the bulbs are found to be faulty. Calculate the probability of finding two faulty bulbs in a single carton.

7. A volleyball team has a 0.25 chance of losing a game. Calculate the probability of the team achieving:
   a) two consecutive wins,
   b) three consecutive wins,
   c) 10 consecutive wins.

Student assessment I

1. A bag contains 12 white counters, 7 black counters and 1 red counter.
   a) If, when a counter is taken out, it is not replaced, calculate the probability that:
      i) the first counter is white,
      ii) the second counter removed is red, given that the first was black.
   b) If, when a counter is picked, it is then put back in the bag, how many attempts will be needed before it is mathematically certain that a red counter will have been picked out?

2. A coin is tossed and an ordinary, fair dice is rolled.
   a) Draw a two-way table showing all the possible combinations.
   b) Calculate the probability of getting:
      i) a head and a six,
      ii) a tail and an odd number,
      iii) a head and a prime number.

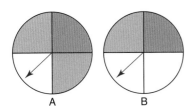

3. Two spinners A and B are split into quarters and coloured as shown. Both spinners are spun.
   a) Draw a fully labelled tree diagram showing all the possible combinations on the two spinners. Write beside each branch the probability of each outcome.
   b) Use your tree diagram to calculate the probability of getting:
      i) two blues,
      ii) two pinks,
      iii) a pink on spinner A and a white on spinner B.

Probability

4. A coin is tossed three times.
   a) Draw a tree diagram to show all the possible outcomes.
   b) Use your tree diagram to calculate the probability of getting:
      i) three tails,
      ii) two heads,
      iii) no tails,
      iv) at least one tail.

5. A goalkeeper expects to save one penalty out of every three. Calculate the probability that he:
   a) saves one penalty out of the next three,
   b) fails to save one or more of the next three penalties,
   c) saves two out of the next three penalties.

6. A board game uses a fair dice in the shape of a tetrahedron. The sides of the dice are numbered 1, 2, 3 and 4. Calculate the probability of:
   a) not throwing a 4 in two throws,
   b) throwing two consecutive 1s,
   c) throwing a total of 5 in two throws.

7. A normal pack of 52 cards is shuffled and three cards picked at random. Calculate the probability that all three cards are picture cards.

Student assessment 2

1. Two normal and fair dice are rolled and their scores added together.
   a) Using a two-way table, show all the possible scores that can be achieved.
   b) Using your two-way table, calculate the probability of getting:
      i) a score of 12,
      ii) a score of 7,
      iii) a score less than 4,
      iv) a score of 7 or more.
   c) Two dice are rolled 180 times. In theory, how many times would you expect to get a total score of 6?

2. A spinner is numbered as shown.
   a) If it is spun once, calculate the probability of getting:
      i) a 1,
      ii) a 2.
   b) If it is spun twice, calculate the probability of getting:
      i) a 2 followed by a 4,
      ii) a 2 and a 4 in any order,
      iii) at least one 1,
      iv) at least one 2.

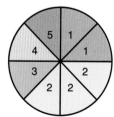

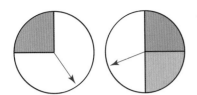

3. Two spinners are coloured as shown (left).
   a) They are both spun. Draw and label a tree diagram showing all the possible outcomes.
   b) Using your tree diagram calculate the probability of getting:
      i) two blues,
      ii) two whites,
      iii) a white and a pink,
      iv) at least one white.

4. Two spinners are labelled as shown:

   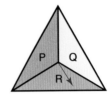

   Calculate the probability of getting:
   a) A and P,
   b) A or B and R,
   c) C but not Q.

5. A vending machine accepts $1 and $2 coins. The probability of a $2 coin being rejected is 0.2. The probability of a $1 coin being rejected is 0.1.

   A sandwich costing $3 is bought. Calculate the probability of getting a sandwich first time if:
   a) one of each coin is used,
   b) three $1 coins are used.

6. A biased coin is tossed three times. On each occasion the probability of getting a head is 0.6.
   a) Draw a tree diagram to show all the possible outcomes after three tosses. Label each branch clearly with the probability of each outcome.
   b) Using your tree diagram calculate the probability of getting:
      i) three heads,
      ii) three tails,
      iii) at least two heads.

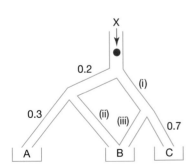

7. A ball enters a chute at X.
   a) What are the probabilities of the ball going down each of the chutes labelled (i), (ii) and (iii)?
   b) Calculate the probability of the ball landing in:
      i) tray A,
      ii) tray C,
      iii) tray B.

Topic 8: Mathematical investigations and ICT

● Probability drop

A game involves dropping a red marble down a chute. On hitting a triangle divider, the marble can bounce either left or right. On completing the drop, the marble lands in one of the trays along the bottom. The trays are numbered from left to right. Different sizes of game exist, the four smallest versions are shown below:

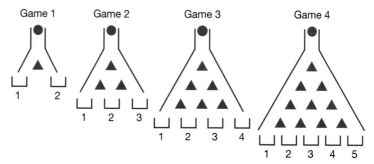

To land in tray 2 in the second game above, the ball can travel in one of two ways. These are: Left − Right or Right − Left.
  This can be abbreviated to LR or RL.

1. State the different routes the marble can take to land in each of the trays in the third game.
2. State the different routes the marble can take to land in each of the trays in the fourth game.
3. State, giving reasons, the probability of a marble landing in tray 1 in the fourth game.
4. State, giving reasons, the probability of a marble landing in each of the other trays in the fourth game.
5. Investigate the probability of the marble landing in each of the different trays in larger games.
6. Using your findings from your investigation, predict the probability of a marble landing in tray 7 in the tenth game (11 trays at the bottom).
7. Investigate the links between this game and the sequence of numbers generated in Pascal's triangle.

The following question is beyond the scope of the syllabus but is an interesting extension.

8. Investigate the links between this game, Pascal's triangle and the binomial expansion.

Dice sum

Two ordinary dice are rolled and their scores added together. Below is an incomplete table showing the possible outcomes:

		Dice 1					
		1	2	3	4	5	6
Dice 2	1	2			5		
	2						
	3				7		
	4				8		
	5				9	10	11
	6						12

1. Copy and complete the table to show all possible outcomes.
2. How many possible outcomes are there?
3. What is the most likely total when two dice are rolled?
4. What is the probability of getting a total score of 4?
5. What is the probability of getting the most likely total?
6. How many times more likely is a total score of 5 compared with a total score of 2?

Now consider rolling two four-sided dice, each numbered 1–4. Their scores are also added together.

7. Draw a table to show all the possible outcomes when the two four-sided dice are rolled and their scores added together.
8. How many possible outcomes are there?
9. What is the most likely total?
10. What is the probability of getting the most likely total?
11. Investigate the number of possible outcomes, the most likely total and its probability when two identical dice are rolled together and their scores added, i.e. consider 8-sided dice, 10-sided dice, etc.
12. Consider two m-sided dice rolled together and their scores added.
    a) What is the total number of outcomes in terms of m?
    b) What is the most likely total, in terms of m?
    c) What, in terms of m, is the probability of the most likely total?

Probability

13. Consider an m-sided and n-sided dice rolled together, where $m > n$.
    a) In terms of m and n, deduce the total number of outcomes.
    b) In terms of m and/or n, deduce the most likely total(s).
    c) In terms of m and/or n, deduce the probability of getting the most likely total.

● ICT activity: Buffon's needle experiment

You will need to use a spreadsheet for this activity.

The French count Le Comte de Buffon devised the following probability experiment.

1. Measure the length of a match (with the head cut off) as accurately as possible.

2. On a sheet of paper draw a series of straight lines parallel to each other. The distance between each line should be the same as the length of the match.

3. Take ten identical matches and drop them randomly on the paper. Count the number of matches that cross or touch any of the lines.

For example in the diagram below, the number of matches crossing or touching lines is six.

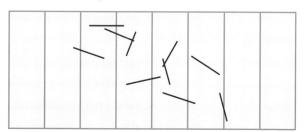

4. Repeat the experiment a further nine times, making a note of your results, so that altogether you have dropped 100 matches.

5. Set up a spreadsheet similar to the one shown below and enter your results in cell B2.

	A	B	C	D	E	F	G	H	I	J	K
1	Number of drops (N)	100	200	300	400	500	600	700	800	900	1000
2	Number of matches crossing/touching lines (n)										
3	Probability of crossing a line ($p = n/N$)										
4	$2/p$										

6. Repeat 100 match drops again, making a total of 200 drops, and enter cumulative results in cell C2.

7. By collating the results of your fellow students, enter the cumulative results of dropping a match 300–1000 times in cells D2–K2 respectively.

8. Using an appropriate formula, get the spreadsheet to complete the calculations in Rows 3 and 4.

9. Use the spreadsheet to plot a line graph of N against $\frac{2}{p}$.

10. What value does $\frac{2}{p}$ appear to get closer to?

Topic 9: Statistics

Syllabus

E9.1
Collect, classify and tabulate statistical data.
Read, interpret and draw simple inferences from tables and statistical diagrams.

E9.2
Construct and read bar charts, pie charts, pictograms, simple frequency distributions, histograms with equal and unequal intervals and scatter diagrams.

E9.3
Calculate the mean, median, mode and range for individual and discrete data and distinguish between the purposes for which they are used.

E9.4
Calculate an estimate of the mean for grouped and continuous data.
Identify the modal class from a grouped frequency distribution.

E9.5
Construct and use cumulative frequency diagrams.
Estimate and interpret the median, percentiles, quartiles and inter-quartile range.

E9.6
Understand what is meant by positive, negative and zero correlation with reference to a scatter diagram.

E9.7
Draw a straight line of best fit by eye.

Contents

Chapter 37   Mean, median, mode and range (E9.3, E9.4)
Chapter 38   Collecting and displaying data (E9.1, E9.2, E9.6, E9.7)
Chapter 39   Cumulative frequency (E9.5)

◯ Statistics in history

The earliest writing on statistics was found in a ninth-century book entitled *Manuscript on Deciphering Cryptographic Messages*, written by the Arab philosopher Al-Kindi (801–873), who lived in Baghdad. In his book, he gave a detailed description of how to use statistics to unlock coded messages.

The *Nuova Cronica*, a 14th-century history of Florence by the Italian banker Giovanni Villani, includes much statistical information on population, commerce, trade and education.

Early statistics served the needs of states — *state-istics*. By the early 19th century, statistics included the collection and analysis of data in general. Today, statistics are widely employed in government, business, and natural and social sciences. The use of modern computers has enabled large-scale statistical computation and has also made possible new methods that are impractical to perform manually.

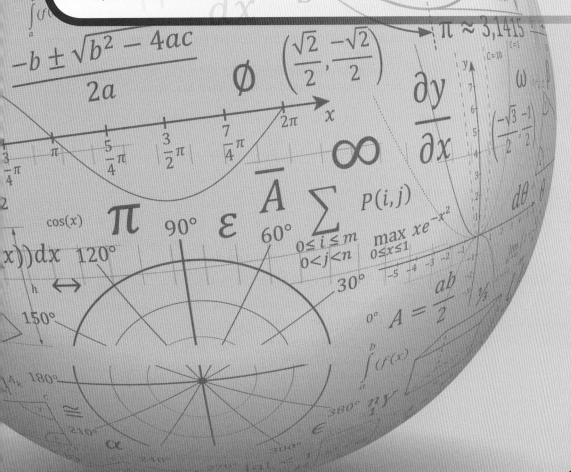

37 Mean, median, mode and range

● Average

'Average' is a word which in general use is taken to mean somewhere in the middle. For example, a woman may describe herself as being of average height. A student may think he or she is of average ability in maths. Mathematics is more exact and uses three principal methods to measure average.

- The **mode** is the value occurring the most often.
- The **median** is the middle value when all the data is arranged in order of size.
- The **mean** is found by adding together all the values of the data and then dividing that total by the number of data values.

● Spread

It is often useful to know how spread out the data is. It is possible for two sets of data to have the same mean and median but very different spreads.

The simplest measure of spread is the **range**. The range is simply the difference between the largest and smallest values in the data.

Another measure of spread is known as the inter-quartile range. This is covered in more detail in Chapter 39.

Worked examples **a)** i) Find the mean, median and mode of the data listed below.

1, 0, 2, 4, 1, 2, 1, 1, 2, 5, 5, 0, 1, 2, 3

$$\text{Mean} = \frac{1+0+2+4+1+2+1+1+2+5+5+0+1+2+3}{15}$$

$$= 2$$

Arranging all the data in order and then picking out the middle number gives the median:

0, 0, 1, 1, 1, 1, 1, ②, 2, 2, 2, 3, 4, 5, 5

The mode is the number which appeared most often. Therefore the mode is 1.

ii) Calculate the range of the data.
Largest value = 5
Smallest value = 0
Therefore the range = 5 − 0 = 5

b) **i)** The frequency chart (below) shows the score out of 10 achieved by a class in a maths test.
Calculate the mean, median and mode for this data.

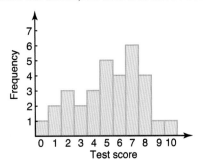

Transferring the results to a frequency table gives:

Test score	0	1	2	3	4	5	6	7	8	9	10	Total
Frequency	1	2	3	2	3	5	4	6	4	1	1	32
Frequency × score	0	2	6	6	12	25	24	42	32	9	10	168

In the total column we can see the number of students taking the test, i.e. 32, and also the total number of marks obtained by all the students, i.e. 168.

Therefore the mean score = $\frac{168}{32}$ = 5.25

Arranging all the scores in order gives:

0, 1, 1, 2, 2, 2, 3, 3, 4, 4, 4, 5, 5, 5, 5, (5, 6,) 6, 6, 6, 7, 7, 7, 7, 7, 7, 8, 8, 8, 8, 9, 10

Because there is an even number of students there isn't one middle number. There is a middle pair. The median is $\frac{(5+6)}{2}$ = 5.5.

The mode is 7 as it is the score which occurs most often.

ii) Calculate the range of the data.
Largest value = 10   Smallest value = 0
Therefore the range = 10 − 0 = 10

Exercise 37.1 In questions 1–5, find the mean, median, mode and range for each set of data.

1. A hockey team plays 15 matches. Below is a list of the numbers of goals scored in these matches.
   1, 0, 2, 4, 0, 1, 1, 1, 2, 5, 3, 0, 1, 2, 2

2. The total scores when two dice are thrown 20 times are:
   7, 4, 5, 7, 3, 2, 8, 6, 8, 7, 6, 5, 11, 9, 7, 3, 8, 7, 6, 5

Statistics

3. The ages of a group of girls are:

   14 years 3 months, 14 years 5 months,
   13 years 11 months, 14 years 3 months,
   14 years 7 months, 14 years 3 months,
   14 years 1 month

4. The numbers of students present in a class over a three-week period are:

   28, 24, 25, 28, 23, 28, 27, 26, 27, 25, 28, 28, 28, 26, 25

5. An athlete keeps a record in seconds of her training times for the 100 m race:

   14.0, 14.3, 14.1, 14.3, 14.2, 14.0, 13.9, 13.8, 13.9, 13.8, 13.8, 13.7, 13.8, 13.8, 13.8

6. The mean mass of the 11 players in a football team is 80.3 kg. The mean mass of the team plus a substitute is 81.2 kg. Calculate the mass of the substitute.

7. After eight matches a basketball player had scored a mean of 27 points. After three more matches his mean was 29. Calculate the total number of points he scored in the last three games.

Exercise 37.2

1. An ordinary dice was rolled 60 times. The results are shown in the table below. Calculate the mean, median, mode and range of the scores.

Score	1	2	3	4	5	6
Frequency	12	11	8	12	7	10

2. Two dice were thrown 100 times. Each time their combined score was recorded. Below is a table of the results. Calculate the mean score.

Score	2	3	4	5	6	7	8	9	10	11	12
Frequency	5	6	7	9	14	16	13	11	9	7	3

3. Sixty flowering bushes are planted. At their flowering peak, the number of flowers per bush is counted and recorded. The results are shown in the table below.

Flowers per bush	0	1	2	3	4	5	6	7	8
Frequency	0	0	0	6	4	6	10	16	18

   a) Calculate the mean, median, mode and range of the number of flowers per bush.
   b) Which of the mean, median and mode would be most useful when advertising the bush to potential buyers?

● The mean for grouped data

The mean for grouped data can only be an estimate as the position of the data within a group is not known. An estimate is made by calculating the mid-interval value for a group and then assigning all of the data within the group that mid-interval value.

Worked example   The history test scores for a group of 40 students are shown in the grouped frequency table below.

Score, S	Frequency	Mid-interval value	Frequency × mid-interval value
$0 \leqslant S \leqslant 19$	2	9.5	19
$20 \leqslant S \leqslant 39$	4	29.5	118
$40 \leqslant S \leqslant 59$	14	49.5	693
$60 \leqslant S \leqslant 79$	16	69.5	1112
$80 \leqslant S \leqslant 99$	4	89.5	358

i) Calculate an estimate for the mean test result.

$$\text{Mean} = \frac{19 + 118 + 693 + 1112 + 358}{40} = 57.5$$

ii) What is the modal class?

This refers to the class with the greatest frequency, if the class width is constant. Therefore the modal class is $60 \leqslant S \leqslant 79$.

Exercise 37.3   **1.** The heights of 50 basketball players attending a tournament are recorded in the grouped frequency table.
Note: 1.8− means $1.8 \leqslant H < 1.9$.

Height (m)	Frequency
1.8−	2
1.9−	5
2.0−	10
2.1−	22
2.2−	7
2.3−2.4	4

a) Copy the table and complete it to include the necessary data with which to calculate the mean height of the players.
b) Estimate the mean height of the players.
c) What is the modal class height of the players?

Statistics

Hours of overtime	Frequency
0–9	12
10–19	18
20–29	22
30–39	64
40–49	32
50–59	20

2. The number of hours of overtime worked by employees at a factory over a period of a month is given in the table (left).
   a) Calculate an estimate for the mean number of hours of overtime worked by the employees that month.
   b) What is the modal class?

3. The length of the index finger of 30 students in a class is measured. The results were recorded and are shown in the grouped frequency table.

Length (cm)	Frequency
5.0–	3
5.5–	8
6.0–	10
6.5–	7
7.0–7.5	2

   a) Calculate an estimate for the mean index finger length of the students.
   b) What is the modal class?

Student assessment 1

1. A rugby team scores the following number of points in 12 matches:

   21, 18, 3, 12, 15, 18, 42, 18, 24, 6, 12, 3

   Calculate for the 12 matches:
   a) the mean score,
   b) the median score,
   c) the mode,
   d) the range.

2. The bar chart (left) shows the marks out of 10 for an English test taken by a class of students.
   a) Calculate the number of students who took the test.
   b) Calculate for the class:
      i) the mean test result,
      ii) the median test result,
      iii) the modal test result,
      iv) the range of the test results.

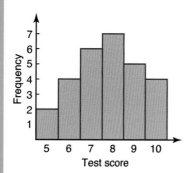

3. Fifty sacks of grain are weighed as they are unloaded from a truck. The mass of each is recorded in the grouped frequency table.
   a) Calculate the mean mass of the 50 sacks.
   b) State the modal class.

Mass (kg)	Frequency
$15 \leq M < 16$	0
$16 \leq M < 17$	3
$17 \leq M < 18$	6
$18 \leq M < 19$	14
$19 \leq M < 20$	18
$20 \leq M < 21$	8
$21 \leq M < 22$	1

Student assessment 2

1. A javelin thrower keeps a record of her best throws over ten competitions. These are shown in the table below.

Competition	1	2	3	4	5	6	7	8	9	10
Distance (m)	77	75	78	86	92	93	93	93	92	89

Find the mean, median, mode and range of her throws.

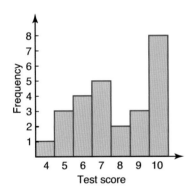

2. The bar chart shows the marks out of 10 for a Maths test taken by a class of students.
   a) Calculate the number of students who took the test.
   b) Calculate for the class:
      i) the mean test result,
      ii) the median test result,
      iii) the modal test result,
      iv) the range of the test results.

3. A hundred sacks of coffee with a stated mass of 10 kg are unloaded from a train. The mass of each sack is checked and the results are presented in the table.
   a) Calculate an estimate for the mean mass.
   b) What is the modal class?

Mass (kg)	Frequency
$9.8 \leq M < 9.9$	14
$9.9 \leq M < 10.0$	22
$10.0 \leq M < 10.1$	36
$10.1 \leq M < 10.2$	20
$10.2 \leq M < 10.3$	8

38 Collecting and displaying data

● **Tally charts and frequency tables**

The figures in the list below are the numbers of chocolate buttons in each of twenty packets of buttons.

35 36 38 37 35 36 38 36 37 35
36 36 38 36 35 38 37 38 36 38

The figures can be shown on a tally chart:

Number	Tally	Frequency
35	IIII	4
36	⋕ II	7
37	III	3
38	⋕ I	6

When the tallies are added up to get the frequency, the chart is usually called a **frequency table**. The information can then be displayed in a variety of ways.

● **Pictograms**

@ = 1 packet of chocolate buttons

Buttons per packet	
35	@@@@
36	@@@@@@@
37	@@@
38	@@@@@@

● **Bar charts**

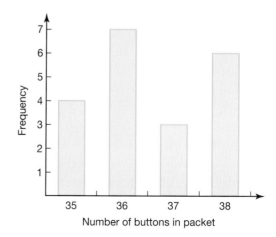

Grouped frequency tables

If there is a big range in the data it is easier to group the data in a **grouped frequency table**.

The groups are arranged so that no score can appear in two groups.

The scores for the first round of a golf competition are shown below.

71 75 82 96 83 75 76 82 103 85 79 77 83 85
88 104 76 77 79 83 84 86 88 102 95 96 99 102
75 72

This data can be grouped as shown:

Score	Frequency
71–75	5
76–80	6
81–85	8
86–90	3
91–95	1
96–100	3
101–105	4
Total	30

Note: it is not possible to score 70.5 or 77.3 at golf. The scores are said to be **discrete**. If the data is **continuous**, for example when measuring time, the intervals can be shown as 0–, 10–, 20–, 30– and so on.

Pie charts

Data can be displayed on a **pie chart** – a circle divided into sectors. The size of the sector is in direct proportion to the frequency of the data. The sector size does not show the actual frequency. The actual frequency can be calculated easily from the size of the sector.

Worked examples a)

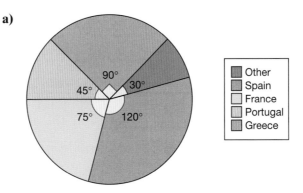

Statistics

In a survey of 240 English children were asked to vote for their favourite holiday destination. The results are shown on the pie chart above (previous page). Calculate the actual number of votes for each destination.

The total 240 votes are represented by 360°.
It follows that if 360° represents 240 votes:

There were $240 \times \frac{120}{360}$ votes for Spain
so, 80 votes for Spain.
There were $240 \times \frac{75}{360}$ votes for France
so, 50 votes for France.
There were $240 \times \frac{45}{360}$ votes for Portugal
so, 30 votes for Portugal.
There were $240 \times \frac{90}{360}$ votes for Greece
so, 60 votes for Greece.
Other destinations received $240 \times \frac{30}{360}$ votes
so, 20 votes for other destinations.

Note: it is worth checking your result by adding them:

$$80 + 50 + 30 + 60 + 20 = 240 \text{ total votes}$$

b) The table shows the percentage of votes cast for various political parties in an election. If a total of 5 million votes were cast, how many votes were cast for each party?

Party	Percentage of vote
Social Democrats	45%
Liberal Democrats	36%
Green Party	15%
Others	4%

The Social Democrats received $\frac{45}{100} \times 5$ million votes
so, 2.25 million votes.
The Liberal Democrats received $\frac{36}{100} \times 5$ million votes
so, 1.8 million votes.
The Green Party received $\frac{15}{100} \times 5$ million votes
so, 750 000 votes.
Other parties received $\frac{4}{100} \times 5$ million votes
so, 200 000 votes.

Check total:

$$2.25 + 1.8 + 0.75 + 0.2 = 5 \text{ (million votes)}$$

c) The table shows the results of a survey among 72 students to find their favourite sport. Display this data on a pie chart.

Sport	Frequency
Football	35
Tennis	14
Volleyball	10
Hockey	6
Basketball	5
Other	2

72 students are represented by 360°, so 1 student is represented by $\frac{360}{72}$ degrees. Therefore the size of each sector can be calculated as shown:

Football $\quad 35 \times \frac{360}{72}$ degrees $\quad$ i.e., 175°

Tennis $\quad 14 \times \frac{360}{72}$ degrees $\quad$ i.e., 70°

Volleyball $\quad 10 \times \frac{360}{72}$ degrees $\quad$ i.e., 50°

Hockey $\quad 6 \times \frac{360}{72}$ degrees $\quad$ i.e., 30°

Basketball $\quad 5 \times \frac{360}{72}$ degrees $\quad$ i.e., 25°

Other sports $\quad 2 \times \frac{360}{72}$ degrees $\quad$ i.e., 10°

Check total:

$$175 + 70 + 50 + 30 + 25 + 10 = 360$$

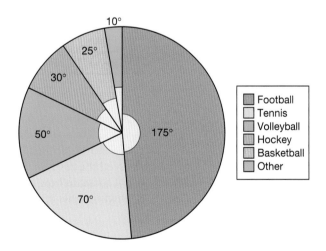

Statistics

Exercise 38.1

1. The pie charts below show how a girl and her brother spent one day. Calculate how many hours they spent on each activity. The diagrams are to scale.

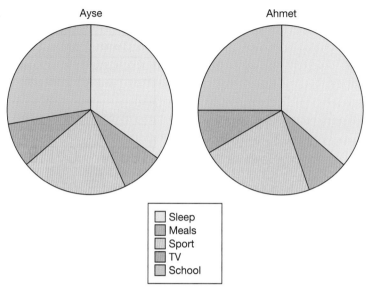

2. A survey was carried out among a class of 40 students. The question asked was, 'How would you spend a gift of $15?'. The results are shown below:

Choice	Frequency
Music	14
Books	6
Clothes	18
Cinema	2

Illustrate these results on a pie chart.

3. A student works during the holidays. He earns a total of $2400. He estimates that the money earned has been used as follows: clothes $\frac{1}{3}$, transport $\frac{1}{5}$, entertainment $\frac{1}{4}$. He has saved the rest.

Calculate how much he has spent on each category, and illustrate this information on a pie chart.

4. A research project looking at the careers of men and women in Spain produced the following results:

Career	Male (percentage)	Female (percentage)
Clerical	22	38
Professional	16	8
Skilled craft	24	16
Non-skilled craft	12	24
Social	8	10
Managerial	18	4

a) Illustrate this information on two pie charts, and make two statements that could be supported by the data.
b) If there are eight million women in employment in Spain, calculate the number in either professional or managerial employment.

● **Surveys**

A survey requires data to be collected, organised, analysed and presented.

A survey may be carried out for interest's sake, for example to find out how many cars pass your school in an hour.
A survey could be carried out to help future planning – information about traffic flow could lead to the building of new roads, or the placing of traffic lights or a pedestrian crossing.

Exercise 38.2
1. Below are a number of statements, some of which you may have heard or read before.
   Conduct a survey to collect data which will support or disprove one of the statements. Where possible, use pie charts to illustrate your results.
   a) Women's magazines are full of adverts.
   b) If you go to a football match you are lucky to see more than one goal scored.
   c) Every other car on the road is white.
   d) Girls are not interested in sport.
   e) Children today do nothing but watch TV.
   f) Newspapers have more sport than news in them.
   g) Most girls want to be nurses, teachers or secretaries.
   h) Nobody walks to school any more.
   i) Nearly everybody has a computer at home.
   j) Most of what is on TV comes from America.

Statistics

2. Below are some instructions relating to a washing machine in English, French, German, Dutch and Italian.

Analyse the data and write a report. You may wish to comment upon:
i) the length of words in each language,
ii) the frequency of letters of the alphabet in different languages.

ENGLISH

ATTENTION
Do not interrupt drying during the programme.
This machine incorporates a temperature safety thermostat which will cut out the heating element in the event of a water blockage or power failure. In the event of this happening, reset the programme before selecting a further drying time.

For further instructions, consult the user manual.

FRENCH

ATTENTION
N'interrompez pas le séchage en cours de programme.
Une panne d'électricité ou un manque d'eau momentanés peuvent annuler le programme de séchage en cours. Dans ces cas arrêtez l'appareil, affichez de nouveau le programme et après remettez l'appareil en marche.

Pour d'ultérieures informations, rapportez-vous à la notice d'utilisation.

GERMAN

ACHTUNG
Die Trocknung soll nicht nach Anlaufen des Programms unterbrochen werden.
Ein kurzer Stromausfall bzw. Wassermangel kann das laufende Trocknungsprogramm annullieren. In diesem Falle Gerät ausschalten, Programm wieder einstellen und Gerät wieder einschalten.

Für nähere Angaben beziehen Sie sich auf die Bedienungsanleitung.

DUTCH

BELANGRIJK
Het droogprogramma niet onderbreken wanneer de machine in bedrijf is.
Door een korte stroom-of watertoevoeronderbreking kan het droogprogramma geannuleerd worden. Schakel in dit geval de machine uit, maak opnieuw uw programmakeuze en stel onmiddellijk weer in werking.

Verdere inlichtingen vindt u in de gebruiksaanwijzing.

ITALIAN

ATTENZIONE
Non interrompere l'asciugatura quando il programma è avviato.
La macchina è munita di un dispositivo di sicurezza che può annullare il programma di asciugaturea in corso quando si verifica una temporanea mancanza di acqua o di tensione.

In questi casi si dovrà spegnere la macchina, reimpostare il programma e poi riavviare la macchina.

Per ulteriori indicazioni, leggere il libretto istruzioni.

● Scatter diagrams

Scatter diagrams are particularly useful if we wish to see if there is a **correlation** (relationship) between two sets of data. The two values of data collected represent the coordinates of each point plotted. How the points lie when plotted indicates the type of relationship between the two sets of data.

Worked example   The heights and masses of 20 children under the age of five were recorded. The heights were recorded in centimetres and the masses in kilograms. The data is shown below.

Height	32	34	45	46	52
Mass	5.8	3.8	9.0	4.2	10.1
Height	59	63	64	71	73
Mass	6.2	9.9	16.0	15.8	9.9
Height	86	87	95	96	96
Mass	11.1	16.4	20.9	16.2	14.0
Height	101	108	109	117	121
Mass	19.5	15.9	12.0	19.4	14.3

i) Plot a scatter diagram of the above data.

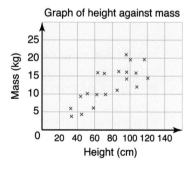

ii) Comment on any relationship you see.

The points tend to lie in a diagonal direction from bottom left to top right. This suggests that as height increases then, in general, mass increases too. Therefore there is a **positive correlation** between height and mass.

iii) If another child was measured as having a height of 80 cm, approximately what mass would you expect him or her to be?

We assume that this child will follow the trend set by the other 20 children. To deduce an approximate value for the mass, we draw a **line of best fit**. This is a solid straight line which passes through the points as closely as possible, as shown.

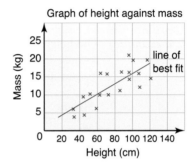

The line of best fit can now be used to give an approximate solution to the question. If a child has a height of 80 cm, you would expect his or her mass to be in the region of 13 kg.

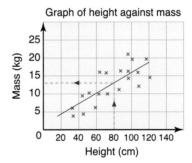

● Types of correlation

There are several types of correlation, depending on the arrangement of the points plotted on the scatter diagram.

A strong positive correlation between the variables x and y.
The points lie very close to the line of best fit.
As x increases, so does y.

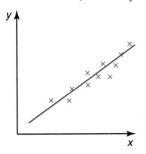

A weak positive correlation.
Although there is direction to the way the points are lying, they are not tightly packed around the line of best fit.
As x increases, y tends to increase too.

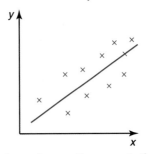

A strong negative correlation.
The points lie close around the line of best fit.
As x increases, y decreases.

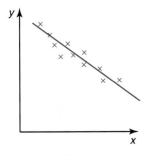

A weak negative correlation.
The points are not tightly packed around the line of best fit.
As x increases, y tends to decrease.

No correlation.
As there is no pattern to the way in which the points are lying, there is no correlation between the variables x and y. As a result there can be no line of best fit.

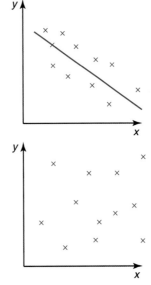

Exercise 38.3

1. State what type of correlation you might expect, if any, if the following data was collected and plotted on a scatter diagram. Give reasons for your answer.
   a) A student's score in a maths exam and their score in a science exam.
   b) A student's hair colour and the distance they have to travel to school.
   c) The outdoor temperature and the number of cold drinks sold by a shop.
   d) The age of a motorcycle and its second-hand selling price.

e) The number of people living in a house and the number of rooms the house has.
f) The number of goals your opponents score and the number of times you win.
g) A child's height and the child's age.
h) A car's engine size and its fuel consumption.

2. A website gives average monthly readings for the number of hours of sunshine and the amount of rainfall in millimetres for several cities in Europe. The table below is a summary for July.

Place	Hours of sunshine	Rainfall (mm)
Athens	12	6
Belgrade	10	61
Copenhagen	8	71
Dubrovnik	12	26
Edinburgh	5	83
Frankfurt	7	70
Geneva	10	64
Helsinki	9	68
Innsbruck	7	134
Krakow	7	111
Lisbon	12	3
Marseilles	11	11
Naples	10	19
Oslo	7	82
Plovdiv	11	37
Reykjavik	6	50
Sofia	10	68
Tallinn	10	68
Valletta	12	0
York	6	62
Zurich	8	136

a) Plot a scatter diagram of the number of hours of sunshine against the amount of rainfall. Use a spreadsheet if possible.
b) What type of correlation, if any, is there between the two variables? Comment on whether this is what you would expect.

3. The United Nations keeps an up-to-date database of statistical information on its member countries.
The table below shows some of the information available.

Country	Life expectancy at birth (years, 2005–2010)		Adult illiteracy rate (%, 2009)	Infant mortality rate (per 1000 births, 2005–2010)
	Female	Male		
Australia	84	79	1	5
Barbados	80	74	0.3	10
Brazil	76	69	10	24
Chad	50	47	68.2	130
China	75	71	6.7	23
Colombia	77	69	7.2	19
Congo	55	53	18.9	79
Cuba	81	77	0.2	5
Egypt	72	68	33	35
France	85	78	1	4
Germany	82	77	1	4
India	65	62	34	55
Israel	83	79	2.9	5
Japan	86	79	1	3
Kenya	55	54	26.4	64
Mexico	79	74	7.2	17
Nepal	67	66	43.5	42
Portugal	82	75	5.1	4
Russian Federation	73	60	0.5	12
Saudi Arabia	75	71	15	19
South Africa	53	50	12	49
United Kingdom	82	77	1	5
United States of America	81	77	1	6

a) By plotting a scatter diagram, decide if there is a correlation between the adult illiteracy rate and the infant mortality rate.
b) Are your findings in part a) what you expected? Explain your answer.
c) Without plotting a graph, decide if you think there is likely to be a correlation between male and female life expectancy at birth. Explain your reasons.
d) Plot a scatter diagram to test if your predictions for part c) were correct.

Statistics

4. A gardener plants 10 tomato plants. He wants to see if there is a relationship between the number of tomatoes the plant produces and its height in centimetres.

   The results are presented in the scatter diagram below. The line of best fit is also drawn.

   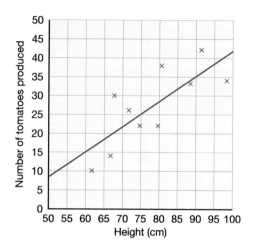

   a) Describe the correlation (if any) between the height of a plant and the number of tomatoes it produced.
   b) The gardener has another plant grown in the same conditions as the others. If the height is 85 cm, estimate from the graph the number of tomatoes he can expect it to produce.
   c) Another plant only produces 15 tomatoes. Deduce its height from the graph.

● Histograms

A histogram displays the frequency of either continuous or grouped discrete data in the form of bars. There are several important features of a histogram which distinguish it from a bar chart.

- The bars are joined together.
- The bars can be of varying width.
- The frequency of the data is represented by the area of the bar and not the height (though in the case of bars of equal width, the area is directly proportional to the height of the bar and so the height is usually used as the measure of frequency).

Worked example

The table (left) shows the marks out of 100 in a maths test for a class of 32 students. Draw a histogram representing this data.

Test marks	Frequency
1–10	0
11–20	0
21–30	1
31–40	2
41–50	5
51–60	8
61–70	7
71–80	6
81–90	2
91–100	1

All the class intervals are the same. As a result the bars of the histogram will all be of equal width, and the frequency can be plotted on the vertical axis. The histogram is shown below.

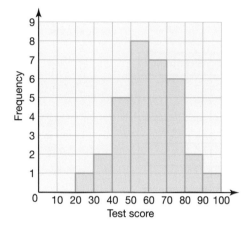

Exercise 38.4

1. The table (below) shows the distances travelled to school by a class of 30 students. Represent this information on a histogram.

Distance (km)	Frequency
$0 \leqslant d < 1$	8
$1 \leqslant d < 2$	5
$2 \leqslant d < 3$	6
$3 \leqslant d < 4$	3
$4 \leqslant d < 5$	4
$5 \leqslant d < 6$	2
$6 \leqslant d < 7$	1
$7 \leqslant d < 8$	1

2. The heights of students in a class were measured. The results are shown in the table (below). Draw a histogram to represent this data.

Height (cm)	Frequency
145–	1
150–	2
155–	4
160–	7
165–	6
170–	3
175–	2
180–185	1

Note that both questions in Exercise 38.4 deal with **continuous data**. In these questions equal class intervals are represented in different ways. However, they mean the same thing.
In question 2, 145– means the students whose heights fall in the range $145 \leqslant h < 150$.

Statistics

So far the work on histograms has only dealt with problems in which the class intervals are of the same width. This, however, need not be the case.

Worked example The heights of 25 sunflowers were measured and the results recorded in the table (below left).

If a histogram were drawn with frequency plotted on the vertical axis, then it could look like the one drawn below.

Height (m)	Frequency
$0 \leqslant h < 1.0$	6
$1.0 \leqslant h < 1.5$	3
$1.5 \leqslant h < 2.0$	4
$2.0 \leqslant h < 2.25$	3
$2.25 \leqslant h < 2.50$	5
$2.50 \leqslant h < 2.75$	4

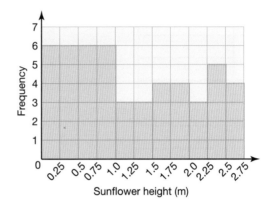

This graph is misleading because it leads people to the conclusion that most of the sunflowers were under 1 m, simply because the area of the bar is so great. In actual fact only approximately one quarter of the sunflowers were under 1 m.

When class intervals are different it is the area of the bar which represents the frequency, not the height. Instead of frequency being plotted on the vertical axis, **frequency density** is plotted.

$$\text{Frequency density} = \frac{\text{frequency}}{\text{class width}}$$

The results of the sunflower measurements in the example above can therefore be written as:

Height (m)	Frequency	Frequency density
$0 \leqslant h < 1.0$	6	$6 \div 1 = 6$
$1.0 \leqslant h < 1.5$	3	$3 \div 0.5 = 6$
$1.5 \leqslant h < 2.0$	4	$4 \div 0.5 = 8$
$2.0 \leqslant h < 2.25$	3	$3 \div 0.25 = 12$
$2.25 \leqslant h < 2.50$	5	$5 \div 0.25 = 20$
$2.50 \leqslant h < 2.75$	4	$4 \div 0.25 = 16$

The histogram can therefore be redrawn as shown below giving a more accurate representation of the data.

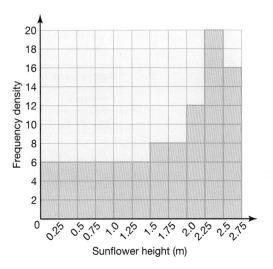

Exercise 38.5

1. The table below shows the time taken, in minutes, by 40 students to travel to school.

Time (min)	0–	10–	15–	20–	25–	30–	40–60
Frequency	6	3	13	7	3	4	4
Frequency density							

   a) Copy the table and complete it by calculating the frequency density.
   b) Represent the information on a histogram.

2. On Sundays Maria helps her father feed their chickens. Over a period of one year she kept a record of how long it took. Her results are shown in the table below.

Time (min)	Frequency	Frequency density
$0 \leqslant t < 30$	8	
$30 \leqslant t < 45$	5	
$45 \leqslant t < 60$	8	
$60 \leqslant t < 75$	9	
$75 \leqslant t < 90$	10	
$90 \leqslant t < 120$	12	

   a) Copy the table and complete it by calculating the frequency density. Give the answers correct to 1 d.p.
   b) Represent the information on a histogram.

Statistics

3. Frances and Ali did a survey of the ages of the people living in their village. Part of their results are set out in the table below.

Age (years)	0–	1–	5–	10–	20–	40–	60–90
Frequency	35			180	260		150
Frequency density		12	28			14	

   a) Copy the table and complete it by calculating either the frequency or the frequency density.
   b) Represent the information on a histogram.

4. The table below shows the ages of 150 people, chosen randomly, taking the 06 00 train into a city.

Age (years)	0–	15–	20–	25–	30–	40–	50–80
Frequency	3	25	20	30	32	30	10

The histogram below shows the results obtained when the same survey was carried out on the 11 00 train.

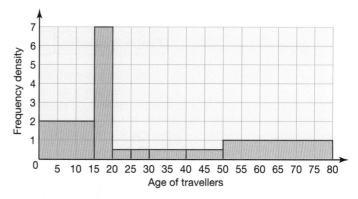

   a) Draw a histogram for the 06 00 train.
   b) Compare the two sets of data and give two possible reasons for the differences.

Student assessment 1

1. The areas of four countries are shown below. Illustrate this data as a bar chart.

Country	Nigeria	Republic of the Congo	South Sudan	Kenya
Area in 10 000 km²	90	35	70	57

2. The table below gives the average time taken for 30 pupils in a class to get to school each morning, and the distance they live from the school.

Distance (km)	2	10	18	15	3	4	6	2	25	23	3	5	7	8	2
Time (min)	5	17	32	38	8	14	15	7	31	37	5	18	13	15	8
Distance (km)	19	15	11	9	2	3	4	3	14	14	4	12	12	7	1
Time (min)	27	40	23	30	10	10	8	9	15	23	9	20	27	18	4

   a) Plot a scatter diagram of distance travelled against time taken.
   b) Describe the correlation between the two variables.
   c) Explain why some pupils who live further away may get to school more quickly than some of those who live nearer.
   d) Draw a line of best fit on your scatter diagram.
   e) A new pupil joins the class. Use your line of best fit to estimate how far away from school she might live if she takes, on average, 19 minutes to get to school each morning.

3. A golf club has four classes of member based on age. The membership numbers for each class are shown below.

	Age	Number	Frequency density
Juniors	0–	32	
Intermediates	16–	80	
Full members	21–	273	
Seniors	60–80	70	

   a) Calculate the frequency density for each class of member.
   b) Illustrate your data on a histogram.

Student assessment 2

1. The table below shows the population (in millions) of the continents:

   Display this information on a pie chart.

Continent	Asia	Europe	America	Africa	Oceania
Population (millions)	4140	750	920	995	35

2. A department store decides to investigate whether there is a correlation between the number of pairs of gloves it sells and the outside temperature. Over a one-year period the store records, every two weeks, how many pairs of gloves are sold and the mean daytime temperature during the same period. The results are given in the table below.

Mean temperature (°C)	3	6	8	10	10	11	12	14	16	16	17	18	18
Number of pairs of gloves	61	52	49	54	52	48	44	40	51	39	31	43	35
Mean temperature (°C)	19	19	20	21	22	22	24	25	25	26	26	27	28
Number of pairs of gloves	26	17	36	26	46	40	30	25	11	7	3	2	0

   a) Plot a scatter diagram of mean temperature against number of pairs of gloves.
   b) What type of correlation is there between the two variables?
   c) How might this information be useful for the department store in the future?
   d) The mean daytime temperature during the next two-week period is predicted to be 20°C. Draw a line of best fit on your graph and use it to estimate the number of pairs of gloves the department store can expect to sell.

3. The grouped frequency table below shows the number of points scored by a school basketball player.

Points	0–	5–	10–	15–	25–	35–50
Number of games	2	3	8	9	12	3
Frequency density						

   a) Copy and complete the table by calculating the frequency densities. Give your answers to 1 d.p.
   b) Draw a histogram to illustrate the data.

39 Cumulative frequency

● Cumulative frequency

Calculating the **cumulative frequency** is done by adding up the frequencies as we go along. A cumulative frequency diagram is particularly useful when trying to calculate the median of a large set of data, grouped or continuous data, or when trying to establish how consistent a set of results are.

Worked example The duration of two different brands of battery, A and B, is tested. 50 batteries of each type are randomly selected and tested in the same way. The duration of each battery is then recorded. The results of the tests are shown in the table below.

Type A		
Duration (h)	Frequency	Cumulative frequency
$0 \leqslant t < 5$	3	3
$5 \leqslant t < 10$	5	8
$10 \leqslant t < 15$	8	16
$15 \leqslant t < 20$	10	26
$20 \leqslant t < 25$	12	38
$25 \leqslant t < 30$	7	45
$30 \leqslant t < 35$	5	50

Type B		
Duration (h)	Frequency	Cumulative frequency
$0 \leqslant t < 5$	1	1
$5 \leqslant t < 10$	1	2
$10 \leqslant t < 15$	10	12
$15 \leqslant t < 20$	23	35
$20 \leqslant t < 25$	9	44
$25 \leqslant t < 30$	4	48
$30 \leqslant t < 35$	2	50

i) Plot a cumulative frequency diagram for each brand of battery.

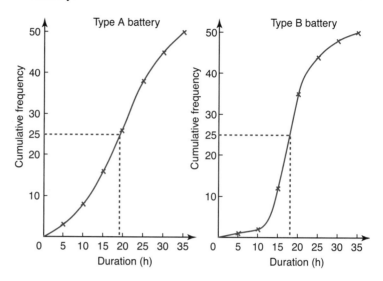

Both cumulative frequency diagrams are plotted above.

Notice how the points are plotted at the upper boundary of each class interval and *not* at the middle of the interval.

ii) Calculate the median duration for each brand.

The median value is the value which occurs half-way up the cumulative frequency axis. Therefore:

Median for type A batteries ≈ 19 h
Median for type B batteries ≈ 18 h

This tells us that the same number of batteries are still working as have stopped working after 19 h for A and 18 h for B.

Exercise 39.1

1. Sixty athletes enter a long-distance run. Their finishing times are recorded and are shown in the table below:

Finishing time (h)	0–	0.5–	1.0–	1.5–	2.0–	2.5–	3.0–3.5
Frequency	0	0	6	34	16	3	1
Cumulative freq.							

a) Copy the table and calculate the values for the cumulative frequency.
b) Draw a cumulative frequency diagram of the results.
c) Show how your graph could be used to find the approximate median finishing time.
d) What does the median value tell us?

Cumulative frequency

2. Three mathematics classes take the same test in preparation for their final exam. Their raw scores are shown in the table below:

Class A	12, 21, 24, 30, 33, 36, 42, 45, 53, 53, 57, 59, 61, 62, 74, 88, 92, 93
Class B	48, 53, 54, 59, 61, 62, 67, 78, 85, 96, 98, 99
Class C	10, 22, 36, 42, 44, 68, 72, 74, 75, 83, 86, 89, 93, 96, 97, 99, 99

a) Using the class intervals $0 \leq x < 20$, $20 \leq x < 40$ etc. draw up a grouped frequency and cumulative frequency table for each class.
b) Draw a cumulative frequency diagram for each class.
c) Show how your graph could be used to find the median score for each class.
d) What does the median value tell us?

3. The table below shows the heights of students in a class over a three-year period.

Height (cm)	Frequency 2007	Frequency 2008	Frequency 2009
150–	6	2	2
155–	8	9	6
160–	11	10	9
165–	4	4	8
170–	1	3	2
175–	0	2	2
180–185	0	0	1

a) Construct a cumulative frequency table for each year.
b) Draw the cumulative frequency diagram for each year.
c) Show how your graph could be used to find the median height for each year.
d) What does the median value tell us?

Statistics

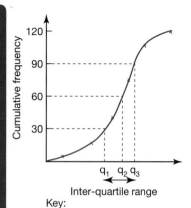

Key:
q_1 Lower quartile
q_2 Median
q_3 Upper quartile

● Quartiles and the inter-quartile range

The cumulative frequency axis can also be represented in terms of **percentiles**. A percentile scale divides the cumulative frequency scale into hundredths. The maximum value of cumulative frequency is found at the 100th percentile. Similarly the median, being the middle value, is called the 50th percentile. The 25th percentile is known as the **lower quartile**, and the 75th percentile is called the **upper quartile**.

The **range** of a distribution is found by subtracting the lowest value from the highest value. Sometimes this will give a useful result, but often it will not. A better measure of spread is given by looking at the spread of the middle half of the results, i.e. the difference between the upper and lower quartiles. This result is known as the **inter-quartile range**.

The graph (left) shows the terms mentioned above.

Worked example   Consider again the two types of batteries A and B discussed earlier (page 491).

i) Using the graphs, estimate the upper and lower quartiles for each battery.

   Lower quartile of type A ≈ 13 h
   Upper quartile of type A ≈ 25 h
   Lower quartile of type B ≈ 15 h
   Upper quartile of type B ≈ 21 h

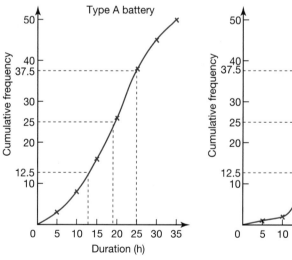

ii) Calculate the inter-quartile range for each type of battery.

   Inter-quartile range of type A ≈ 12 h
   Inter-quartile range of type B ≈ 6 h

iii) Based on these results, how might the manufacturers advertise the two types of battery?

Type A: on 'average' the longer-lasting battery
Type B: the more reliable battery

Exercise 39.2

1. Using the results obtained from question 2 in Exercise 39.1:
   a) find the inter-quartile range of each of the classes taking the mathematics test,
   b) analyse your results and write a brief summary comparing the three classes.

2. Using the results obtained from question 3 in Exercise 39.1:
   a) find the inter-quartile range of the students' heights each year,
   b) analyse your results and write a brief summary comparing the three years.

3. Forty boys enter for a school javelin competition. The distances thrown are recorded below:

Distance thrown (m)	0–	20–	40–	60–	80–100
Frequency	4	9	15	10	2

   a) Construct a cumulative frequency table for the above results.
   b) Draw a cumulative frequency diagram.
   c) If the top 20% of boys are considered for the final, estimate (using the graph) the qualifying distance.
   d) Calculate the inter-quartile range of the throws.
   e) Calculate the median distance thrown.

4. The masses of two different types of orange are compared. Eighty oranges are randomly selected from each type and weighed. The results are shown below.

Type A		Type B	
Mass (g)	Frequency	Mass (g)	Frequency
75–	4	75–	0
100–	7	100–	16
125–	15	125–	43
150–	32	150–	10
175–	14	175–	7
200–	6	200–	4
225–250	2	225–250	0

a) Construct a cumulative frequency table for each type of orange.
b) Draw a cumulative frequency diagram for each type of orange.
c) Calculate the median mass for each type of orange.
d) Using your graphs estimate:
   i) the lower quartile,
   ii) the upper quartile,
   iii) the inter-quartile range
   for each type of orange.
e) Write a brief report comparing the two types of orange.

5. Two competing brands of battery are compared. A hundred batteries of each brand are tested and the duration of each is recorded. The results of the tests are shown in the cumulative frequency diagrams below.

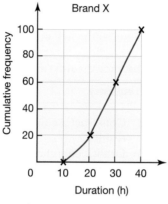
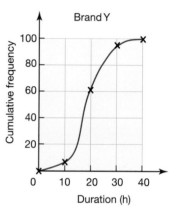

a) The manufacturers of brand X claim that on average their batteries will last at least 40% longer than those of brand Y. Showing your method clearly, decide whether this claim is true.
b) The manufacturers of brand X also claim that their batteries are more reliable than those of brand Y. Is this claim true? Show your working clearly.

Student assessment I

1. Thirty students sit a Maths exam. Their marks are given as percentages and are shown in the table below.

Mark	20–	30–	40–	50–	60–	70–	80–	90–100
Frequency	2	3	5	7	6	4	2	1

   a) Construct a cumulative frequency table of the above results.
   b) Draw a cumulative frequency diagram of the results.
   c) Using the graph, estimate a value for:
      i) the median,
      ii) the upper and lower quartiles,
      iii) the inter-quartile range.

2. 400 students sit an IGCSE exam. Their marks (as percentages) are shown in the table below.

Mark (%)	Frequency	Cumulative frequency
31–40	21	
41–50	55	
51–60	125	
61–70	74	
71–80	52	
81–90	45	
91–100	28	

   a) Copy and complete the above table by calculating the cumulative frequency.
   b) Draw a cumulative frequency diagram of the results.
   c) Using the graph, estimate a value for:
      i) the median result,
      ii) the upper and lower quartiles,
      iii) the inter-quartile range.

3. Eight hundred students sit an exam. Their marks (as percentages) are shown in the table below.

Mark (%)	Frequency	Cumulative frequency
1–10	10	
11–20	30	
21–30	40	
31–40	50	
41–50	70	
51–60	100	
61–70	240	
71–80	160	
81–90	70	
91–100	30	

a) Copy and complete the above table by calculating the cumulative frequency.
b) Draw a cumulative frequency diagram of the results.
c) An 'A' grade is awarded to a student at or above the 80th percentile. What mark is the minimum requirement for an 'A' grade?
d) A 'C' grade is awarded to any student between and including the 55th and the 70th percentile. What marks form the lower and upper boundaries of a 'C' grade?
e) Calculate the inter-quartile range for this exam.

Topic 9: Mathematical investigations and ICT

● Heights and percentiles

The graphs below show the height charts for males and females from the age of 2 to 20 years in the United States.

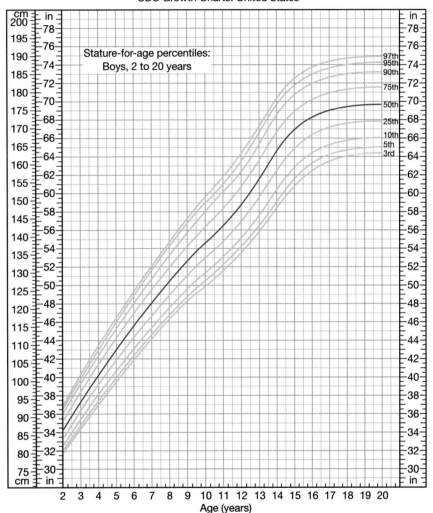

Note: Heights have been given in both centimetres and inches.

Statistics

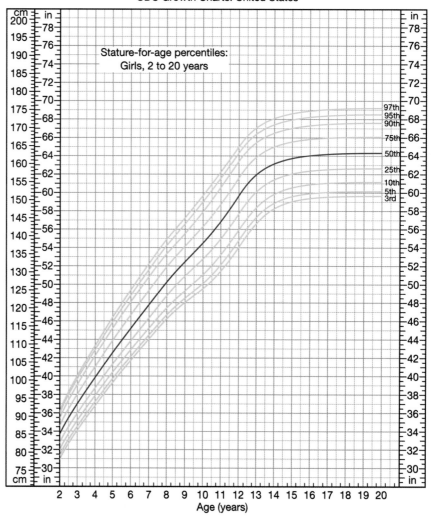

1. From the graph find the height corresponding to the 75th percentile for 16 year-old girls.
2. Find the height which 75% of 16 year-old boys are taller than.
3. What is the median height for 12 year-old girls?
4. Measure the heights of students in your class. By carrying out appropriate statistical calculations, write a report comparing your data to that shown in the graphs.
5. Would all cultures use the same height charts? Explain your answer.

● Reading ages

Depending on their target audience, newspapers, magazines and books have different levels of readability. Some are easy to read and others more difficult.

1. Decide on some factors that you think would affect the readability of a text.
2. Write down the names of two newspapers which you think would have different reading ages. Give reasons for your answer.

There are established formulae for calculating the reading age of different texts.
   One of these is the Gunning Fog Index. It calculates the reading age as follows:

$$\text{Reading age} = \frac{2}{5}\left(\frac{A}{n} + \frac{100L}{A}\right) \text{ where}$$

A = number of words
n = number of sentences
L = number of words with 3 or more syllables

3. Choose one article from each of the two newspapers you chose in question 2. Use the Gunning Fog Index to calculate the reading ages for the articles. Do the results support your predictions?
4. Write down some factors which you think may affect the reliability of your results.

● ICT activity

In this activity you will be collecting the height data of all the students in your class and plotting a cumulative frequency diagram of the results.

1. Measure the heights of all the students in your class.
2. Group your data appropriately.
3. Enter your data into graphing software such as Excel or Autograph.
4. Produce a cumulative frequency diagram of the results.
5. From your graph find:
   a) the median height of the students in your class.
   b) the inter-quartile range of the heights.
6. Compare the cumulative frequency diagram from your class with one produced from data collected from another class in a different year group. Comment on any differences/similarities between the two.

Index

A
acceleration 178
accuracy, appropriate 14
acute angles 214
addition
   of algebraic fractions 115
   of fractions 35–6
   of matrices 397–9
   of vectors 384
adjacent side, of a right-angled triangle 349
al-Karaji 103
al-Khwarizmi 103
al-Kindi 465
algebraic fractions
   addition and subtraction 115
   simplifying 114, 115–18
angles
   alternate 246
   at the centre of a circle 257
   at a point and on a line 245
   between a line and a plane 374–6
   between a tangent and a radius of a circle 254–5
   bisecting 223–5
   calculating size by trigonometry 359
   corresponding 246
   in a quadrilateral 128, 249–51
   in a semi-circle 253–4
   in a triangle 128, 248–9
   in opposite segments of a circle 258
   in the same segment of a circle 257–8
   obtuse 214
   of elevation and depression 357–9
   of irregular polygons 256
   of rotation 412
   reflex 214
   right 214
   supplementary 258
   types 214
   vertically opposite 246
   within parallel lines 246
answers, estimating 14–15
apex, of a pyramid 298
Apollonius of Perga 213
approximation 12
Arab mathematicians 103
arc 216, 292
Archimedes 213
area
   converting from one unit to another 278–80
   of a circle 286–7
   of a parallelogram 283–5
   of a rectangle 281
   of a sector 294–6
   of a trapezium 284–5
   of a triangle 281–3, 370–1
   of similar shapes 235–7
   under a speed–time graph 181–3
area factor 231, 235
arithmetic sequences 155–7
Aryabhata 3
averages 466–8

B
back bearings 346–7
bar charts 472
base, of a triangle 282
bearings 346
Bernoulli family of mathematicians 345
Bhascara 3
bonus 76
bounds, upper and lower 16–21
brackets
   and order of operations 26, 33–4
   expanding 104–5, 108
Brahmagupta 3

C
calculations
   estimating answers 14–15
   order of operations 26
   with fractions 34–7
   with upper and lower bounds 17–21
calculator calculations
   appropriate accuracy 14
   order of operations 26–8
capacity, metric units 276, 278
Cardano, Girolamo 381
Cartesian coordinates 315
centre
   of a circle 216
   of enlargement 415–17
   of rotation 410, 412
chaos theory 441
chord, of a circle 216, 242
circle properties 242–3
circles
   angle properties 253–5, 257–9, 273
   area 286–7
   circumference 286–7
   tangents from an external point 243
   terminology 216
circular prism *see* cylinder
circumcircle, of a triangle 226
circumference, of a circle 286–7
class intervals 485
column matrix 394
column vectors, translation 386
common denominators 116, 117
common difference, in an arithmetic sequence 155
common ratio, in a geometric sequence 162
compass points 346
complement of a set 90
completing the square 142
composite functions 207–8

Index

compound interest 81–3, 163–4
concyclic points 258
cone
  rotational symmetry 241
  surface area 305–6
  volume 301–5
constant of proportionality 167
constant of variation 167
conversion graphs 174–5
correlation 479–84
cosine 352–3
cosine curve 360–1
cosine rule 367–70
cost price 77
cube, rotational symmetry 241
cube numbers 5
cube roots 9–10
cubic function 193
cubic sequences 157–61
cuboid
  planes of symmetry 240
  rotational symmetry 241
  surface area 288–9
  volume 290
cumulative frequency 491–3
currency conversions 75, 174
curves, gradients 195–7
cyclic quadrilateral 258
cylinder
  rotational symmetry 241
  surface area 288–9
  volume 290

D

data
  continuous or discrete 473, 485
  correlation 479–84
decagon 218
deceleration 178
decimal places (d.p.) 12–13
decimals 31–2
  changing to fractions 38, 39–41
  terminating and recurring 4, 38–41
deductions from gross earnings 76
denominator of a fraction 30
depreciation 78
depression, angle 357–9
Descartes, René 315
determinant, of a matrix 403–4
diameter, of a circle 216

difference of two squares 109
direct proportion 52–4, 167
direct variation 167–9
directed numbers 10
direction of rotation 412
discount 77
distance see speed, distance and time
division
  of a quantity in a given ratio 54–6
  of fractions 36–7
  rules 33
dodecagon 218
double time 76

E

earnings 76–7
Egyptian mathematics 275
elements
  of a matrix 394
  of a set 87–8
elevation, angle 357–9
elimination, in solving simultaneous equations 131–2
empty set 88
enlargement 415–21
  negative 419
equations
  constructing 128–31, 136–8
  of a line through two points 332–3
  of a straight line 320–5
  simple linear 127
  solving by graphical methods 197–200
Euclid of Alexandria 213
Eudoxus of Asia Minor 213
Euler, Leonhard 345
evaluation
  of functions 203–5
  of numerical expressions 110
events, combined 452
exchange rates 75
exponential equations 65
  graphical solution 211
exponential functions, graphs 195

F

factorisation
  and evaluation of numerical expressions 110
  by difference of two squares 109–10
  by grouping 108–9
  in solving quadratic equations 138–9
  of quadratic expressions 110–11
  simple 105–6
factors 5–6
Fermat, Pierre de 315, 441
Fermat's Last Theorem 315
Fibonacci (Leonardo Pisano) 385
formulae
  for the terms of an arithmetic sequence 155–6
  transformation (rearrangement) 107, 112–14
fraction, proper 30
fractional indices 69–71
  algebraic 124–5
fractions 30–1
  addition and subtraction 35–6
  algebraic 114–18
  changing to decimals 37–8
  equivalent 34
  improper 30–1
  multiplication and division 36–7
  simplest form (lowest terms) 34–5
French mathematicians 315
frequency
  cumulative 491–3
  relative 447–9
frequency density 486
frequency tables 472
functions
  definition 203
  evaluating 203–5
  inverse 205–6

G

Gauss, Carl Friedrich 441
geometric figures, constructing 221–3
geometric sequences 162–5
gradient
  of a distance–time graph 176
  of a straight line 195, 316–19
  of curves 195–6
  of parallel lines 326–7
  of perpendicular lines 333–6

Index

positive/negative 318
gradient–intercept form 324, 326
graphical solution
   of equations 197–200
   of exponential equations 211
   of inequalities 149–52
   of quadratic equations 191–2
   of simultaneous equations 329–30
graphs
   exponential functions 195
   for finding square roots 9
   in practical situations 174
   of quadratic functions 189–91
   of reciprocal function 192–3
   types 193–4
Greek mathematicians 213
grouped data, mean 469
grouped frequency tables 473

H
height (altitude), of a triangle 214, 282
heptagon 218
hexagon 218, 251
hexagonal-based pyramid 298
highest common factor (HCF) 7, 35
Hindu mathematicians 3
histograms 484–8
Huygens, Christiaan 441
hyperbola 192, 193
hypotenuse 349

I
identity matrix 403
image 408
increase and decrease
   by a given ratio 57–9
   percentage 46–7
index (indices) 62–4
   algebraic 123
inequalities 16
   graphing 149–52
   in linear programming 152–3
   manipulation 143
   representing on a number line 24–5, 143–4
   solving 143, 149
   symbols 24, 143, 149
integers 4

inter-quartile range 466, 494–6
interest 79–83
intersection of sets 90
inverse, of a matrix 404–6
inverse functions 205–6
inverse proportion 56–7, 169–70
inverse variation 169–70
irrational numbers 4–5, 7–8
Italian mathematicians 381

K
kite, properties 217

L
Laghada 3
Laplace, Pierre Simon de 441
laws of indices 62, 123
length, metric units 276
line
   bisecting 223–5
   of best fit 480
   of symmetry 240
   see also straight line
line segment
   calculating length 330–1
   midpoint 331–2
linear equations, solving 127–8
linear function, straight-line graph 193
linear inequalities 143
linear programming 152–3
locus (loci) 265–8
lowest common multiple (LCM) 7, 116

M
maps see scale drawings
mass, metric units 276, 278
matrix (matrices)
   addition and subtraction 397–9
   determinant 403–4
   elements 394
   inverse 404–6
   and transformations 426–7
   multiplication by a scalar quantity 400
   multiplication by another matrix 400–2
   order 394–6
   representing a transformation 423–5

   see also column matrix; identity matrix; row matrix; square matrix; transformation matrix; zero matrix
mean
   definition 466
   for grouped data 469
median, definition 466
metric units 276
   converting from one unit to another 277–8
midpoint, of a line segment 331–2
mirror line 408
mixed numbers 30–1, 35
modal class 469
mode, definition 466
multiples 7
multiplication
   in solving simultaneous equations 134
   of a matrix by a scalar 400
   of a matrix by another matrix 400–2
   of a vector by a scalar 384
   of fractions 36–7
   rules 33

N
natural numbers 4
negative indices 64, 68–9, 124
net pay 76
nets 219
Newtonian universe 441
numbers, types 4
numerator of a fraction 30

O
object and image 408
octagon 218
Omar Khayyam 103
opposite side, of a right-angled triangle 349
order
   of a matrix 394–6
   of operations 26, 33
   of rotational symmetry 240
ordering 24
outcomes 442
overtime 76

Index

P

π (pi)
  approximations for 3
  irrational number 5
parabola 189, 193
parallel lines 214
  angle properties 246
  equations 326–7
parallelogram
  area 283–5
  constructing 222–3
  properties 217
Pascal, Blaise 315, 441
Pascal's Triangle 315
pentagon 218, 251
percentage increases and decreases 46–7
percentage interest 79
percentage of a quantity, calculating 44–5
percentage profit and loss 78–9
percentages 32
  fraction and decimal equivalents 43–4
  reverse 48
percentiles 494
perimeter, of a rectangle 281
perpendicular bisector 223, 242, 265
perpendicular lines 214, 333–6
pictograms 472
pie charts 473–7
piece work 76
place value 33
plane of symmetry 240
plans see scale drawings
polygons
  angle properties 251–3, 256
  regular 218, 251
  similar 219, 231–4
  types 218
position vectors 387
positive indices 62, 66, 123
positive or negative number see directed numbers
prime factors 6–7
prime numbers 5, 100
principal 79
prism 290–2
probability
  definition 442
  of combined events 452

practical and theoretical 442–6
profit and loss 77–8
  percentage 78–9
pyramid 298–301
Pythagoras of Samos 213
Pythagoras' theorem 353–6

Q

quadratic equations
  graphical solution 191–2
  solving by completing the square 142
  solving by factorising 138–9
quadratic expressions, factorisation 110–11
quadratic formula 141
quadratic functions
  graphs 189–91
  parabola 193
quadratic sequences 157–61
quadrilaterals 218, 251
  angle properties 249–51
  types 217
quantity
  dividing in a given ratio 54–6
  expressing as a percentage of another quantity 45–6
quantum mechanics 441
quartiles 494–6

R

radius, of a circle 216
range 466, 494
ratio, enlargement/reduction 57–9
ratio method
  for direct proportion 52, 53
  for dividing a quantity in a given ratio 54
rational numbers 4, 7–8
real numbers 4–5
reciprocal function, graph 192–3
reciprocal of a number 36–7
rectangle, properties 217, 281
rectangular prism see cuboid
reflection 408–10
reflective symmetry 240
rhombus, properties 217
rotation 410–13
rotational symmetry 240
rounding 12
row matrix 394

S

scale drawings 227–9
scale factor 231, 235
  of enlargement 415–17
  of negative enlargement 419
scatter diagrams 479–84
scientific notation see standard form
sector
  area 294–6
  definition 293
  of a circle 216
segment, of a circle 216
selling price 77
semi-circle, angle in 253–4
sequences
  arithmetic 155–7
  definition 155
  geometric 162–5
  quadratic and cubic 157–61
  terms 155
sets
  complement 90
  elements 87–8
  empty 88
  intersection 90
  notation 87, 88, 90
  problems involving 93–4
  union 91
  universal 90
shapes, similar 231–4
significant figures (s.f.) 13
similar shapes, area and volume 235–7
simple interest 79–80
simplification, using indices 62–3, 123, 125–6
simultaneous equations 131–6
  graphical solution 329–30
sine 351–2
sine curve 359–60
sine rule 366–7
solids see three-dimensional shapes
speed, distance and time 175–6
speed–time graphs 178–83
sphere
  rotational symmetry 241
  surface area 297–8
  volume 296–7
spread, measures 466
square, properties 217
square matrix 394
square numbers 5, 100

Index

square roots 7, 8–9
square-based pyramid 298
   rotational symmetry 241
standard form 65–9
statistics, historical development 465
straight line
   equation 320–5
   shortest distance between two points 214
straight-line graphs 316
   drawing 327–8
subsets 88–9
substitution 106–7
   in solving simultaneous equations 131, 132
subtraction
   of algebraic fractions 115
   of fractions 35–6
   of matrices 397–9
   of vectors 384
surface area
   of a cone 305–6
   of a cuboid 288–9
   of a cylinder 288–9
   of a pyramid 300–1
   of a sphere 297–8
surveys 477–9
Swiss mathematicians 345
symmetry 240–1

T

tally charts 472
tangent
   to a circle 243
   to a curve, gradient 196
   trigonometric ratio 349–51
temperature scale 10
terms of a sequence 155
Thales of Alexandria 213
Theano (Greek mathematician) 213
three-dimensional shapes
   nets 219
   symmetry 240–1
three-figure bearing system 346
time 85–6
   see also speed, distance and time
time and a half 76
timetables 85
transformation
   and inverse matrices 426–7
   combinations 422, 427–30
   definition 408
   of formulae 107, 112–14
transformation matrix (matrices) 423–5
   combination 427–30
translation 414–15
   of a column vector 386
trapezium 217, 284–5
travel graphs 176–8
tree diagrams 453–7
triangles
   acute-angled 215
   angle properties 128, 248–9
   area 281–3, 370–1
   base 282
   circumcircle 226
   congruent 215
   constructing 221–2, 272–3
   equilateral 215, 251
   height (altitude) 214, 282
   isosceles 215
   obtuse-angled 215
   properties 218
   right-angled 215, 349
   scalene 215
   similar 216, 231–4
   types 215–16
   vertex 282
triangular prism
   rotational symmetry 241
   volume 290
trigonometric ratios 349
trigonometry, in three dimensions 371–3
two-way table 452

U

union of sets 91
unitary method
   direct proportion 52, 53
   for dividing a quantity in a given ratio 55
universal set 90

V

vector geometry 387–9
vectors
   addition and subtraction 384
   magnitude 386
   multiplying by a scalar 384–5
   see also column vectors; position vectors
Venn diagrams 90–3
vertex, of a triangle 282
Villani, Giovanni 465
volume
   converting from one unit to another 278–80
   of a cone 301–5
   of a prism 290–2
   of a pyramid 298–300
   of a sphere 296–7
   of similar shapes 235–7

W

Wiles, Andrew 315

Z

zero index 64, 123–4
zero matrix 403